fundamentals of finite mathematics

fundamentals of finite mathematics

r. w. negus
rio hondo college
department of mathematics
whittier, california

john wiley & sons, inc.
new york london sydney toronto

Negus, R W 1916-
Fundamentals of finite mathematics.
1. Mathematics – 1961- I. Title.
QA39.2.N43 1974 510 73-17469
ISBN 0-471-63121-3

Printed in the United States of America

10 9 8 7 6 5 4 3 2 1

PREFACE

This book grew out of my need for a textbook to use in a finite mathematics course for students majoring in business and social science. Some parts of the book were written originally to explain material from texts using the usual terse, concise style of mathematicans—a style that students find very difficult to understand. Other sections were originally written to supplement texts that had been simplified by the omission of the more difficult concepts. This book provides a reasonable depth of material, presented in language that the student comprehends. It was written for students, not mathematicians. The inferences that a mathematician makes when he reads a concise, carefully worded definition or statement do not occur to the student. Thus, here, the most important meanings and consequences of the basic statements are stated, restated, and discussed to make sure that the student understands them.

The material was developed over several years of actual classroom use. The students provided critical comment on the extent of the discussion and its depth. The material, in essentially its present form, has been used successfully for more than two years by different instructors at two different colleges.

As is usual in a mathematics text, the exercises at the end of each section are a vital part of the material. In each set there are a few mechanical exercises to give the student practice in working with basic concepts and symbols. There are also problems that require the student to use the concepts in constructing mathematical models for described situations. In doing this, the student must place himself in the situation, analyze the information available, determine what he needs to know, and perceive how the mathematical models discussed in the text can be applied to the situation. This is the process that transforms mathematics from an abstract intellectual exercise into a useful tool for solving everyday problems.

Each chapter is, as much as possible, independent of the others. There are a few exceptions. The concepts and notation relating to sets (Chapter 1) are used throughout the text. Chapter 6 is a necessary preparation for Chapter 7. The material in Chapters 4 and 7 is used in Chapter 8. The Simplex Method was purposely omitted from the chapter on linear programming (Chapter 5). I think that a mathematics course at this level should build a foundation of understanding for later study. The Simplex Method is a special technique, which the student will promptly forget at the end of the course. If he should need it in his later work, he can learn it then. Since learning the method requires a sizable part of a semester course (and space here is a problem), it has been omitted in favor of more basic material.

I am indebted to many people for assistance and encouragement in this project. I thank my colleagues at Rio Hondo College, Professors Gerald E. Bruce and J. Gordon Eversole for their cooperation and constructive suggestions while using the material; and Professors Louis Wilson, Willis Powers and Ken Lingren of Cerritos College for using the text and providing constructive comments. I thank also the people who reviewed the book, Professors Samuel W. Kodis of Cabrillo College, Kenneth Goldstein of Miami-Dade Junior College, and George C. Dorner of Loyola University of Chicago, for their many useful suggestions. I am especially grateful to my wife for her help in typing and proofreading and for her cooperation during the writing. And I certainly appreciate the candid and reasoned suggestions of the many students in my classes, who made this book possible.

R. W. Negus

CONTENTS

FOREWORD TO THE STUDENT 1

CHAPTER 1 SETS 5

1.1 INTRODUCTION 5

1.1.1 Definitions and Undefinitions 6
1.1.2 Set Specifications 7
1.1.3 Symbology 8
1.1.4 Set Relationships 9
Subset 10
Equal Sets 11
Proper Subset 11
Exercise 1.1 12

1.2 SET OPERATIONS 14

1.2.1 Union 14
1.2.2 Intersection 15
1.2.3 Null Set 16
1.2.4 Universal Set 16
1.2.5 Complement of a Set 17
1.2.6 Relative Complement 17
1.2.7 Disjoint Sets 18
1.2.8 Summary 20
Exercise 1.2 20

1.3 ADDITIONAL SET OPERATIONS: VENN DIAGRAMS 23

1.3.1 Venn Diagrams 26
1.3.2 Identifying Compound Sets 29
Exercise 1.3 30

1.4 COUNTING THE ELEMENTS IN SETS 32

1.4.1 Partitions and Cross Partitions 37
Exercise 1.4 39

CHAPTER 2 DEDUCTIVE REASONING AND SYMBOLIC LOGIC 43

2.0 INTRODUCTION 43

2.1 STATEMENTS 45

2.1.1 Compound Statements 46
2.1.2 Truth Tables 48
2.1.3 Definition of Conjunction 49
2.1.4 Definition of Disjunction 50
2.1.5 Negation 51
Exercise 2.1 51

2.2 CONDITIONAL AND BICONDITIONAL 53

2.2.1 Conditional 53
2.2.2 Biconditional 55
2.2.3 Variations of the Conditional 55
2.2.4 Converse 57
2.2.5 Contrapositive 57
2.2.6 Inverse 57
2.2.7 Verbal Expressions for the Conditional 58
2.2.8 Necessary and Sufficient Conditions 59
Exercise 2.2 60

2.3 TRUTH TABLES FOR COMPLEX STATEMENTS 62

2.3.1 Compound Statements with More Than Two Components 65
2.3.2 Logical Equivalents 67
2.3.3 Tautologies 68
2.3.4 Logical Absurdity 69
Exercise 2.3 69

2.4 DEDUCTION AND ARGUMENT 70

2.4.1 Truth-Table Method for Checking Argument Validity 72
Exercise 2.4 75

2.5 ARGUMENT FORMS 76

2.5.1 Direct Proof 78
2.5.2 Use of Logical Equivalents 79

2.5.3 Formal Proof 80
2.5.4 Indirect Proof 83
Exercise 2.5 85

2.6 QUANTIFIED PROPOSITIONS 87

2.6.1 Universal Propositions 87
2.6.2 Existential Propositions 89
2.6.3 Negations of Quantified Propositions 90
Exercise 2.6 92

CHAPTER 3 RELATIONS AND FUNCTIONS 95

3.1 GENERAL DESCRIPTION 95

3.1.1 Variables—Independent and Dependent 96
3.1.2 Functions 99
Exercise 3.1 101

3.2 GRAPHS 103

3.2.1 Line Graphs 103
3.2.2 Cartesian Plane 105
3.2.3 Graphs of Relations 108
Exercise 3.2 113

3.3 LINEAR FUNCTIONS 115

3.3.1 Alternate Forms for Linear Functions 119
3.3.2 Equation of Line from Two Points 120
Exercise 3.3 122

3.4 APPLICATIONS 123

3.4.1 Straight-Line Depreciation 123
3.4.2 Simple Interest and Loan Discounting 124
3.4.3 Break-Even Analysis 126
3.4.4 Linear Correlation and Projections 128
Exercise 3.4 131

CHAPTER 4 MATRICES AND VECTORS 134

4.0 INTRODUCTION 134

4.1 BASIC OPERATIONS WITH MATRICES 135

4.1.1 Equality of Matrices 136
4.1.2 Sum of Two Matrices 137
4.1.3 Multiplication of a Matrix by a Scalar 138
4.1.4 Associative Property of Matrix Addition 138
Exercise 4.1 139

4.2 MULTIPLICATION OF MATRICES 142

4.2.1 Product of Two Vectors 142
4.2.2 Product of Two Matrices 143
4.2.3 Associative Property of Matrix Multiplication 146
4.2.4 Distributive Property of Matrices 147
4.2.5 Some Additional Symbology 147
Exercise 4.2 148

4.3 SYSTEMS OF LINEAR EQUATIONS 150

Exercise 4.3 153

4.4 SOLUTIONS TO SYSTEMS OF LINEAR EQUATIONS 155

Exercise 4.4 161

4.5 USE OF MATRICES IN SYSTEMS OF LINEAR EQUATIONS 162

4.5.1 Matrix Solutions to Systems of Linear Equations 164
4.5.2 Additional Forms of Solutions 167
Exercise 4.5 173

4.6 INVERSE MATRIX 174

4.6.1 Identity Matrix 174
4.6.2 The Inverse of a Matrix 174
4.6.3 Use of Inverse Matrix in Solutions of Linear Systems 178
4.6.4 Matrix Equations 180
Exercise 4.6 182

CHAPTER 5 LINEAR INEQUALITIES AND LINEAR PROGRAMMING 184

5.1 LINEAR INEQUALITIES 184

5.1.1 Terminology 184
5.1.2 The Algebra of Inequalities 185
5.1.3 Linear Inequalities 187
Exercise 5.1 189

5.2 GRAPHING INEQUALITIES 191

5.2.1 Graphs for Systems of Inequalities 195
Exercise 5.2 200

5.3 LINEAR PROGRAMMING—PART I 201

5.3.1 Linear Programming—Graphical Solutions 204
Exercise 5.3 207

5.4 LINEAR PROGRAMMING—PART II 209

5.4.1 Graphical Solutions for More Than Two Variables 209
5.4.2 Linear Programming—Algebraic Solutions 214
Exercise 5.4 216

CHAPTER 6 COUNTING PATTERNS 219

6.0 INTRODUCTION 219

6.1 FUNDAMENTAL COUNTING PRINCIPLES 220

6.1.1 Fundamental Principle of Counting 221
Exercise 6.1 224

6.2 PERMUTATIONS 227

6.2.1 Factorials 228
6.2.2 Applications of Permutations to Counting 229
6.2.3 Circular Permutations 231
Exercise 6.2 232

6.3 COMBINATIONS 234
6.3.1 A Simple Problem in Sampling 236
6.3.2 Sampling with Replacement 238
6.3.3 Selection of a Counting Pattern 239
Exercise 6.3 239

6.4 COMPLEX COUNTING PATTERNS 241
Exercise 6.4 245

6.5 PARTITIONS 247
6.5.1 Ordered Partitions 247
6.5.2 Unordered Partitions 249
6.5.3 Ordered Listings—Agreements and Disagreements 251
Exercise 6.5 257

CHAPTER 7 PROBABILITY 259
7.0 INTRODUCTION 259
7.1 SIMPLE PROBABILITY 260
7.1.1 Terms and Definitions 260
7.1.2 Basic Properties of Probability 263
7.1.3 Equally Probable Outcomes 264
7.1.4 The Urn Model 266
Exercise 7.1 267

7.2 COMPOUND EVENTS 268
Exercise 7.2 273

7.3 CONDITIONAL PROBABILITY 275
Exercise 7.3 283

7.4 INDEPENDENT EVENTS AND UNION OF EVENTS 286
7.4.1 Probability of the Union of Events 289
Exercise 7.4 293

7.5 REPEATED TRIALS 296
7.5.1 Bernoulli Trials (Binomial Probabilities) 301
Exercise 7.5 304

7.6 STOCHASTIC PROCESSES; BAYES' THEOREM 307

Exercise 7.6 316

7.7 RANDOM VARIABLES AND EXPECTED VALUE 319

7.7.1 Random Variables 319
7.7.2 Expected Value 320
7.7.3 Use of Expected Value in Decision Making 324
Exercise 7.7 326

7.8 PROBABILITIES AND CERTAINTIES; ODDS 329

7.8.1 Effect of Knowledge on Probability 329
7.8.2 Odds 334
Exercise 7.8 337

CHAPTER 8 MARKOV CHAINS 339

8.1 SIMPLE MARKOV PROCESSES 339

Exercise 8.1 343

8.2 PROBABILITY VECTORS 345

Exercise 8.2 352

8.3 ABSORBING MARKOV CHAINS 353

Exercise 8.3 359

8.4 RANDOM WALK 361

8.4.1 Gambler's Ruin 365
8.4.2 Expected Duration of Walk 367
Exercise 8.4 368

8.5 MARKOV PROCESSES IN GENETICS 369

Exercise 8.5 373

ANSWERS TO SELECTED EXERCISES 375

INDEX 399

foreword to the student

The mathematics in this text is probably quite different from that which you have studied previously. In your early work in mathematics you became familiar with numbers and the operations and relationships used with them. The first courses in arithmetic and algebra are directed toward this end. It is assumed that before you start this text you have developed a reasonable facility in manipulating mathematical symbols and expressions, and in using these manipulations in "solving" simple algebraic equations.

In this text the emphasis is on the application of mathematics to certain types of situations frequently encountered in business analysis and in some phases of the behavioral sciences. Our attention is focused on those mathematical concepts and ideas that are particularly useful in these applications. In our work we shall use two aspects of mathematics that may be new to you: mathematics as a **language,** and mathematics as a **method** for **organizing our thoughts.**

Language. Mathematics is, in part, a **language.** In ordinary language **words,** made up of sequences of letters, serve as **abstract symbols** to represent ideas and concepts. Meanings are conveyed by the letters chosen and the order in which they are arranged. In mathematics there is a greater variety of basic symbols—letters are used along with numerals and other symbols. They are arranged above, below, to one side or the other, as well as in horizontal sequence, to convey the meaning desired.

It is essential that you learn these symbols and their meanings. It is no more possible to use mathematics without knowing mathematical symbols than it is possible to read and write without knowing words and their meanings. In this text, however, we shall express and discuss each mathematical idea in words before introducing the mathematical symbol that represents it.

Organization. Much of the utility of mathematics stems from its organization. Mathematical systems are devised so that ideas can be applied to a wide variety of situations. In fact, mathematical ideas and concepts are formed by abstracting the inherent structure of these situations. Mathematical development consists of exploring the nature and properties of the structure and making generalizations about them.

As a simple illustration, consider the process of subtraction as an adjunct to **counting.** If we have five apples and remove two of them, we have three left. We can determine the number remaining either by counting them, or by using the mathematical operation of subtraction. Two subtracted from five leaves three. The same process applies to removing two from five of anything. The process of subtraction is generalized in mathematics, so that we can compute the number remaining regardless of the quantity we start with and the quantity removed. The development of the process of subtraction is a part of the development of the **structure** of our number system.

In this text we shall study the basic structure of sets, vectors, and matrices. We shall also take a brief look at functions and relations. These concepts are particularly useful in business analysis and the behavioral sciences. They are used to form **mathematical models** for situations encountered in these fields of activity.

A **mathematical model** is a set of one or more mathematical expressions that state the relationships existing among the components in a situation. As a simple example, suppose you are going to have some posters printed for a club. The printer tells you it will cost \$5 to set the type for your poster, and he will print posters for 25¢ each, including the materials used. If you need 10 posters, you compute the cost as $\$5 + 10 \cdot (25¢)$ or \$7.50. When you report back to the club, however, someone suggests that you get 25 posters. The treasurer says that \$12 has been budgeted for posters, and wants to know, for instance, how many posters you can get for that amount. To answer all these questions a **mathematical model** of the way to find the cost of posters is useful. If n represents the number of posters, the cost, C, is given by

$$\$C = 5 + (.25)n$$

With this statement of the relationship between the number of posters and the cost, you can compute the value of either of these quantities for any value of the other.

This model is a particular case of a general model for situations of this type, whether the item involved is a poster or something else that is produced by first setting up a machine and then making the item. If C represents the cost; S, the "set-up" charge; p, the price for making each item; and n, the number of items,

$$C = S + p \cdot n$$

A mathematical model that takes a relatively simple form, as in the foregoing example, is often called a **formula**. The mathematical models we shall study are more complex, and they apply to more sophisticated situations.

The task of obtaining a working understanding of the mathematical concepts in this text is not an easy one. It requires serious concentration for you to grasp the concepts and to see how they can be applied. The following tips on how to study are based on my experience as a student and with students.

They are intended to help you avoid the pitfalls of the "pseudostudy" that is widely practiced in mathematics courses.

Pseudostudy results from the fact that assignments in most mathematics courses include problems to be solved. The **answers obtained** for the problems are usually the only tangible evidence that you have "done" the assignment. Thus, the temptation may be strong for you to attempt the problems first when starting the assignment. If you can find the answers without studying the text, you can avoid a lot of dull reading. Mathematics texts usually contain many examples. The problems are similar to the examples. You find an example in the text that resembles the assigned problem, follow the same procedures used in the example, and behold! you obtain the answer. The trap is that you have done so without **understanding** what you have done. It is possible therefore for you to "get through" your math courses with satisfactory grades without learning anything about mathematics that can be used outside class.

Following the suggestions listed below will take considerably longer and involve much more work than pseudostudy. The reward is **understanding** and the **ability to reason through situations** that heretofore were completely baffling.

STUDY SUGGESTIONS

First (before looking at the problems), study the text material carefully—this means reading the material and **thinking it through**. It may be necessary to read the material several times to get the idea.

The text material includes many examples. These are **illustrations** of the **concepts being discussed**. (They are not samples of how to get the answers to the problems in the exercises.) The examples should be studied along with the text material as an aid to understanding the concepts involved.

Important: Particular attention should be paid to the **definitions**. These are the basis for the ideas discussed. Many of the problems encountered can be solved merely by applying definitions.

Second, after the text material begins to make sense, try to work the problems in the exercises. These are intended as a self-administered quiz to determine whether the material is understood. If you have trouble with a problem, restudy the sections containing the relevant material. (The answers to about half the problems can be found at the back of the book.) **Do not** use these answers to suggest modifications to what you have done in trying to solve a problem. Do not make any steps in the solution unless you know **why** you are making them. If you can not understand the material, seek an explananation either from your instructor or another text.

These suggestions are not "mathematics made easy." They are the way to "mathematics understood." There are aspects of our lives that inherently involve certain mathematical concepts. In fact, mathematics was developed as a part of the study of these aspects of our lives. To cope with them we must incorporate mathematical methods into our way of thinking. This is far less unpleasant than it may sound. I hope that this text will show you some of the ways that you can use mathematics and even enjoy it.

1

sets

1.1 INTRODUCTION

In organizing a large quantity of material it is almost automatic to start by grouping the things that are alike in one or more ways. We perform such groupings, consciously or subconsciously, regularly in our thoughts. In the latter part of the nineteenth century a German mathematician, Georg Cantor (1845–1918), in connection with some work in higher mathematics, made a study of groupings and collections and their properties. He called the collections **"sets,"** and he developed a formalized theory of sets as a mathematical system, in many ways analogous to the theory of numbers. At first he used sets only as a tool in working with more complex mathematical ideas. The applicability of set concepts to simpler situations became more apparent as the theory developed, and now the basic notions are used in the very first formalized training in mathematics in elementary school.

The development of set theory is a beautiful example of what mathematics is all about: the establishment, through carefully constructed definitions, of a precise language in which ideas can be expressed and communicated clearly; the establishment of operations for "working with" the ideas; and the formulation of a body of facts (usually called theorems), which enable the ideas and concepts to be used and applied in a variety of ways.

Here, we shall study only the basic concepts of set theory to obtain an indication of how these concepts can be applied to a variety of situations encountered in business, psychology, and sociology.

1.1.1
Definitions and Undefinitions

Before we discuss the language of sets, let us examine some of the general aspects of a language. By a language we mean a system of abstract symbols, each of which is used to convey or express an idea. Those using the language agree on the meaning to be assigned to each symbol. Such assignment of meaning is called a definition. If, however, we try to define **every** term and symbol we are to use, where do we start? What symbols do we use to define the first one? Our spoken language makes no attempt to define words with the precision that we need for defining the symbols in mathematics. The words in our everyday language acquire meaning from the way they are used. They are defined primarily by experience. The words are **related formally to each other** in a dictionary, but they are not **defined** in the sense that mathematical terms are defined. Look up a word in a dictionary. Pick one of the synonyms given, and look that up. Pick one of its synonyms, and look that up. Note how few times you can do this before the synonyms given are all words you have already looked up. This is a satisfactory situation in ordinary language, since the main purpose of the dictionary is to systematize and standardize language. In learning the meaning of a word from a dictionary you read the dictionary's definition. If that explanation of the word does not make the meaning clear, you choose one of the synonyms listed, and try that. You continue the process until you find an explanation that relates to your own experience, that is, that you can understand.

In mathematics, however, we seek to **construct** a language. Each new symbol is defined using those symbols **already defined**. This means that we need some symbols to start with. Mathematicians call such symbols **undefined terms**. This is not to say that these terms have no meaning, but, instead, that these are the terms we agree to define from our experience or "intuition." **All the other terms and symbols we use will then be defined precisely**, using the **undefined terms** as a **starting point**. To keep the language precise, of course, it is desirable to have as few undefined terms as possible.

In our work with sets, the term **set** itself is undefined. It is intended to represent a group or collection – for example, the concept our experience associates with a **set** of dishes, a chess **set**, and a **set** of rules. We make one qualification on this concept of set, however. To qualify as a set, a collection, or group, must be characterized so that there is no doubt about whether any "object" is or is not included in the set.

For example, we could specify as a set the books in our college library. There is no doubt about whether or not a given book is in the set – the library catalog is a listing of all the members of the set.

If we were to specify the set as all the **good** books in the library, we could run into trouble. Good by what criterion? Such a grouping does not qualify as a set under our rules, since we cannot determine with certainty whether a given book is "good," and is to be included in the set.

In our discussions we use several more terms that must remain undefined. **Element** and **member** are undefined terms used to refer to one of the objects, ideas, or whatever, that is included in a set. **Belongs to** is another undefined term, as is the term **contains**. We say that an **element belongs to a set**, or a **set contains** an **element**.

1.1.2 Set Specifications

We shall speak of "defining" a set and mean the **specifications for a particular set**. (This is a slight deviation from "defining" a term. Although the use of the same word for two different meanings occurs in language, we try to avoid it in mathematics.) **A set is defined in terms of the members that it contains.** The membership list for a set is specified in one of two ways:

a. A **listing** of the elements of the set, called the "roster method"; or

b. A **rule**, or **description** of the elements provided that there can be no ambiguity in the interpretation of the rule.

Example 1.1.1

a. The set of letters, a, e, i, o, u.

b. The set of men who have served as President of the United States.

c. The set of numbers greater than 3.

The set in **(a)** is specified by listing the elements. There can be no question about what is and what is not included in the set. Note that if the set had been defined by the rule, "The vowels in the English alphabet," there would be a question of whether or not "*y*" is to be included. In our uses of sets, we shall avoid definitions subject to such questions.

The set in **(b)** is specified by a rule. As new Presidents are elected, the actual membership of the set will change. But at any given time there is no question of who the members of the set are. It is possible to define this set also by listing the names of the members.

The set of **(c)** is defined by a rule, and there is no doubt about whether or not a given number is an element of the set. In this case, however, it is not possible to list all the elements because of the nature of our number system. There is an unlimited quantity of numbers greater than three.

The sets in Example 1.1.1, **(a)** and **(b)** are called **finite** sets, because each has a **finite** number of elements. The set of Example 1.1.1**(c)** is called an **infinite** set. Most of our attention will be directed toward finite sets, although occasionally we shall use an infinite set.

The following conventions apply to the specifications of sets by listing the members:

Repeated Elements: The concept of set membership involves only the fact that an item satisfies the specifications that define the set. Thus, in listing the members of a set, listing the same item twice does not mean that there are two of these items in the set. The second listing has no significance and does not increase the membership.

Identical Elements: If we specify a set of several ostensibly identical objects, each object is considered to have a separate identity. A set of six plates, for example, is not complete if one is missing. Although all of the plates look alike, they do have separate identities; and it is assumed that some means exists for establishing their identities. On the other hand, if we were to list the kinds of utensils to be found in a set of dishes, "plate" would be listed once as a **type** of dish.

Order of Elements: The order in which the elements of a set are listed is of no significance. The set $\{a,e,i,o,u\}$ is the same set as $\{e,i,u,o,a\}$.

1.1.3 Symbology

As is customary in mathematics the ideas and concepts we shall work with are represented by symbols of various kinds. There is a fair degree of uniformity in the symbology used for sets, and we shall use those symbols most frequently found in other books and articles.

Braces, { }, are used to designate the establishment or existence of a set.

When a set is defined by listing its members, the list is enclosed by braces to indicate the formation of a set. The set listed in Example 1.1.1**a** is designated $\{a,e,i,o,u\}$.

When a set is defined by a rule, braces enclose the statement of the rule, to indicate a set is designated. Example 1.1.1**b** becomes {The men who have served as the President of the United States}.

A notation, known as **set-builder notation**, is frequently used, particularly when a set of numbers is designated. In set-builder notation the set of Example 1.1.1**c** is designated as

$$\{x \mid x \text{ is a number greater than } 3\}$$

If it is clear from the context that x is a number, this might be shortened to

$$\{x \mid x > 3\}$$

(The symbol ">" is from number algebra and means "greater than.")

The first designation is translated into English as, "The set of all x such that x is a number greater than 3"; and the second is, "The set of all x such that x is greater than 3."

The vertical line between the x's is translated "such that." The first x specifies the symbol that will be used to represent a "typical" element of the set in an accompanying discussion. The statement following the vertical line is the specification by which x qualifies as a member of the set. It can also be considered as a property that is represented by the x.

Capital letters are used to represent sets in discussions about them. The statement, $V = \{a,e,i,o,u\}$, means that in discussing this particular set the letter, V, is going to be used to represent it. The V plays the role of a **name** for the set. Such representations are used regularly in mathematics to simplify the wording in discussions.

The symbol "$\in$" is used to mean "is an element of." In the example above we could say, "$a \in V$," to mean "a is an element of the set represented by V."

This is usually shortened to "a is an element of the set V." The symbol "$\notin$" means "is not an element of." Thus, "$b \notin V$" is "b is not an element of V."

Example 1.1.2

Let us reprise the use of set-builder notation to designate the set of numbers greater than 10, and call this set, D.

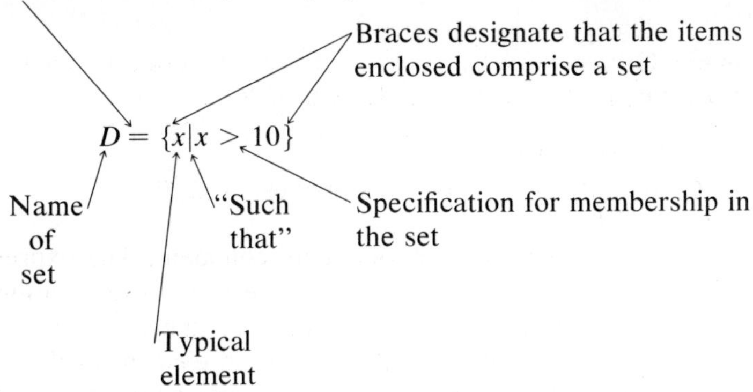

We can say $12 \in D$, "Twelve is an element of D"
"Twelve belongs to the set D"
"Twelve is contained in the set D"
"D contains the number 12."

or $3 \notin D$ "Three is not an element of D"
"Three does not belong to the set D"
"Three is not contained in D"
"D does not contain three."

1.1.4 Set Relationships

With a concept as general as that of a set, there are certain to be many interrelationships among collections that qualify as sets. Part of the great utility of the algebra of sets is that they can all be expressed in terms of a few simple relationships and operations. These are defined and discussed in this and the next section. Our discussions in later chapters use these relationships and operations, and are presented in the language and format of set algebra.

Subsets

When working with sets it is sometimes desirable to use the elements contained in one set in forming other sets. This process leads to the concept of a **subset**. A **subset** is a set whose elements are **all** elements also of another set (sometimes called the **parent** set). Note that the term "subset" defines a relationship between two sets. If we have two sets, A and B, and **every** element of A is also an element of B, then A is called a **subset** of B.

The symbol "$\subseteq$" is used to denote the subset relationship. "$A \subseteq B$" is a **mathematical sentence** whose translation into English is "A is a subset of B."

(Although we shall not use it in this text, the subset symbol is sometimes reversed and used to mean "contains." "$A \supseteq B$" means that the set A contains all of the elements of set B. That is, set B is a subset of set A.)

All of the ramifications of the subset relationship are summed up concisely in the following definition:

Definition 1.1.1

Subset: Given two sets, A and B. A is a **subset** of B if, and only if, every element of A is also an element of B.

In mathematical symbols:

$A \subseteq B$ if, and only if, $x \in A$ implies $x \in B$.

The form of this definition merits some comment. The expression* "if, and only if" is used in mathematics to denote a two-way relationship. (We shall study this relationship, called a biconditional, in greater detail in the next chapter.) If each element of the set A is also an element of the set B, then A is called a subset of B. Conversely, if A is specified to be a subset of B, it means that every element of A is an element of B. Both of these statements are condensed, in the terse manner of mathematical expression, into the single statement in which "if, and only if" couples the two parts. Notice that many, but by no means all, definitions in mathematics use "if, and only if" as a connective.

Note also how the statement of the definition in mathematical symbols is both precise and succinct. This is characteristic of mathematical statements. They must be read and pondered to obtain the full meaning.

Let us summarize a few of the implications of the definition of a subset.

If we start with the declaration, or knowledge, that $A \subseteq B$, this implies that every element of A is also an element of B. On the other hand, if we can somehow determine that every element of A is also contained in B, we have established that $A \subseteq B$.

The definition of subset includes the possibility that a subset could include all the elements of the other set, that is, the set itself. Thus, every set is a subset of itself. In the language of mathematics this is stated, "For all sets A, $A \subseteq A$."†

Equality of Sets

The situation will arise in which two sets specified in different ways actually turn out to be the same set. In accordance with the fundamental precepts in mathematics one designation for the set could be substituted for the other

* Many of the expressions and symbols used in mathematics are confusing to students, especially those whose principal interests lie in nontechnical areas. We shall give explanations of these symbols and expressions as they are encountered in subsequent sections. When the symbol or expression is only incidental to the discussion, the explanation is placed in a footnote.

† In this case the expression "For all sets A" is a mathematical way of saying "for **any** set." The A is the symbol used in this case as the representative for **any** set.

without affecting the truth of whatever statements are made. As we shall see, this is often a convenient way of simplifying statements to make them more easily understood. We, therefore. define what we mean by two equal sets:

Definition 1.1.2

Equal Sets: Two sets are **equal** if, and only if, they contain exactly the same elements.

As discussed in the preceding section this definition is two statements in one. The "if, and only if" phrase makes the definition mean, "If two sets contain exactly the same elements, then they are equal," **and also**, "If two sets are known to be equal, they must contain exactly the same elements." The significance of this dual statement will become apparent in later discussions.

An implication of the definition leads to an alternate definition for set equality that is often useful. Note that if two sets, A and B, contain the same elements, every element of A is an element of B. As we have seen, this means $A \subseteq B$. Also, since every element of B is a member of A, $B \subseteq A$. We have as an alternate definition:

Definition (Alternate) 1.1.2A

Equal Sets: Two sets, A and B, are **equal** if, and only if, A is a subset of B **and** B is a subset of A.

In symbols,

$A = B$ iff* $A \subseteq B$ **and** $B \subseteq A$.

Proper Subset

A **proper subset** is a subset that does not contain all of the elements of the parent set. In more precise terms:

Definition 1.1.3

Proper Subset: Given two sets, A and B. A is a **proper subset** of B if, and only if, A is a subset of B and there is at least one element in B that is not in A.

Another way of expressing the concept of proper subset is to observe that if A is **completely contained** in B, but B is **not** completely contained in A, then A is a **proper** subset of B.

The symbol "$\subset$" is used to denote "is a proper subset of." $A \subset B$ is read "A is a proper subset of B." Note that this symbol is the same as the symbol for subset "$\subseteq$" with the bar omitted. The lower bar is intended to imply the **possibility** that the two sets are equal. We could translate "$A \subseteq B$" as, "A is a proper subset of, or equal to B." This is analogous to the symbol "$\leq$" used in number algebra to mean, "less than, or equal to."

* "iff" is a frequently used abbreviation for "if, and only if."

Using our previous definition for equal sets, we note that in the proper subset relationship, the two sets cannot be equal. We have, therefore, an alternate definition for proper subset:

Definition (Alternate) 1.1.3A

Proper Subset: Given two sets, A and B, A is a **proper subset of** B if, and only if, A is a **subset** of B, **and** A is **not equal** to B.

In symbols,

$A \subset B$ iff, $A \subseteq B$ **and** $A \neq B$.

Observe that the symbol $\subset$ contains more information than the symbol $\subseteq$. When we know that all the elements of A are contained also in B, we can say $A \subseteq B$. If we know **in addition** that not all of the elements of B are contained in A, we can say $A \subset B$. Conversely, the statement $A \subset B$ tells us more about the sets A and B than does the statement $A \subseteq B$.

EXERCISE 1.1

1. Which of the following can be used to define a set?
   a. The counting numbers.
   b. The young people holding elective office.
   c. The major oil companies.
   d. The people eligible for social security benefits.
   e. The leading money winners in professional golf last year.
   f. The animals in the San Diego Zoo.
   g. The products marketed by General Motors Corporation.
   h. The brand names of quality soup products.
   i. The symptoms of malaria.

2. Given that $A = \{x | x \text{ is a counting number and } x > 10\}$. Which of the following are true and which false?
   a. $11 \in A$ b. $5 \in A$ c. $x \in A$
   d. $\{15\} \in A$ e. $15 \notin A$ f. $12\frac{1}{2} \in A$

3. Given the sets
   $A = \{x | x \text{ is less than } 10\}$
   $B = \{x | x \text{ is greater than } 3\}$
   $C = \{x | x \text{ is a letter of the alphabet}\}$
   $D = \{a, b, c, d, e, 1, 2, 3, 4, 5\}$
   Supply elements in the spaces to make the following statements true.
   ____ $\in A$.
   ____ $\in D$.
   ____ is an element of A, but not of D.
   ____ is an element of both C and D.
   ____ is an element of both A and B.
   ____ is an element of B, but not of A.

____ is an element of B, but not of C.
____ is an element of both A and B, but not of D.

4. Give two examples of sets by listing the members.

5. Give two examples of sets using a statement to define the set.

6. Give two examples of sets using set-builder notation.

7. Define sets that could have these members:
  a. Set, element, member, contains, belongs to.
  b. George, John, Thomas, James, Andrew, William, Millard.
  c. 3, 5, 7, 9.
  d. Lions, Bears, Rams, Eagles.

8. Indicate which of the following are finite and which are infinite sets.
  a. All fractions with 5 as the denominator.
  b. All the words ever written.
  c. All the counting numbers divisible by 7.
  d. The people who voted in the 1968 national election.

9. Express the sets listed below in set-builder notation. Make sure that your rule for membership includes all of the elements listed and only those elements.
  a. $\{a, b, c, d, e, f\}$
  b. $\{3, 4, 5, 6, 7, 8\}$
  c. $\{3, 6, 9, 12, 15\}$
  d. $\{b, c, d, f, g, h, j, k, l, m\}$

10. Choose the most appropriate symbol, $\subset$, $\subseteq$, $\in$, or $=$, to complete the following:
  a. x . . . {letters of the alphabet}
  b. $\{x\}$ . . . {letters of the alphabet}
  c. {The employees of the Postal Service who earn $10,000 per year or more} . . . {employees of the Postal Service}
  d. $\{c, d\} \ldots \{\{a, b\}, \{b, c\}, \{c, d\}\}$
  e. {Students in this class who correctly answer this question} . . . {students in this class}

11. Determine whether or not the following sets are equal; and if not, why not.

a.	$\{m, e, a, l\}$	$\{l, a, m, e\}$
b.	$\{d, e, w, a, r\}$	$\{r, e, w, a, r, d\}$
c.	$\{n, o, m, a, d\}$	$\{d, o, m, a, i, n\}$
d.	$\{a, b, c, d, e\}$	{The first five letters of the alphabet}
e.	$\{3, 6, 9, 12, 15\}$	{The counting numbers divisible by 3}
f.	{The counting numbers less than 10}	{The one-digit counting numbers}

12. Formulate the following statements into symbols:
  a. x is an element of the set A.
  b. B is not a subset of A.
  c. a is not contained in set B.
  d. C is contained in B, but it is not equal to B.
  e. All of the elements of set B are in set A.

13. Formulate the following symbol statements into words.
    a. $B \subseteq C$ b. $a \notin A$
    c. $A \subset C$ d. $B \not\subset C$
    e. $A \in B$

14. Let S be the set of students enrolled in this college. List at least five proper subsets of S that can be formed.

15. A survey was made of the people attending a baseball game. They were asked to answer yes or no to the following questions.
    a. Are you currently unemployed?
    b. Did your income last year exceed $5000?
    c. Do you own an automobile?

    Let S represent the set of people answering the questions. List at least five different subsets of S that can be formed on the basis of the responses to the questions.

1.2 SET OPERATIONS

In the previous section we dealt with the two primary set **relationships:** the subset and set equality. Each of these relationships involves **two** sets, and each makes a statement about the elements of one set with respect to the elements of another. If we say $A \subseteq B$, we are stating that each element of set A is also an element of set B. Note that the statement, $A \subseteq B$, **may, or may not, be true,** depending on what the elements of set A and B are.

We turn now to the **operations** that are performed with sets. The operations are ways in which two sets can be **combined to make a new set.** The particular operation to be performed is specified by a symbol. It specifies additional conditions to be met by the elements of the component sets to qualify them for membership in a **compound set. Compound set** is the term applied to a set formed as the result of one or more set operations.

1.2.1 Union

The **union** of two sets is a set formed by combining all of the elements of the two sets into a single set. Thus, an element becomes a member of the union of two sets by being a member of either one, or both, of the two sets.

The symbol used to designate the union of two sets is "$\cup$." "$A \cup B$" means the union of sets A and B.

Definition 1.2.1

Union: The **union** of two sets A and B, $A \cup B$, is the set whose elements are either elements of A, **or** elements of B, or both.

In symbols:

Set-membership notation:

$x \in (A \cup B)$ iff $x \in A$ **or** $x \in B$ or both

Set-builder notation:

$A \cup B = \{x \mid x \in A$ **or** $x \in B$ or both$\}$

Example 1.2.1

Consider the sets $A = \{a, b, c, d\}$ and $B = \{c, d, e, f\}$. The union of A and B, $A \cup B = \{a, b, c, d, e, f\}$.

To illustrate the comments at the start of this section we note that performing the operation of union on the two sets in the example produces another **set**. The **name** of the set produced is $A \cup B$. There is no relationship between A and B involved or implied; a union can be formed of any two sets by taking all of the elements of one of the sets and combining them with the elements of the other to form a new set.

1.2.2 Intersection

The **intersection** of two sets is the set of elements common to both sets.

The symbol used to represent the intersection is "$\cap$." "$A \cap B$" means the intersection of the sets A and B.

Definition 1.2.2

Intersection: The **intersection** of two sets A and B, $A \cap B$, is the set formed from those elements of set A that are also elements of B.

In symbols:

Set-membership notation:

$x \in A \cap B$ iff $x \in A$ **and** $x \in B$

Set-builder notation:

$A \cap B = \{x \mid x \in A$ **and** $x \in B\}$

Example 1.2.2

Let $A = \{a, b, c, d\}$, and $B = \{c, d, e, f\}$. The intersection of A and B is the set $\{c, d\}$. If we write this in symbols, we have

$$A \cap B = \{a, b, c, d\} \cap \{c, d, e, f\} = \{c, d\}$$

As with the union, the intersection of two sets, A and B, $A \cap B$, names a new set containing those elements of the one set that also are elements of the other. The operation of intersection can be performed on **any** two sets, and the result is another **set**.

We note the possibility that the two sets whose intersection is formed have no elements in common; that is, none of the elements of either set is contained in the other. In such a case we say that the intersection of the two sets is the **empty set**; that is, it contains no elements. The "empty set," more frequently called the "null set," is a special set defined specifically to be used in situations of this kind.

1.2.3 Null Set

The **null set** is a set that contains no elements. It performs a function in set algebra analogous to that of "0" in arithmetic. "0" has no magnitude, but **it is a number**. The null set contains no elements, but **it is a set**. It plays an important role in the formulation of definitions and in performing operations on sets. It makes many mathematical expressions much simpler and more convenient to use.

As an illustration of the null set—the set with no elements—consider the following:

The set of all numbers less than 10 and larger than 100.
The set of all living eighteenth century poets.
The set of all dairy farms on Manhattan Island.

We observe in the above examples that a null set results when two or more incompatible requirements are made for membership in a set.

$\emptyset$ is the symbol usually used to designate the null set; but empty braces, { }, are also used.

For convenience, and to keep the structure of set algebra neater, an arbitrary agreement is made that the **null set** is a **subset** of **every set, including itself.**

1.2.4 Universal Set

In defining sets, set relationship, and set operations we have thus far directed our attention to those elements that are contained **in** the sets. It is often useful to agree to restrict the things, or entities, relevant to a situation and form them into a set. Such a set is called a **universal set**, and all of the sets in discussions of the situation are subsets of the universal set. In this way we can form sets of those elements **not** in a particular set without being hampered by irrelevant material.

For example, in making a study of corporations, the universal set is the set of all corporations. There is no need to include public institutions, numbers, books and other such irrelevant items.

If a sociologist is studying the hierarchy in labor unions, a universe that includes corporate executives would contain many elements not involved in the study. These would be excluded by an appropriate selection of a universal set.

In general, the set that defines the "universe" for a discussion should be restricted as much as possible. We establish the concept of **universal set** by the following definition:

Definition 1.2.3

> **Universal Set:** The set that contains **all** of the elements relevant to a situation in which sets are used, but only those elements, is called the **universal set** for that situation.

The universal set is usually represented by the letter U.

1.2.5 Complement of a Set

Now that we have restricted our universe to manageable size, we can consider the elements that do not belong to a set. If we consider a set in a given situation, those elements of the appropriate universal set that are not elements of this set form a set called the **complement** of the set under consideration.

Definition 1.2.4

> **Set Complement:** For a set A, the set of all the elements in an appropriate universal set that are **not** elements of A is called the **complement** of set A.

The complement of a set is denoted by "′." A' is the complement of A.*

If A is a set, then **any element** of the appropriate universal set must be a member **either of** A **or of** A'. The following relationships will always hold:

If $a \in A$, then $a \notin A'$. If $a \notin A$, then $a \in A'$
If $a \in A'$, then $a \notin A$. If $a \notin A'$, then $a \in A$

1.2.6 Relative Complement

In Section 1.2.5 we defined the complement of a set A as the set of elements in the universal set that are not in A. This is the complement of a set relative to its universal set. The **relative complement** of a set A with respect to a set B (not the universal set) is the set of elements that are in B, but not in A.

The relative complement of A with respect to B is represented by the symbols $B - A$.

The formal statement of the definition is:

Definition 1.2.5

> **Relative Complement:** The relative complement of a set A with respect to a set B, $B - A$, is the set of all elements of B that are not elements of A.

*The tilde, "~," is also used to denote a set complement. $\sim A$, or $\tilde{A}$, each specifies the complement of the set A.

Set-membership notation:

$x \in B - A$ iff $x \in B$ **and** $x \notin A$

Set-builder notation:

$B - A = \{x | x \in B$ **and** $x \notin A\}$

Example 1.2.3

Let $A = \{a, b, c, d\}$ and $B = \{c, d, e, f\}$. The relative complement of A with respect to B is $B - A = \{e, f\}$.

We note from the definition that the elements of $B - A$ are not elements of A. From the definition of A's complement, A', the elements of $B - A$ must be elements of A'. Using the definition of intersection we see that the relationship $B - A = B \cap A'$ is true for all sets A and B. That is, the set $B \cap A'$ can be used as an alternative designation of the relative complement of A with respect to B.

1.2.7 Disjoint Sets

Sets that have no common elements are of special interest in many applications. This relationship is defined as follows:

Definition 1.2.6

Disjoint Sets: Two nonempty* sets, A and B, are said to be **disjoint** if, and only if, their interesection is the null set.

In symbols:

A and B are disjoint iff $A \cap B = \emptyset$ **and** A and B are both nonempty.

Example 1.2.4

If $A = \{a, b, c, d\}$ and $B = \{e, f, g, h\}$. A and B are disjoint because each set has elements and none of the elements of A is also an element of B; that is, $A \cap B = \emptyset$.

Note that any nonempty set and its complement are disjoint. We consider now an example of the use of set symbology as a convenience in working with information.

Example 1.2.5

Suppose we are making a study of the job attitudes of the employees of a large company to see if there are any data about employees that correlate with degree of job satisfaction.

As a partial listing of the data on the employees, we have sex,

*The term **nonempty** is frequently used in set algebra to specify that a set has at least one element; that is, a set A is nonempty iff $A \neq \emptyset$.

amount of education, and department of the company in which the employee works. Each of the classifications of data determines a set of employees, which is a subset of the universal set of all the employees of the company. We have, for instance, the set of women, the set of men, the set of college graduates, and the set of employees in the production department. We list these sets and assign each a symbol to represent it.

Symbol	**Set Represented**
M	Men
W	Women
G	College graduate
C	People with high school plus some college
H	High school graduate, but no college
L	Less than high school education
P	Works in production
R	Works in research
S	Works in sales
A	Works in accounting

In this study we are concerned with people with various combinations of these characteristics, such as the men in production, the women with college degrees, and the people in the accounting department who have less than a high school education. They are represented by various combinations of the sets listed. The men in production are the members of the set of men who are also members of the set of people who work in the production department; that is, $M \cap P$. We list a few of these and the symbols which represent them.

Women in sales	$W \cap S$
The college graduates in accounting	$A \cap G$
The women with a high school diploma plus some college	$W \cap C$

Thus, the symbols provide a convenient shorthand for expressing cumbersome descriptions of the various groupings of employees. As we shall see later, set algebra provides a means for additional combinations and simplifications of the sets, and these provide still further information.

We have also the means of stating information about these groups of people. If, for example, the company does not employ any women with less than a high school education, the set of women with less than a high school education is the empty set; that is, $W \cap L = \emptyset$.

Other information that might be obtained includes:

"Everyone with less than a high school education works in production." That is, every member of the set of people with less than a high school education is also a member of the set of people working in production; this is translated into symbols as $L \subseteq P$.

"All of the women in research have college degrees." This statement is translated into set symbols as $W \cap R \subseteq G$.

"All of the college graduates in sales are men;" becomes $S \cap G \subseteq M$.

When the symbols are understood (when the set language is learned), they provide a convenient and useful way to express ideas and facts. The algebra for working with these symbols provides a way to obtain new insights into the information.

1.2.8 Summary

The three basic set operations are **set union**, **set intersection**, and **set complement**.

The set operations are ways to **form** sets. The operation, together with the names of the sets involved, is the **name** of the set formed. $A \cap B$ is a **set named** the **intersection of** A **and** B. It **exists** whether or not A and B have elements in common.

We use a slash with the element symbol, $\in$, the subset symbols, $\subset$ and $\subseteq$, and the equal sign to denote "not." For example, $A \not\subseteq B$ means A is not a subset of B. However, we do not use the slash with the intersection or union symbols. For example, the fact that A and B have no elements in common is symbolized $A \cap B = \emptyset$, **not** $A \not\cap B$.

The following **relationships** involving set unions and intersections are **always true**:

The union of any set with its complement is the universal set; that is, $A \cup A' = U$ for all A.

The union of any set with the null set is the set itself; that is, $A \cup \emptyset = A$, for all A.

The union of a set with itself is the set itself; that is, $A \cup A = A$, for all A.

The intersection of a set with the universal set is the set itself; that is, $A \cap U = A$, for all A.

The intersection of a set with the null set is the null set; that is, $A \cap \emptyset = \emptyset$, for all A.

The intersection of a set with itself is the set itself; that is, $A \cap A = A$, for all A.

The intersection of a set with its complement is the null set; that is, $A \cap A' = \emptyset$, for all A.

EXERCISE 1.2

1. Let $A = \{a,b,c,d,e\}$, $B = \{c,d,e,f,g\}$, $C = \{f,g,h,i,j\}$, and $U =$ {first 12 letters of the alphabet}. Define the following sets by listing their members:

   a. $A \cup B$ b. $A \cap B$ c. B'
   d. $B \cup C$ e. $B \cap C$ f. $A \cap B'$

g. $A \cup C$ h. $A \cap C$ i. $A \cup C'$
j. $A - B$ k. $C - A$ l. $A' \cap B'$

2. Indicate whether each of the following statements is always true (AT), sometimes true (ST), or never true (NT).

a. If $a \in A$, $a \in A \cup B$.
b. If $a \in A$, $a \in A \cap B$.
c. If $a \in A'$, $a \in A \cup B$.
d. If $a \in B'$, $a \in A \cap B$.
e. If $a \in B - A$, $a \in A'$.
f. If $a \in A \cup B$, $a \in A'$.
g. If $a \in (A \cup B)'$, $a \in A'$.
h. If $a \in (A \cap B)$, $a \in B$.
i. If $a \in (A \cap B)'$, $a \in A$.
j. If $A \subset B'$, $A \cap B = \emptyset$.
k. If $A \subseteq B$, $A \cup B = A$.
l. If $A \subset B$, $A \cap B = B$.

3. Use the symbols $\subset$, $\subseteq$, or $=$, to make the following statements true, and as informative as possible.

a. If $A \subset B'$, then $B \quad A'$.
b. If $A \subseteq B$, then $B' \quad A'$.
c. If $A \subset B$, then $B' \quad A'$.
d. If $A' \subseteq B$, then $B' \quad A$.
e. If $A \not\subset U$, then $\emptyset \quad A'$.

4. Designate each of the following statements as always true (AT), sometimes true (ST), or never true (NT). Assume the sets A, B, and C, are subsets of the same universal set.

a. If $A \not\subset B$, then $A = B$.
b. If $A \subset U$, then $A' = \emptyset$.
c. If $A \not\subset B'$, then $B \subseteq A'$.
d. If $A \subseteq B$ and $B \subseteq C$, then $A \subseteq C$.
e. If $A \subset U$ and $B \subset U$, then $A' = B'$.

5. Sets A and B are subsets of a universal set U. Arrange the following sets in a sequence so that each set in the sequence is a subset of the next one:

$A \cap B$, $\emptyset$, U, $A \cup B$, B

6. Let U be the students in a math class. Define two subsets of U, A and B, that could be formed on the basis of the results of the first examination, subject to the following conditions: (Part *a* and part *b* are different sets of conditions.)

a. A and B are disjoint, (and nonempty), and $A \cup B = U$.
b. A and B are disjoint, and (nonempty), and $A \cup B \subset U$.

Be sure you consider **all** the requirements carefully in formulating the sets.

7. A survey team was given the job of gathering information on the educational backgrounds of a group of people and how this affected their earning ability. They made up the following two sets of categories into which each of the people was to be placed.

For earnings they had

A. Less than \$10,000 per year
B. \$10,000 to \$20,000 per year
C. More than \$20,000 per year

For education they had

D. High school only
E. Some college, not graduate
F. College graduate

Describe in words the following sets:

a. $A \cap F$ b. $B \cap D$ c. $A \cup B$
d. $C \cup D$ e. $C \cap D$

Translate into symbols the following:

f. The people with some college, but no degree, who earn less than \$10,000 per year.
g. The people earning between \$10,000 and \$20,000 per year who graduated from college.
h. No college graduate earns less than \$10,000 per year.
i. The people who earn \$20,000 per year or less.
j. All of the high school graduates with some college, but no degree, earn \$10,000 per year or more.

8. All the applicants for clerical positions in a company are given a placement examination on which they can score between zero and 100 points. Let A represent the set of those who score less than 25; B, those who score between 25 and 50; C, those who score between 50 and 75; and D, those who score more than 75. Let H represent the set of applicants who are hired. There are three job classifications that a new clerical employee can be assigned: file clerk, represented by F; typist, represented by T; and stenographer, represented by S.

Translate the following into symbols:

a. The applicants scoring between 25 and 50 who are hired.
b. The applicants scoring between 50 and 75 who become typists.
c. No one scoring less than 25 points on the placement examination is hired.
d. All of those scoring more than 75 points on the placement exam are hired.
e. Some of those scoring between 50 and 75 on the placement exam become typists.

Translate into words the following symbols:

f. $B \cap F$
g. $C \cup D$
h. $B \cap H \neq \emptyset$
i. $B \cap T = \emptyset$
j. $B \cup C \subseteq F$

9. Make a list of *all* the subsets of the following sets:

a. $\{a, b\}$
b. $\{a, b, c\}$
c. $\{a, b, c, d\}$

10. Consider the list of Exercise 9*c*:
    a. How many subsets will there be if another element is added to the set?
    b. Analyze the sequence of the number of subsets for 9*a*, 9*b*, 9*c*, and 10*a*. Consider the manner in which the list of subsets is changed as more elements are added. Formulate an expression for the number of subsets of a set with *n* elements.

1.3 ADDITIONAL SET OPERATIONS: VENN DIAGRAMS

In set algebra as in the algebra of numbers, the operations performed on sets are defined for either one or two sets. An example of an operation defined for one set is the complement. (In number algebra operations involving only one number include raising to a power and extracting a root.) The operations we have defined for two sets are the union, the intersection, and the relative complement. (In number algebra we have addition, subtraction, multiplication, and division.) Since we frequently deal with more than two sets, we need rules to govern set operations in these circumstances.

All of the operations we have defined thus far have the property of **closure**. That is, the **result** of the **operation** in each case is **another** set. The **union** of two sets is the **set** that contains all the elements in either of the sets. The **intersection** of two sets is the **set** that contains the elements common to both. The **complement** of a set is the **set** of elements not in the set. We have also defined a special set, the null set, that contains no elements. Thus, we can **always** treat the **result** of a **set operation** as another **set** and **perform additional operations with it**. Recall from the previous section that a set formed by the combination of two or more sets is called a **compound set**.

When we wish to form a compound set from more than two sets, we are faced with the question of which **two** sets to start with. The answer is given by the meaning that we intend to convey by the set symbols; that is, what elements are to be included in the compound sets. We shall use grouping symbols, as they are used in number algebra, to specify the order in which the operations are to be performed. In some cases it does not matter in which order the operations are performed. In other cases the results will be quite different, and we must follow the grouping instructions carefully.

For example, if we are forming the union of three sets, A, B, and C, we can form first the union of A and B and combine the resulting set with C. The symbolic representation for this process is $(A \cup B) \cup C$. Another possibility it to form the union of B and C and then combine A with this set. The symbolic representation in this case is $A \cup (B \cup C)$. As we expect from the definition of set union, the resulting set in each case has the same elements. That is,

$$(A \cup B) \cup C = A \cup (B \cup C) \tag{1.3.1}$$

The relationship expressed in Equation 1.3.1 is called the **associative property for set union**. It is directly analogous to the **associative property for addition**

for numbers. Since the order of performing successive unions is of no consequence, we usually omit the grouping symbols and refer to the union of the three sets, A, B, and C, as $A \cup B \cup C$.

If, however, we wish to use both the intersection and the union combining the three sets into a single compound set, we must specify the order in which the operations are to be performed. The result of forming the intersection of A with the union of B and C, $A \cap (B \cup C)$, can be quite different from the union of the intersection of A and B with C, $(A \cap B) \cup C$. The student will have the opportunity of checking an example of the difference in the exercises.

Occasionally we encounter complex expressions involving set operations on compound sets specified in inconvenient ways. Table 1.3.1 lists some equivalent expressions that can be used to simplify or modify a compound set into a more convenient form. In accordance with a fundamental axiom in mathematics, any mathematical expression can be replaced by its equal at any time without affecting the meaning or truth of the statement in which the substitution is made. Since all the relationships in the table are statements of equality between sets, one of the forms in any statement can replace the other in any expression involving sets.

Table 1.3.1
Properties of Set Operations

Identity Laws	
(1) $A \cup \emptyset = A$	(2) $A \cap \emptyset = \emptyset$
(3) $A \cup U = U$	(4) $A \cap U = A$
Laws on Complements	
(5) $A \cup A' = U$	(6) $A \cap A' = \emptyset$
(7) $(A')' = A$	(8) $U' = \emptyset$
Idempotent Laws	
(9) $A \cup A = A$	(10) $A \cap A = A$
Commutative Property	
(11) $A \cup B = B \cup A$	(12) $A \cap B = B \cap A$
Associative Property	
(13) $(A \cup B) \cup C = A \cup (B \cup C)$	(14) $(A \cap B) \cap C = A \cap (B \cap C)$
Distributive Property	
(15) $A \cup (B \cap C) = (A \cup B) \cap (A \cup C)$	(16) $A \cap (B \cup C) = (A \cap B) \cup (A \cap C)$
De Morgan's Laws	
(17) $(A \cup B)' = A' \cap B'$	(18) $(A \cap B)' = A' \cup B'$

Example 1.3.1

Consider the compound set $A \cap (B \cup A')$. It can be shown that $A \cap (B \cup A')$ can be simplified to $A \cap B$.

1. $A \cap (B \cup A') = (A \cap B) \cup (A \cap A')$ — Property 16
2. Since $A \cap A' = \emptyset$ — Property 6
3. $A \cap (B \cup A') = (A \cap B) \cup \emptyset$ — Substitution of $\emptyset$ for $(A \cap A')$

4. $(A \cap B) \cup \emptyset = A \cap B$ Property 1

So we have by substitution of $A \cap B$ for $(A \cap B) \cup \emptyset$ in Equation 3:

5. $A \cap (B \cup A') = A \cap B$

Special mention should be made of Properties 17 and 18, of Table 1.3.1, De Morgan's laws. These are very useful relationships for working with complements. Property 17 states that the complement of the **union** of two sets, A and B, is equal to the **intersection** of their complements. Let us consider the meaning of this statement. Any element of $A \cup B$ is by definition an element of A, an element of B, or both. Thus, an element of $(A \cup B)'$ cannot be an element of A, and it cannot be an element of B. (Otherwise it would be an element of $A \cup B$.) That is, an element of $(A \cup B)'$ is an element both of A' and of B'. This is another way of saying it is an element of $A' \cap B'$. Thus, $(A \cup B)'$ is equal to $A' \cap B'$.

Example 1.3.2

Let $A = \{a, b, c, d, e\}$ and $B = \{e, f, g, h, i\}$ with $U = \{a, b, c, d, e, f, g, h, i, j, k, l\}$

$$A \cup B = \{a, b, c, d, e, f, g, h, i\} \text{ and } (A \cup B)' = \{j, k, l\}$$

$$A' = \{f, g, h, i, j, k, l\} \text{ and } B' = \{a, b, c, d, j, k, l\}$$

$$A' \cap B' = \{j, k, l\} = (A \cup B)'$$

Example 1.3.3

Let U be the students in your math class. Let A be the set of students in this class who have jobs; and B, the set of students who own cars. Then, $A \cup B$ is the set of students in the class who have either a job or a car or both. The remaining students (if any) are members of the set $(A \cup B)'$. These students have no jobs and no cars. That is, they are members of $A' \cap B'$, the intersection of the set of students who do not have jobs, A', with the set who do not have cars, B'.

Property 18 has the same form as Property 17 with the roles of the intersection and union reversed. The statement says that those elements that are not in both sets A and B (the elements of $(A \cap B)'$) are those which are either not in A or not in B or neither—they are elements of $A' \cup B'$.

Example 1.3.4

Let $A = \{a, b, c, d, e\}$ and $B = \{e, f, g, h, i\}$ with $U = \{a, b, c, d, e, f, g, h, i, j, k, l\}$.

$$A \cap B = \{e\} \text{ and } (A \cap B)' = \{a, b, c, d, f, g, h, i, j, k, l\}$$

$$A' = \{f, g, h, i, j, k, l\} \quad B' = \{a, b, c, d, j, k, l\}$$

$$A' \cup B' = \{a, b, c, d, f, g, h, i, j, k, l\} = (A \cap B)'$$

1.3.1 Venn Diagrams

Many of the relationships and operations with sets can be clarified by the use of illustrative diagrams developed by the English mathematician, John Venn. A Venn diagram uses a closed plane figure to depict a set, and the elements of the set are considered to be "located" in the enclosed region.

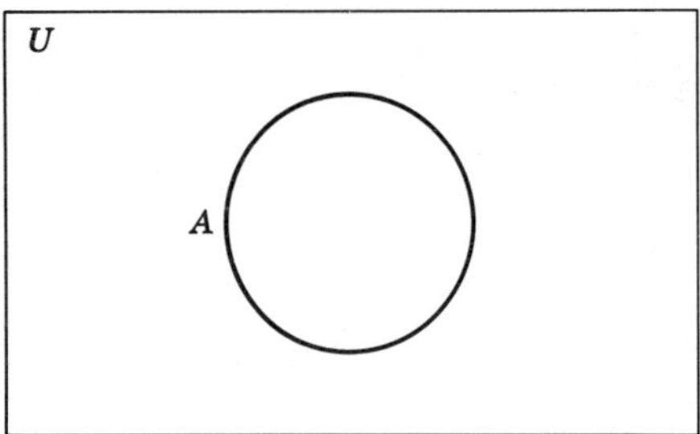

Figure 1.3.1

In Figure 1.3.1 the rectangle represents the universal set U; and the circle represents the set A, which is a subset of the universal set. The fact that the circle is entirely contained in the rectangle depicts the concept of a subset, in which all the elements of A are also contained in U.

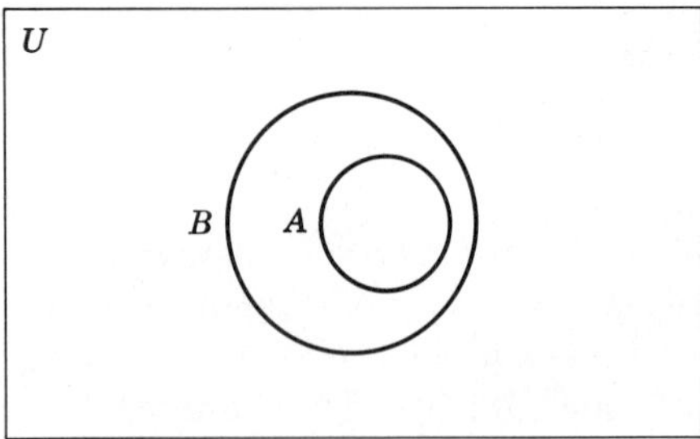

Figure 1.3.2

When two sets are considered, a Venn diagram can be used to illustrate the different relationships possible between the sets. The diagram of Figure 1.3.2 shows the situation in which one set is a subset of the other. In this figure $A \subset B$. Note that the area representing set A is entirely contained in the area representing set B—that is, all of the elements of A are also elements of B.

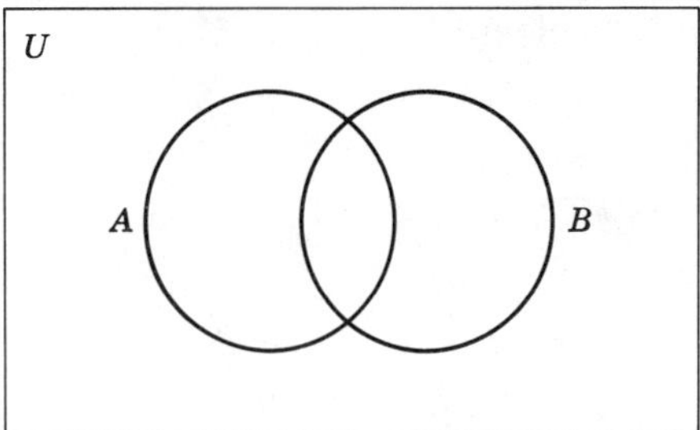

Figure 1.3.3

Figure 1.3.3 depicts the two sets with some elements in common ($A \cap B \neq \emptyset$), but each set contains elements the other does not.

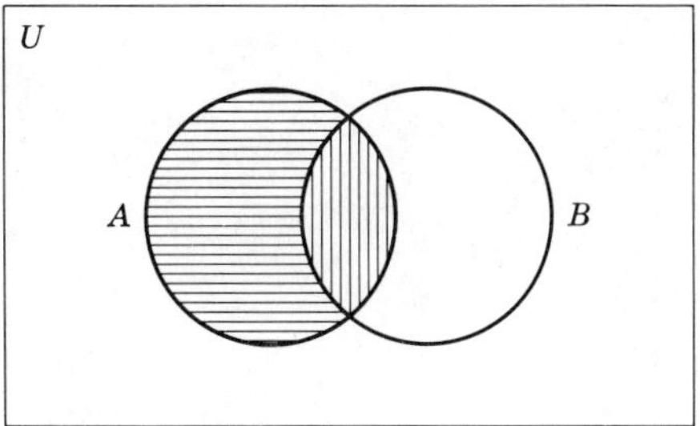

Figure 1.3.4

Shading can be used on a diagram such as Figure 1.3.3 to depict a particular set formed from two components. In Figure 1.3.4 the vertical shading depicts $A \cap B$, whereas the horizontal shading shows $A \cap B'$.

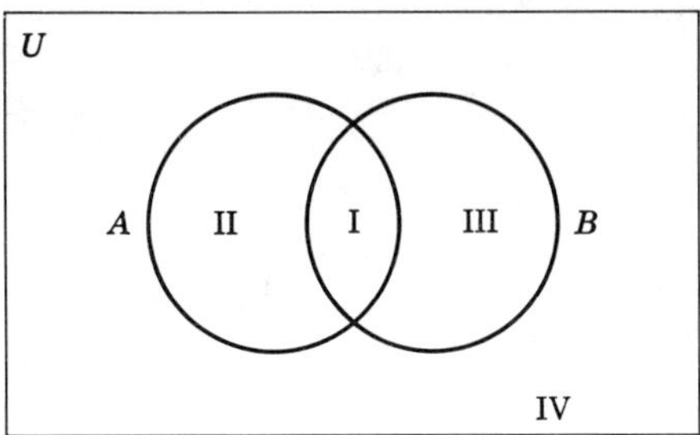

Figure 1.3.5

With appropriate labeling a diagram such as Figure 1.3.5 can be used to show any possible relationship between two sets. Note that the sets A and B, as shown, divide the universal set into four separate, or disjoint, sets.
In the figure

Region I represents $A \cap B$
Region II represents $A \cap B'$
Region III represents $A' \cap B$
Region IV represents $A' \cap B'$

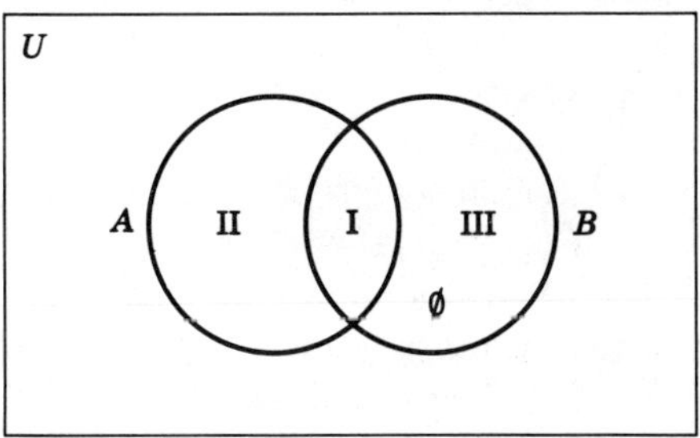

Figure 1.3.6

By labeling Region III as the null set, $\emptyset$, we can use this diagram to illustrate the relationship $B \subset A$, as seen in Figure 1.3.6.

In this diagram set B has no elements that are not also in set A. Thus, $B \subset A$.

By labeling Region I as the null set, as shown in Figure 1.3.7, we depict the situation in which the two sets, A and B, are disjoint.

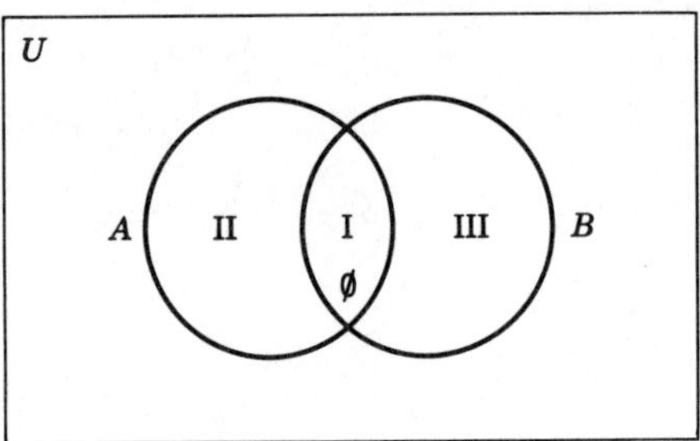

Figure 1.3.7

A more important use of a diagram, such as shown in Figure 1.3.5, is that it enables us to express any compound set as the **union** of **disjoint sets**. As we shall see in the next section, this is very useful in determining the number of elements in a compound set.

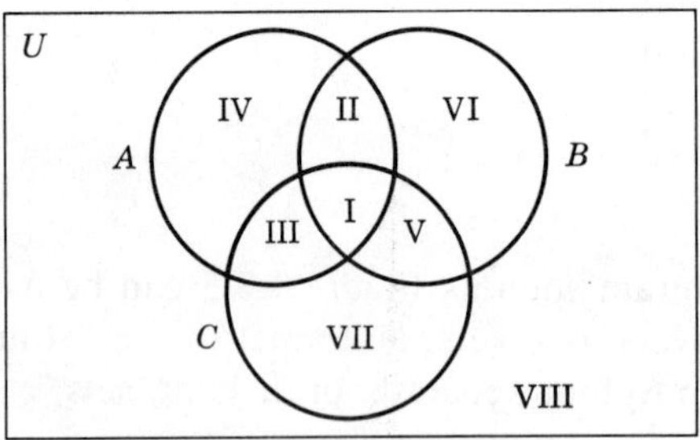

Figure 1.3.8

The most general relationship among three sets, A, B, and C, is shown in the diagram of Figure 1.3.8. In this case there are eight disjoint regions in the universal set.

Region I $= A \cap B \cap C$	Region II $= A \cap B \cap C'$
Region III $= A \cap B' \cap C$	Region IV $= A \cap B' \cap C'$
Region V $= A' \cap B \cap C$	Region VI $= A' \cap B \cap C'$
Region VII $= A' \cap B' \cap C$	Region VIII $= A' \cap B' \cap C'$

Any compound set other than those listed that can be formed from these three sets can be expressed as the union of two or more of these eight disjoint sets.

For example, if we let (I) be the set of Region I, and (II), the set of Region II, and so forth, we have

$(A \cup B) - C = (\text{II}) \cup (\text{IV}) \cup (\text{VI})$
$A \cap (B \cup C) = (\text{I}) \cup (\text{II}) \cup (\text{III})$
$B \cup (A \cap C) = (\text{I}) \cup (\text{II}) \cup (\text{III}) \cup (\text{V}) \cup (\text{VI})$

1.3.2 Identifying Compound Sets

The most frequent direct use of sets and set algebra in business and the behavioral sciences is as a convenient means for identifying groups of people or things of special interest in certain situations. For example, let us suppose a manufacturer of electronic equipment is conducting a market survey in advance of the introduction of a new product. He is interested in whether there is a correlation between owning a house and the purchase of major home entertainment equipment. In the survey he asks the respondents to answer yes or no to three questions:

a. Do you own your house?
b. Do you have a color television set?
c. Do you have a stereo sound system?

He identifies the yes answers as elements of the following sets:

H: Homeowners
C: Owners of color television
S: Owners of stereo equipment

Some of the groups of people he may wish to identify:

–Those who own their own homes and have a color TV, but not a stereo system.

Those who own their homes and have color TV are elements of $H \cap C$. Those who do not have stereo systems are elements of S'. Thus, the set becomes $H \cap C \cap S'$, or $(H \cap C) - S$.

–Those who do not own their homes, but have either a color TV, a stereo system, or both.

Those having either a color TV or a stereo system belong to the set $C \cup S$. Those who do not own their own homes are elements of H'. Thus, the set of interest is $(C \cup S) \cap H'$ or $(C \cup S) - H$.

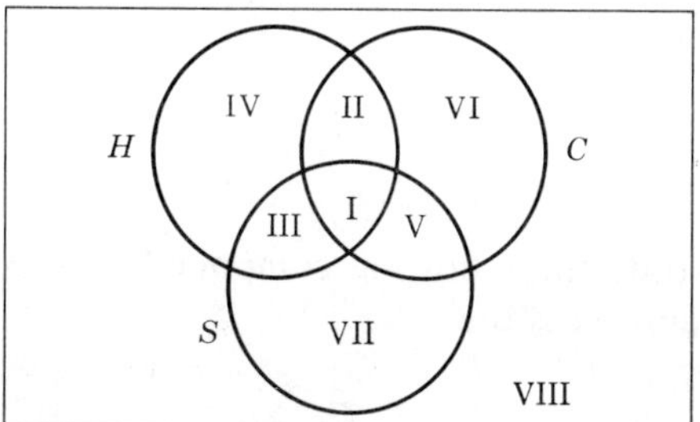

U = set of respondents in survey

Figure 1.3.9

A Venn diagram can be used to help identify these sets. Figure 1.3.9 shows the most general situation possible.

The people in the first set specified must be included in the region labeled H. These are the people who own their own homes. Since they also have color TV, they are elements of the part of H also in C, Regions I and II of the diagram. Since they do not have stereo systems, they are outside S: that is, they are elements of Region II of the diagram.

By a similar process the people who do not own their homes, but who have either a color TV or a stereo system, or both, are seen to be elements of the union of Regions V, VI, and VII of the diagram.

EXERCISE 1.3

1. $A = \{a, b, c, d, e\}$, $B = \{c, d, e, f, g\}$, $C = \{f, g, h, i, j\}$, and $D = \{j, k, l, m, n\}$ with $U = \{\text{alphabet}\}$.
Define the following sets by listing their members.

a. $(A \cup B) \cap C$ b. $A \cup (B \cap C)$
c. $(A \cap B) \cup (A \cap C)$ d. $(A \cup B) \cap (A \cup C)$
e. $(A \cup B) - C$ f. $(A \cap B) - C$
g. $(A \cup B) \cap (C \cup D)$ h. $(A \cap B) \cup (C \cap D)$
i. $(A \cup D) \cap (B \cup C)$ j. $(A \cap C) \cup (B \cap D)$
k. $(A \cup B) \cap C'$ l. $A \cup (B \cap C)'$

2. $U = \{\text{all real numbers}\}$
$A = \{\text{positive integers}\}$
$B = \{\text{negative integers}\}$
$C = \{0\}$
$D = \{\text{real numbers less than } 100\}$
$E = \{\text{odd integers}\}$
Define the following sets by specifying their members:

a. $A \cup B \cup C$ b. $(A \cup C) \cap D$
c. $A \cap D \cap E$ d. $A \cap B \cap C$
e. $(A \cup B) \cap E'$ f. $B \cup (D' \cap E)$

3. Use the properties of set operations listed in Table 1.3.1 to simplify the expressions for the following sets:

a. $(A' \cup B') \cap (A \cap B)$
b. $(A \cup B) \cap (A \cup B')$
c. $(A' \cup B')'$
d. $(A' \cap B')'$
e. $[(A \cup B) \cap C] \cup [C \cap (A \cup B)]$

4. Draw a Venn diagram to illustrate the following compound sets. Shade the area on the diagram that represents the set.

a. $A \cap B'$ b. $(A \cup B) \cap C$
c. $A \cup B \cup C$ d. $(A \cup B)' \cap C$
e. $A - (B \cup C)$ f. $(A \cap B) \cup (A \cap C) \cup (B \cap C)$

5. Using a Venn diagram with regions labeled as in Figure 1.3.8, express the following sets as unions of the appropriate regions:

a. $A \cup (B \cap C)$ b. $(A \cap B) - C$
c. $B - (A \cup C)$ d. $(A \cup C) - B$
e. $[(A \cup B) - (A \cap B)] \cup [C - (A \cup B)]$

6. Referring to Figure 1.3.8, name the compound sets represented by the following regions:

a. (I) ∪ (II) ∪ (III)

b. (III) ∪ (IV) ∪ (VII)
c. (VII) ∪ (VIII)
d. (III) ∪ (IV) ∪ (V) ∪ (VI)
e. (IV) ∪ (VI) ∪ (VII)

7. Using as a universal set the heads of families in a certain medium-sized city, let $A =$ {those who earned more than \$10,000 last year}, $B =$ {those who purchased a new automobile last year}, and $C =$ {those who own a color television set}. Describe in words the members of the following compound sets:

a. $A \cap (B \cup C)$
b. $B' \cup C'$
c. $A' \cap (B \cap C)$
d. $A \cap (B - C)$
e. $A' \cap [(B \cup C) - (B \cap C)]$

(*Hint*: The descriptions are sometimes clearer if an equivalent set expression is used.)

8. Using the same set definitions given in Exercise 7, translate the following sets into symbols:

a. The set of those who earned less than \$10,000 last year who own a color television.
b. The set of those who own a color television, but who did not buy a new car last year.
c. The set of those earning more than \$10,000 last year who bought a new car, but who do not own a color TV.
d. The set of those who either earned more than \$10,000 last year or bought a new car or own a color TV.
e. Those who either bought a new car last year or who own a color TV, but not both.

9. In a study of people's reading habits, a researcher used the following sets:

U = Sample set of adults in a medium-sized city
M = Set of men in the sample
B = {People who read five or more complete books the previous year} (This set was used to define a "regular book reader.")
P = {People who read all or part of each edition of a magazine during the previous year}
(This set was used to define a "regular periodical reader.")

Name the sets whose members meet the following specifications:

a. The men who regularly read either books or periodicals.
b. The women who read both books and periodicals regularly.
c. The people who regularly read either books or periodicals, but not both.
d. The men who do not read either books or periodicals regularly.
e. The women who read books but not periodicals regularly.

Use the set designations above, and describe the members of the following sets:

f. $M' \cap P$
g. $M \cap B' \cap P$
h. $M \cap B \cap P$
i. $M' \cap B' \cap P'$
j. $M \cap (P \cup B) - M \cap (P \cap B)$

10. A counselor at a community college is studying the effect of student activities and interests on their academic performance. Some of the sets of students he has set up are

U = {All students at the college}
M = {Men students}
W = {Women students}
A = {Students active in campus affairs}
C = {Students who own cars}
J = {Students who have jobs}
D = {Students on the dean's list}
P = {Students on academic probation}
T = {Students who plan to transfer to a four-year college or university}
V = {Students enrolled in a vocational training program}

Translate the following into symbols:

a. {The men students who own cars who are on academic probation}
b. {Students with either jobs or cars who are on the dean's list}
c. {Women students enrolled in a vocational program who are active in campus affairs}
d. None of the women students on academic probation have jobs.
e. Some of the men students with jobs are active in campus affairs.
f. All of the women on the dean's list plan to transfer to a four-year school.

Translate the following into words:

g. $(M \cap J) \cap T$
h. $(J \cup C) \cap D$
i. $(T \cup V)' \cap D = \emptyset$
j. $[M \cap (J \cup C)] \cap D \neq \emptyset$
k. $(M \cap T) \cap D \subseteq A$
l. $(M \cap A) \cap T \subseteq (J \cap C)'$

1.4 COUNTING THE ELEMENTS IN SETS

In our subsequent studies we shall frequently be interested in the number of elements in compound sets. The exercises and examples in previous sections have suggested some of the types of sets of interest to businessmen and behavioral scientists. In most cases the interest centers on the number of elements in these sets—the number of people or things having certain characteristics or properties in common. The process of determining these numbers usually goes beyond the simple addition or subtraction of the number of elements in the component sets. For example, if we have two sets A and B, in which $A = \{a, b, c, d, e\}$ and $B = \{c, d, e, f, g\}$, each set contains five elements. If we put all the elements into one set—form $A \cup B$—we might expect the total number of elements to be 10. However, if we count the different elements in $A \cup B$, we find there are only seven.

We observe that if we list the elements of A and the elements of B, we have for $A \cup B$

$$A \cup B = \{a, b, c, d, e, c, d, e, f, g\} \tag{1.4.1}$$

The three elements, c, d, and e, have been listed twice. From our previous agreement about the listing of an element of a set more than once, we can omit the second listing; and the seven members of $A \cup B$ are seen to be

$$A \cup B = \{a, b, c, d, e, f, g\} \tag{1.4.2}$$

Note that in Equation 1.4.1 there are 10 elements listed for $A \cup B$, the five from A and the five from B. Note that the three listed twice are the elements in $A \cap B$. Therefore, in finding the number of elements in $A \cup B$, we add the five elements for A and the five for B, and then subtract the three elements in $A \cap B$. We can write, therefore,

$$^{*}n(A \cup B) = n(A) + n(B) - n(A \cap B) \tag{1.4.3}$$

Equation 1.4.3 is the general equation relating the number of elements in the union of two sets with the number of elements in each of the sets and the number in their intersection. If we know any three of these quantities, we can compute the fourth.

By a reasoning process similar to that above we can show that for three sets we have

$$^{\dagger}n(A \cup B \cup C) = n(A) + n(B) + n(C) - n(A \cap B) - n(A \cap C) - n(B \cap C) + n(A \cap B \cap C) \tag{1.4.4}$$

The pattern of Equation 1.4.4 comes from the fact that the elements in the intersection $A \cap B$ are included in both $n(A)$ and $n(B)$; the elements of $A \cap C$ are included in both $n(A)$ and $n(C)$; and the elements of $B \cap C$ are included in both $n(B)$ and $n(C)$. That is, the number of elements in each of the possible two-set intersections is counted twice, in adding the number of elements in each of the individual sets; and, therefore, one times the number of elements in each of these intersections must be subtracted out of the sum. Also, the number of elements in $A \cap B \cap C$ is included three times—once in each of the terms $n(A)$, $n(B)$, and $n(C)$. It is then subtracted out three times, once for each of the two-set intersections. This number must, therefore, be added back in to give the correct total for the number of elements in $A \cup B \cup C$.

Example 1.4.1

Let $A = \{a, b, c, d, e\}$, $B = \{c, d, e, f, g\}$, and $C = \{e, f, g, h, i\}$.

$$A \cup B \cup C = \{a, b, c, d, e, f, g, h, i\}$$

$$n(A \cup B \cup C) = 9 \quad \text{by direct count}$$

*We shall use the symbol $n(\ \ )$ to mean the number of elements in set ( ). $n(A)$ means the number of elements in set A.

†Since the **associative property for set union** and **set intersection** (See Properties 13 and 14, Table 1.3.1) permit us to perform each of these operations in any order, compound sets consisting of the **union** of several sets, or the **intersection** of several sets, are customarily written without grouping symbols. Note that this convention applies only in case the operation is the **same** for all the sets. Mixed operations require grouping symbols. For example, we can write $A \cup B \cup C \cup D$ without ambiguity, but $(A \cup B) \cap (C \cup D)$ requires that grouping symbols be used.

If we compute $n(A \cup B \cup C)$ with Equation 1.4.4, we need the following information:

$$
\begin{array}{ll}
A \cap B = \{c, d, e\} & n(A \cap B) = 3 \\
A \cap C = \{e\} & n(A \cap C) = 1 \\
B \cap C = \{e, f, g\} & n(B \cap C) = 3 \\
A \cap B \cap C = \{e\} & n(A \cap B \cap C) = 1
\end{array}
$$

Equation 1.4.4 *becomes:*

$$n(A \cup B \cup C) = 5 + 5 + 5 - 3 - 1 - 3 + 1 = 9$$

Patterns of relationships are important in mathematics. Much useful information can be inferred from a careful observation of the patterns followed by formulas. For example, a study of Equations 1.4.3 and 1.4.4 shows that the number of elements in the union of several sets is obtained by adding the number of elements in each individual set, then subtracting the number of elements in each of the possible two-set intersections, then adding the number in the three-set intersections, and so on. An application of this pattern to the union of four sets, $A \cup B \cup C \cup D$, yields the following formula, which turns out to be the correct one:

$$
\begin{aligned}
n(A \cup B \cup C \cup D) = {} & n(A) + n(B) + n(C) + n(D) - n(A \cap B) \\
& - n(A \cap C) - n(A \cap D) - n(B \cap C) \\
& - n(B \cap D) - n(C \cap D) + n(A \cap B \cap C) \\
& + n(A \cap B \cap D) + n(A \cap C \cap D) + n(B \cap C \cap D) \\
& - n(A \cap B \cap C \cap D) \qquad (1.4.5)
\end{aligned}
$$

Equations 1.4.3, 1.4.4, 1.4.5, and their counterparts for larger numbers of sets are used to compute the number of elements in the union of sets. They are rather cumbersome, however, particularly as the number of sets increases. In the previous section we saw how the union of two or more sets can be separated into disjoint sets. This type of separation is of particular use in counting elements in sets because all of the intersections are empty—they contain zero elements. Thus, the number of elements in the union of these disjoint sets is the simple sum of the number of elements in each component set.

In the previous section we separated the union of two sets into four disjoint "regions" as shown in Figure 1.4.1. If we know the number of elements in each of the four "regions," we can readily compute the number of elements in any possible compound set that can be formed from the two sets.

We recall from the preceding section that the union of three sets can be

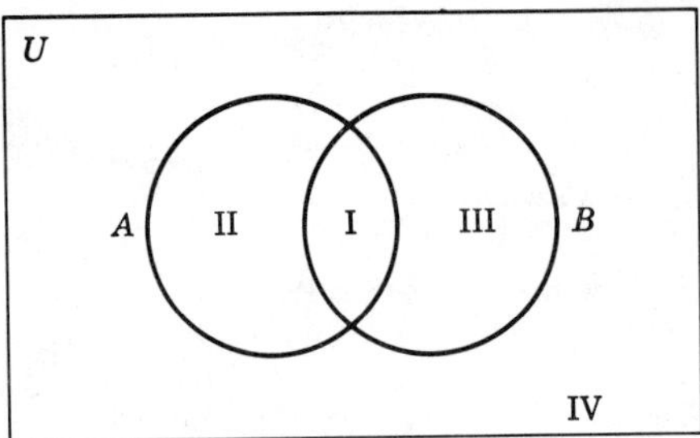

Figure 1.4.1

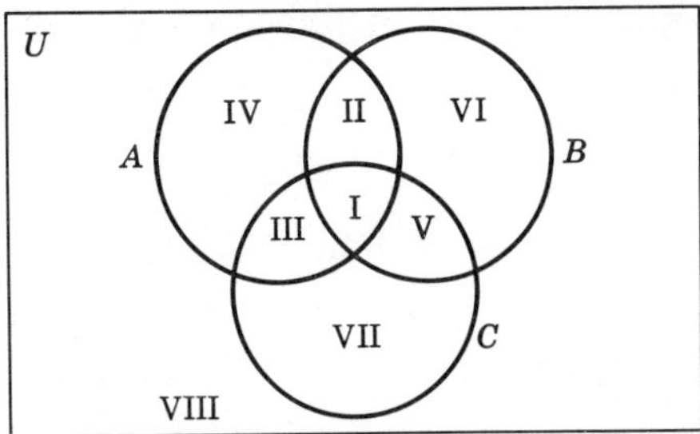

Figure 1.4.2

separated into eight disjoint regions, as shown in Figure 1.4.2. Knowing the number of elements in each of these regions enables us to compute the numbe of elements in any possible combination of three sets. The sum of the number of elements in each of the eight regions is the number of elements in the universal set.

The number of elements in each of the eight regions of Figure 1.4.2 can be obtained if we know the number of elements in the universal set, the number of elements in each of the three sets, and the number of elements in each of the four intersections possible. That is, if the three sets are designated as A, B, and C, we need to know $n(U)$, $n(A)$, $n(B)$, $n(C)$, $n(A \cap B)$, $n(A \cap C)$, $n(B \cap C)$, and $n(A \cap B \cap C)$.

Notice that the above information is easily obtained by the usual counting techniques employed in surveys—the principal application of the set counting techniques we are discussing. A card is made for each element of the universal set—for each person or thing sampled. A box is printed on the card for each of the sets of interest, similar to that shown in Figure 1.4.3. This box is marked if the "element" is a member of that set. When the survey is completed, the cards are counted by machine. The machine records how many cards have Box A marked, $(n(A))$, how many have Box B marked, $(n(B))$, and so on. It also records, for example, how many cards have both A and B marked $(n(A \cap B))$ and how many have all three marked $(n(A \cap B \cap C))$. (For more complex surveys the card has more boxes. The counting process is longer, but follows the same pattern.)

Referring to Figure 1.4.2 we note that the number of elements in Region I is the number in $A \cap B \cap C$—the number of cards with all three boxes marked. The number of elements in Region II is $n(A \cap B) - n(A \cap B \cap C)$. The num-

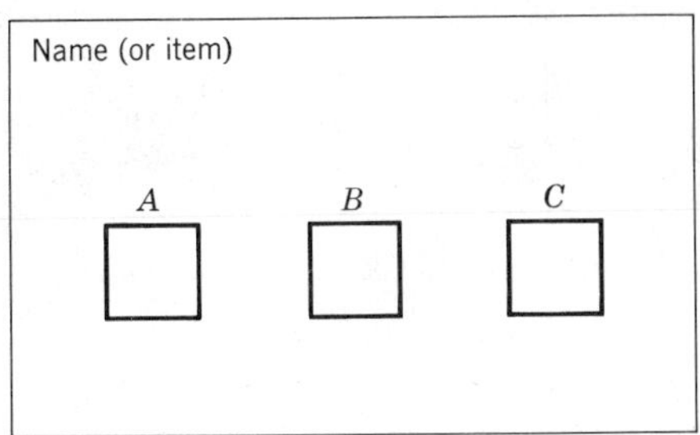

Figure 1.4.3

ber in Region III is $n(A \cap C) - n(A \cap B \cap C)$. The number in Region IV is $n(A)$ less the sum of the numbers of elements in Regions I, II, and III, and so on.

Example 1.4.2

A survey of 150 women is made at a market by a manufacturer seeking information about laundry products. Each woman is asked three questions:

a. Do you have a washing machine at home?
b. Do you use a laundry detergent?
c. Do you use a laundry presoak product?

A card is marked for each woman showing the questions, if any, to which she gave a yes answer. A tabulation of the results showed the following:

Yes answers to question *a*:	$95 = n(A)$
Yes answers to question *b*:	$120 = n(B)$
Yes answers to question *c*:	$38 = n(C)$
Yes answers to questions *a* and *b*:	$78 = n(A \cap B)$
Yes answers to both *a* and *c*:	$32 = n(A \cap C)$
Yes answers to both *b* and *c*:	$30 = n(B \cap C)$
Yes answers to all three questions:	$25 = n(A \cap B \cap C)$

A Venn diagram for this survey is shown in Figure 1.4.4.

To find the number of elements in each of the regions, we start with Region I, $A \cap B \cap C$. The survey results showed 25 elements in this set. $A \cap B$ includes Regions I and II. Thus, the number in Region II $= n(A \cap B) - n(A \cap B \cap C) = 78 - 25 = 53$.

$A \cap C$ includes Regions I and III. Since $n(A \cap C) = 32$, and 25 of these are in Region I, Region III must contain 7 elements.

Set A is made up of Regions I, II, III, and IV. The survey results show it contains 95 elements. $53 + 25 + 7 = 85$ of the elements are in Regions I, II, and III. Thus, Region IV contains 10 elements, as shown in the diagram.

By similar procedures we obtain the number of elements in Regions V, VI, and VII. These are shown in Figure 1.4.4.

The sum of the elements in Regions I through VII (138) is the total number of elements in $A \cup B \cup C$. Thus, $(A \cup B \cup C)'$ – Region VIII – must contain the remainder, 12, in the universal set.

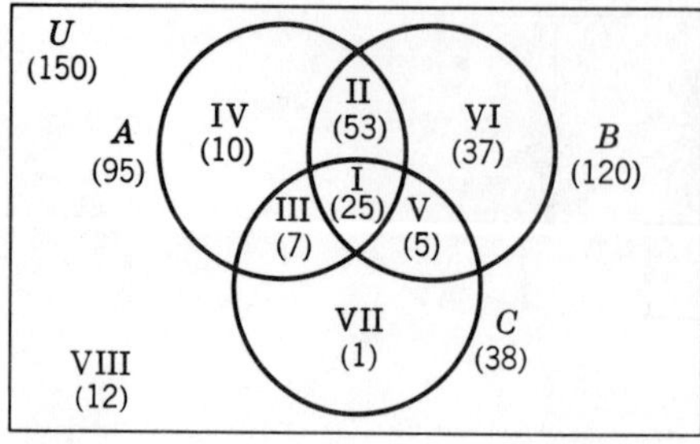

Figure 1.4.4

We now have the information needed to answer any question relating to the number of women in any category covered by the survey. We note, for example, that the number of women who have a washing machine, but who use neither a detergent nor a presoak, is 10 (Region IV). 53 women with washing machines use a detergent, but no presoak (Region II); whereas, 7 with washing machines use a presoak, but no detergent (Region III). There are 43 women who do not have washing machines who use either a detergent or a presoak or both (Regions V, VI, and VII).

1.4.1 Partitions and Cross Partitions

The discussion and examples of the preceding section illustrate the analysis and computations used to extract information in a survey in which there are only two possibilities for each category. That is, the questions or classifications used are such that any element of the survey either does or does not qualify for membership in one of the sets of interest. For instance, in Example 1.4.2 each woman surveyed either did or did not have a washing machine. There are no other possibilities.

There are some types of surveys, notably opinion surveys, in which a respondent has three (and sometimes more) alternative responses. As a simple example, he can be asked a question in which he answers, "Yes," "No," or "No opinion."

We observe that in this process each category, or question, of the survey divides the **entire** universal set into two or more **separate** (disjoint) **subsets.** In a "Yes-No-No opinion" situation, for example, each respondent is placed in one, but no more than one, set on the basis of his response. Such a separation of an entire set into disjoint subsets is called a **partition**. A partition has the following properties:

1. All of the elements of the set are placed into the subsets that make up the partition.
2. No element is placed in more than one subset.

In the symbology of algebra, the definition of a partition becomes

Definition 1.4.1

Let A_1,* A_2, $A_3, \ldots, A_n$† be subsets of a set A. The collection of subsets, $A_1, A_2, A_3, \ldots, A_n$, constitutes a **partition** of A if, and only if,

* The numbers and letters written to the right of and slightly below an algebraic symbol are called **subscripts**. They are used widely in algebra to designate particular units, or elements, in a collection. In this case, for example, A_1 means the first subset of the collection; A_2, the second subset of the collection. The A_n means the nth subset of the collection. The n is used to indicate that there can be any number of subsets. n is customarily used to represent a counting number or an integer.

† The three dots represent the subsets between the 3rd and the nth. This is another algebraic idiom. This one is used to specify a sequence of items if all the items normally in the sequence are to be included. When operation symbols are used in a sequence, as in part 1 of the definition, the operation symbol precedes and follows the three dots.

1. $A_1 \cup A_2 \cup A_3 \cup \ldots \cup A_n = A$; and

2. $A_i \cap A_j^* = \emptyset$, where A_i and A_j are any two different subsets of the collection.

Note that the union of all the subsets, as symbolized in part 1 of Definition 1.4.1, guarantees that all of the elements of set A are included in the subsets. We cannot write a multiple intersection for part 2, however, because this would not guarantee that each **pair** of subsets is disjoint; that is, that each subset is disjoint from each of the others. The iterative symbology of part 2 of the definition is used to make the precise specification that is required. Note that the language of algebra is very concise. It is hard to read because it says so much in so few symbols. Algebraic expressions must be read carefully to appreciate their full meaning. This is hard at first, but after a while the beauty of the precision and conciseness of the language begins to appear.

A set can, of course, be partitioned in many different ways. Partitioning a set in more than one way produces a **cross-partition**. In an opinion survey, for example, the respondents will most likely be asked more than one question. Each question develops a partition of the sample set. Two or more questions considered together produce a cross-partition.

We considered above a survey question that could be answered in one of three ways: Yes, No, No opinion. If there are two such questions in a survey, there are three times three, or nine, separate categories in which the respondents are placed (for example, Yes on both questions; Yes on question 1, and No on question 2; and No on question 1, Yes on question 2). In general, the number of separate subsets produced by successive partitions of a set is the product of the numbers of subsets of each partition.

Notice that there are no simple processes for obtaining the number of elements in each subset of a cross-partition. Each category (subset) possible must be counted separately. In a survey with three questions, each of the Yes-No-No opinion variety, there are three times three times three, or 27, different possible sets of responses. The number of responses in each of these 27 sets is counted separately—usually by a machine tabulator.

The student may recognize that cross-partitioning describes a classification process that occurs regularly in almost every field of activity. In business, we see that inventories, and personnel classification as to pay grade, type of duties, and the department assigned to are two examples of cross-partitions of sets. In sociology, people are classified according to such information as their place of residence, their income, and their employment status. The census is a gigantic cross-partitioning of the population into a large number of different sets of classifications. Psychologists classify people with respect to their temperament, their heredity factors, their environment, and so on.

Multiple partitioning of a set forms the basis for the widely used "punch card" method of record keeping. The objects, or people, about whom the

*The i and the j used as subscripts are index subscripts. Each represents successively all of the possible subscripts in the collection. They are used to avoid writing a statement over and over again for each possibility. They are the algebraic counterpart of the word "each." This statement would be read, "The intersection of **each** pair of subsets is the null set."

records are maintained constitute the elements of the universal set. A different card is used for each element. Each partitioning of the set (each of the different ways in which the elements are to be classified) is assigned an area on the card, and a hole is punched in a specific location in this area to identify the cell of the partition in which the element is to be placed. Each card has a hole punched in each area. The particular collection of holes determines the cell of the cross-partition in which the element belongs. A sorting machine can be set up to respond to holes in a selected location. Thus, the cards for the elements of the set with a specified set of properties (a particular cell of the cross-partition) can be separated out, and the elements readily identified.

Example 1.4.3

Suppose we have an employment service to supply part-time help for various jobs. Every person who registers for the service supplies information that prospective employers are concerned with: for example, sex, age, education, types of experience, wage requirements, and special times available for work. Each of these categories constitutes a partition of the set of clients of the employment service. Thus, when someone wants a man over 50 with a high school education who is an experienced gardener to work week-ends for no more than \$3.50 per hour, a particular cell of the cross-partition has been defined. The cards are put through the sorting machine to determine which, if any, represent people who are the elements of the cell; that is, which represent people with the required qualifications. Note that the same kind of cards and the same sorting machines can be used to identify books on particular subjects, by certain authors, and so on; items of clothing of a particular size, color, material, and style; or any of a multitude of other possibilities. The system involves sets and partitions of sets. The meaning for the various algebraic entities—the sets, partition cells, cell elements, and so on—is supplied by the user.

EXERCISE 1.4

1. Two sets, A and B, are subsets of the same universal set.

a. If set A contains 58 elements and A' contains 37 elements, how many elements are in the universal set?

b. If set A contains 58 elements, set B 43 elements, and the intersection of A and B contains 26 elements, how many elements are in the union of A and B?

c. Use the results of (a) and (b) to find the number of elements that are in neither A nor B.

2. a. If $n(A \cup B) = 50$, $n(A) = 30$, and $n(B) = 32$, how many elements are there in $A \cap B$?

b. If A and B are both subsets of a universal set with 80 elements, how many elements are there in A'? B'? $(A \cup B)'$?

3. Three sets, A, B, and C are subsets of the same universal set U.

$n(A) = 45$
$n(B) = 49$
$n(C) = 55$
$n(A \cap B) = 25$
$n(A \cap C) = 22$
$n(B \cap C) = 28$
$n(A \cap B \cap C) = 10$

a. How many elements are in the following sets?

$(A \cap B) - C$ $\quad B - (A \cup C)$ $\quad A \cup B$
$(A \cup C) - B$ $\quad (A \cup B \cup C)$

b. If $n(A \cup B \cup C)' = 12$, what is $n(U)$?

4. In a universal set with 95 elements, we consider three subsets, A, B, and C.

$n(A) = 40$
$n(B) = 50$
$n(C) = 60$
$n(A \cap B) = 30$
$n(A \cap C) = 20$
$n(B \cap C) = 25$
$n(A \cap B \cap C) = 15$

a. Compute the number of elements in each of the eight regions of U.
b. Draw a Venn diagram to represent these sets.

5. In a survey of public opinion on financing education, the following three questions were asked:

A. Do you favor a state lottery to finance education?
B. Do you favor increased federal financial aid to schools?
C. Do you believe attending school should be compulsory up to age 18?

One hundred fifty people responded to the survey. A tabulation of the survey cards showed:

Yes on question A:	43
Yes on question B:	110
Yes on question C:	88
Yes on questions A and B:	38
Yes on questions A and C:	21
Yes on questions B and C:	61
Yes on all three:	18

a. How many people favored an increase in federal aid to schools, but not a lottery or compulsory attendance?
b. How many favored either an increase in federal aid or a state lottery?
c. How many favored increased federal aid and compulsory school attendance, but not a lottery?
d. How many either answered no to all the questions or answered yes only to the state lottery?

6. $U = \{x | x \text{ is a student at State College}\}$. One partition of U is the collection

$A_1 = \{x | x \text{ is a student less than 18 years of age}\}$.
$A_2 = \{x | x \text{ is a student 18 or older, but less than 21}\}$.

$A_3 = \{x | x$ is a student 21 or older, but less than 25$\}$.
$A_4 = \{x | x$ is a student 25 or older, but less than 30$\}$.
$A_5 = \{x | x$ is a student 30 or older$\}$.

Another partition of this same set is the collection

$B_1 = \{x | x$ is a woman student$\}$.
$B_2 = \{x | x$ is a man student$\}$.

Suppose there are 5000 students at the college, of which 3000 are men and 2000, women. Suppose

$n(A_1) = 300$; $n(A_2) = 2500$; $n(A_3) = 1500$;
$n(A_4) = 500$; and $n(A_5) = 200$

What additional information, if any, do we need to determine the number of elements in each cell of the cross-partition between the partition, A, and the partition, B? Explain.

7. a. Given a set containing eight elements. Show that by forming three appropriate partitions of two subsets each, any element of the set can be identified uniquely from the answers to three yes-or-no questions about its membership in the subsets of the partitions.

b. How many elements could the set contain and have the identity of an element determined by four yes-or-no questions about four partitions?

8. In an opinion survey the respondents were asked to identify themselves as "Republicans," "Democrats," or "Independents." They were then asked if they favored a year of compulsory public service, either military or social, for every nineteen-year-old. Two hundred people were interviewed.

Thirty percent said they were Republicans, 45% said they were Democrats, and the remainder said they were Independents.

Sixty percent said they favored the compulsory service, 30% opposed it, and 10% had no opinion. Of the Republicans, 50% were in favor, 40% opposed, and 10% had no opinion. Of the Democrats, 70% were in favor; 20% opposed; and 10% had no opinion.

How many Independents favored the public service; how many opposed it; and how many had no opinion?

9. One hundred men who thought they might be color-blind were given a test in which they viewed for a few seconds a card on which the outline of a letter was printed in a pattern of various colored dots.

Card 1 had an A printed in red and green dots;
Card 2 had a B in yellow and blue dots; and
Card 3 had a C in orange and brown dots.
38 men correctly identified the A.
75 men correctly identified the B.
47 men correctly identified the C.
30 men correctly identified the A and the B.
18 men correctly identified the A and the C.
26 men correctly identified the B and the C.
12 men correctly identified all three.

a. How many identified only the A?
b. How many identified only the C?
c. How many identified none of the three?
d. How many identified the A and the B, but not the C?

10. A field worker for an opinion poll submitted the following report of his day's work.
Number of people interviewed: 80.
72 yes on question 1
67 yes on question 2
59 yes on question 3
55 yes on questions 1 and 2
46 yes on questions 1 and 3
51 yes on questions 2 and 3
37 yes on questions 1, 2, and 3
His supervisor recommended that he be fired for not understanding his job and for falsifying his report. What evidence, if any, can the supervisor produce to substantiate his charge?

2

deductive reasoning and symbolic logic

2.0 INTRODUCTION

The scientific method of study and research involves two types of reasoning: inductive and deductive. Inductive reasoning is used to formulate an "explanation" for things that are observed to happen. A phenomenon is observed to occur in a particular set of circumstances. A scientist proposes a model—an explanation—that produces this phenomenon under these circumstances. This model is a tentative theory for understanding and describing the situation.

The next step in the scientific process is to use the theory to **deduce**, or **predict**, what will happen if the circumstances are altered slightly. The theory is then tested by performing an experiment under the altered conditions. If the results of the experiment match the prediction, the theory gains a measure of acceptance. If they do not, inductive reasoning is again used to modify the theory until it satisfactorily explains both the original observations and the new, unpredicted one. In either case, further deductions are made and tested in the same manner.

As the process continues, the theory is developed to explain more completely the nature of the situations being studied. It becomes more nearly the "true" model of the situation. Notice, however, that we can never be certain that the explanation is, indeed, complete. There always remains the possibility that a new variation of the conditions in the situation will produce unpredicted results that will require another modification to the theory. This is one of the lures that make scientific research a never-ending process.

Each of us uses the process as we gain experience. Consider a young child learning about electric lights. The child observes older members of the family push a little lever on the wall and as they do, the room is suddenly filled with light. The child forms a tentative theory that light is produced inside a house by pushing a little lever on the wall. He will probably experiment and find that he, too, can make light. He now "knows" how light is produced inside a house. He may test his knowledge by pushing little levers in all the rooms of the house. But one day he pushes the little lever on the wall of his room, and nothing happens. He pushes it again and again, and still nothing happens. He faces a new situation, which his "theory" does not explain. After consultation with an older member of the family, he learns that the light bulb is burned out and must be replaced with a new one. His theory now reads, "If the little lever on the wall is pushed, there will be light if the light bulb is good. If there is no light, a new light bulb is needed." The second part of his theory makes the deduction that the reason no light appears when the lever is pushed is that the light bulb needs replacement. Later experience will probably extend his "theory of indoor light" to include the role of circuit breakers, fuses, and the power company.

Procedures in scientific research are more highly organized and on a more sophisticated level than our everyday learning experience. The reasoning processes involved, however, are basically the same. We usually do not express the results of our experiences in formal, explicit statements as researchers do; and sometimes we make deductions that are not valid (as sometimes researchers do also). In this chapter we shall study the form usually used to make explicit statements of theories and the process by which we can determine the validity of deductions made from them. We shall see that the validity of a deduction does not depend on the validity of the theory it is based on, but instead on the process used to form it. The validity of the deductive process can be checked independently of any test of the theory.

The study of the deductive reasoning process, or "logic," has been going on for centuries. Aristotle (384–322 B.C.) is generally credited with being the first to analyze logic systematically and to develop a "method" of logical thought. He developed most of the forms of logical argument* in use today. Although we use Aristotle's methods in our everyday "logical thinking," logic has been considered until relatively recently as primarily a tool of philosophers.

The organization and formalization of the deductive reasoning process as a branch of mathematics was done by two nineteenth-century logicians, Augustus De Morgan and George Boole. De Morgan and Boole believed that the use of language was a major hindrance to the reasoning process. Too often the meaning of the statements diverted attention from the validity of the conclusions being drawn. They believed that the validity of deductions depended entirely on the **form**, or **structure**, of the thought process by which they were reached, instead of on the meaning of the sentences being used to express them. They devised a system of **symbolic logic**, in which ideas are

* In the study of logic the process of establishing the validity of a deduction is called an **argument**. In mathematics the deduced statement whose truth is being established is called a **theorem**, and the process of establishing its validity is called a **proof**.

represented by abstract symbols rather than words. Conclusions are reached by working with these symbols according to an agreed upon set of rules. The validity of the conclusions depends on the way the symbols are manipulated, instead of the meanings they represent. Thus, logical thought is converted into an algebra, in which the same set of symbols and operations can be used to represent countless different ideas about many different subjects.

We now discuss some of the basic characteristics of the algebra of logic. A thorough study of this subject, usually called Boolean algebra, is beyond our scope here. We shall, however, study the fundamental concepts and consider some examples of their use.

2.1 STATEMENTS

The declarative sentence is the standard format in our language for the communication of ideas and observations. We use this same type of sentence also to express emotion, opinion, feeling, and the products of our sometimes rich and vivid imaginations. Since our present purpose is to develop a system for use in scientific reasoning, which deals with "facts" and "objective observations," we need a screening process to remove from our operations the nonscientific sentences. We do this by agreeing to use only those sentences that can be characterized as either "true" or "false." Such sentences are called **statements**, and their properties are stated formally as follows:

Properties of Statements A declarative sentence is a **statement** if, and only if

1. It can be adjudged to be "true" or "false."
2. In the context of a situation it is either "true" or "false," but not both.

We need some explanations and comments about these properties. Notice first that they are stated as **properties** and not as a **definition**. "Statement" is an undefined term, just as "set" is. A statement has the properties, or characteristics, listed above, and these are analogous to the properties of a set. We shall also leave "true" and "false" as undefined terms, but our experience and intuition should provide an adequate idea of their meaning.

The two properties of statements might well be stated as one; they are stated separately here to emphasize that the **form** and **content** of a statement makes it **capable** of being labeled as "true" or "false," and second, that this judgment must be made. The restriction that statements cannot be true and false simultaneously rules out paradoxical sentences such as, "This statement is false." Such declarations are assigned to discussions of the philosophy of communications; our purpose here is to provide machinery primarily for use with scientific concepts and "proofs."

Example 2.1.1

The following sentences are statements:

a. The consumer price index rose by more than 4% in 1970.

b. The federal government runs the banks in the United States.

c. Mozambique is a city in Canada.

d. Performing experiments is a basic process of scientific research.

e. The population of our 10 largest cities is decreasing.

The following sentences are not statements:

f. Mathematics is dark brown.

g. A large city is the best place to live.

h. Our tax structure needs a complete revision.

i. He is a member of Congress.

In Example 2.1.1 statements **a** and **d** are true while **b** and **c** are false. Statement **e** may be true or false, depending on the situation at the time the statement is made. In the **context of that situation**, however, it will have a specific truth value, and it qualifies as a statement for that situation. Contrast this with sentence **g**, which is not a statement. This sentence is too vague. No criteria exist for what constitutes the "best" place to live. Any such criteria would certainly be different for different people. The sentence states an opinion instead of a fact.

Sentence **h** does not qualify as a statement because of the word "complete." In this context "complete" means different things to different people, and no single truth value could be assigned.

Sentence **i** is not a statement, in a strict interpretation of the properties of statements, because "he" is not identified. In a particular situation, if it were known who was meant, the sentence would be a statement. In this example, without any context to supply the identity for "he," it is called an **open statement**, which can be made either true or false by letting the "he" represent different people. In a later section we shall make a distinction between statements and open statements and call the latter **propositions**. Note that the same distinction is made in number algebra. We have the statement that $(x+a)^2 = x^2 + 2ax + a^2$. This is a true statement no matter what numerical values are assigned to "x" or "a." Such statements are often called **identities**. We have also open statements, such as $3x^2 + 5x = 8$, which are true only when certain specific numbers are used to replace the x.

2.1.1 Compound Statements

The statements in Example 2.1.1 are **simple statements**; that is, each expresses a single idea. It is possible to combine simple statements, with words called **connectives**, to form **compound statements**. Compound statements have all the properties of simple statements and can be combined to form other compound statements. The process is analogous to forming compound sets using set operations.

In symbolic logic, compound sentences are dissected into simple sentences, each of which is assigned a symbol (usually a letter) to represent it. These symbols are connected by symbolic connectives to form the symbolic representation for the compound statements.

There are five basic connectives used in symbolic logic. We shall consider first the simplest three of these, the conjunction, the disjunction, and the negation.

Conjunction (symbol, $\wedge$): The name given to the connective used to symbolize the concept "and."

Disjunction (symbol, $\vee$): The name given to the connective used to symbolize the concept "or."

Negation (symbol, $-$): The name given to the connective used to symbolize "not."

(**Note**: The negation is not strictly a **connective** since it is not used to join two statements. It is, however, a necessary modifier for use with a single statement, and it is usually grouped together with the more conventional connectives.)

Example 2.1.2

"The leaf is green and smooth" is a compound statement combining "The leaf is green" with "The leaf is smooth." If we use the letter p to represent "The leaf is green," and q to represent "The leaf is smooth," the compound statement is symbolized, $p \wedge q$.

If we change the statement to read "The leaf is green, but not very smooth," the second part of the statement is now a negation. It is symbolized $-q$. The modifier "very" is not translated into the symbols, since it does not change the essential meaning of the sentence. The entire sentence is symbolized $p \wedge -q$. Note that we use the "and" connective where the language says "but." In language we use "but" in place of "and" to show that the second part of the sentence runs contrary to our expectations. We need such nuance in our communications, but in logic we are concerned only with the essential meaning, which is that the leaf is both green **and** not smooth.

Example 2.1.3

"The Senate will consider either the welfare proposal or the tax revision bill next week." This statement signifies that an alternative exists between the statements, "The Senate will consider the welfare proposal," and "The Senate will consider the tax revision bill." If we let p represent the first of the alternatives and q, the second, the statement is symbolized using the disjunction: $p \vee q$.

2.1.2 Truth Tables

In accordance with the properties of statements each statement in logic must have a truth value, and there are exactly two possibilities for this value, true and false. In a logical analysis of a situation we can dissect the statements about the situation into their basic components, some of which will be true and some of which will be false. But when these basic components are combined to form compound statements, are the resulting statements true, or are they false? If symbolic logic is to be of any value to us, there must be a standard procedure for finding the answer to this question.

For example, consider the two simple statements, "John is intelligent," and "John studies hard." Suppose both statements are true. We have little difficulty in deciding that the compound statement, "John is intelligent, and he studies hard," is true. A question arises, however, for the statement, "Either John is intelligent, or he studies hard." Is this statement true, or does the use of "or" imply that the combined statement is true only if one or the other of the components is false? It is desirable to have a consistent, orderly procedure to resolve such questions and to assign truth values that take into account both the connectives used and the truth values of the simple components.

In developing such a procedure we start with the two "logical possibilities" for each simple statement. If the statement is true, we assign it a "truth value" of T. If it is false, its truth value is F. In ensuing discussions we shall use the term "logical possibility" to mean one assignment of truth values to the **basic components** of a statement. If a compound statement has two components, there are four logical possibilities—both components are T; the first is T, and the second is F; the first is F, and the second is T; and both components are F. For each of these four logical possibilities the compound statement is assigned a truth value determined by the connective used to form the compound statement. Thus, it is not necessary to puzzle over a complicated statement trying to decipher from its meaning whether it is true or false. It can be broken apart into its simple components, whose truth value is more easily determined; and then the **form of the statement**—the connectives used to combine the components—will determine the truth value.

When working with compound statements, we must, of course, make sure to provide for all of the logical possibilities of different truth values for all of the components. A convenient way for tabulating these possibilities is a truth table. If a compound statement has two simple components, say p and q, the table listing all of the logical possibilities has four rows, as shown in Figure 2.1.1

p	q
T	T
T	F
F	T
F	F

Figure 2.1.1

Any compound statement with two simple components will have four truth values, one for each of the logical possibilities shown in the four rows of the table.

To use the system we must now take a typical statement made with each connective, and decide upon its truth value for each of its logical possibilities. Then any compound statements formed from the given connectives will always receive this same set of truth values. We use intuition and expected meaning whenever possible in deciding upon these truth values. There are situations, however, in which an arbitrary assignment of truth value must be made. In these cases the truth value is chosen to make the system as useful as possible.

The system can be used in two ways: first, to obtain truth values for a statement formed with a given connective; and second, to decide which connective to use when we want to form a statement with a given set of truth values. Thus, we can consider the choice of a truth value for each logical possibility as a **definition** for a connective. We shall define a conjunction, for example, as that connective having a particular assignment of truth values. Definitions of this type for the conjunction, the disjunction, and the negation follow.

2.1.3 Definition of Conjunction

Let us proceed step by step through the definition of the conjunction. The task is to decide upon the truth value of a compound statement, comprised of two simple statements joined by "and" for each of the four possible combinations of truth values for the components. If p and q represent the two components, $p \wedge q$ will represent the compound statement. We refer to the truth values listed in Figure 2.1.1. We shall use this table and add a third column for the truth values of $p \wedge q$.

We have no problem in deciding that if p and q are both true, $p \wedge q$ should be true also. Similarly, if p and q are both false, $p \wedge q$ is also false. Thus, we place a T in column 3, row 1 of the table; and, an F in column 3, row 4. In rows 2 and 3, one of the components is true; and the other false. It is useful, and reasonably consistent with the expected meaning of "and" that the compound statement should be false in both these cases. That is, a statement formed with a conjunction is true only when both its components are true; it is false otherwise. Accordingly, we have the following definition for the conjunction, the connective $\wedge$.

Definition 2.1.1

Conjunction:

p	q	$p \wedge q$
T	T	T
T	F	F
F	T	F
F	F	F

In any given situation the conjunction has **one** of the truth values listed in the third column of the table. The truth values of the components determines in which row the appropriate truth value is listed. This same truth value is used without regard to the meaning or the complexity of the component statements represented by p and q. That is, either p or q, or both, could represent a compound statement. If so, the truth values would have to be determined from their components before the truth value could be obtained for $p \wedge q$ from the table. Combinations of compound statements are studied in greater detail in the next section.

2.1.4 Definition of Disjunction

To develop a truth-table definition for the disjunction we consider the expected meaning of the connective "or." Certainly, a compound statement joined by "or" would be true if either of the components were true and the other false. Also, the compound statement would be false if both components were false. If we use the same format as before for the truth table, the third column, representing $p \vee q$, has a T in rows 2 and 3, and an F in row 4. A question arises concerning the truth value when both components are true. It is useful and not inconsistent with the meaning of "or" to assign the truth value T in this case. That is, the disjunction forms a compound statement that is false only when both component statements are false, and true in all other cases. We have, therefore, the following definition for the disjunction, the connective $\vee$.

Definition 2.1.2

Disjunction:

p	q	$p \vee q$
T	T	T
T	F	T
F	T	T
F	F	F

As with the conjunction, a compound statement formed with a disjunction will have **one** truth value, which is determined by the truth values of the components. The components, of course, can be compound statements themselves.

Although the truth value T was chosen for the case in which both components are true, a choice of F could have been made without contradicting the meaning of "or." In fact, a connective incorporating this choice is sometimes used. It is called an **exclusive disjunction** and is usually symbolized "$\underline{\vee}$." It is the connective of "forced choice." "Joe will join either the Air Force or the Navy." "Either a statement is true, or it is false." In our subsequent discussions we shall not need this connective, and we shall use only the "standard" disjunction defined above.

2.1.5 Negation

The negation is that "connective" which changes the truth value of a statement. It can be formed in words by preceding the statement with the phrase, "It is not the case that. . . ." In verbal statements the negation is called the **denial** of a statement. It is not strictly a connective, since it does not combine two statements. It is included in this section, however, because it is an important and widely used modifier for statements.

Definition 2.1.3

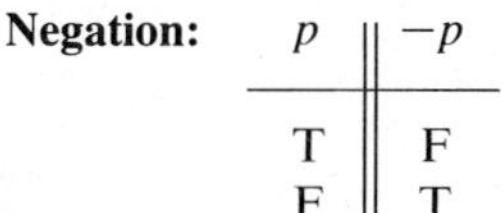

Negation:

p	$-p$
T	F
F	T

Note that the definition for the negation contains only two logical possibilities, since the negation involves only one component statement. The "–" used to symbolize the negation* is placed immediately to the left of the statement it modifies. Grouping symbols are used when needed. $-(p \wedge q)$ is the negation of the entire statement, $p \wedge q$.

EXERCISE 2.1

1. Which of the following are statements of logic?

a. The sun is shining.
b. The tax rate will increase next year.
c. No one under 18 will be admitted.
d. Please help me with this problem.
e. You get a 10% discount if you pay your income tax in cash.
f. Do not enter.
g. The speed limit is 25 miles per hour.
h. I never tell the truth.
i. Bill never tells the truth.
j. The price index will never go up by more than two points in one month.
k. The world is coming to an end.
l. The business outlook for next year appears brighter.

*There are other symbols in use to denote the negation. p', $\sim p$, and $\tilde{p}$ are all used in place of our $-p$. When grouping symbols are used, we have $-(p \wedge q)$, $(p \wedge q)'$, $\sim(p \wedge q)$, $(\widetilde{p \wedge q})$; all are used to symbolize the negation of $(p \wedge q)$.

2. Write five sentences that are statements of logic, and five that are not.

3. Construct a truth table for each of the following.
   a. $-(-p)$ (the double negation)
   b. $p \wedge -p$ (the law of contradiction)
   c. $p \vee -p$ (the law of excluded middle)

4. "The school colors are red and white." This sentence uses the connective "and." Is this a compound statement in the form of a conjunction? Explain.

5. Given the compound statements, r, s, t, u, and v, formed from the simple statements p and q. The truth values for r, s, t, u, and v are shown in the table.

p	q	r	s	t	u	v
T	T	T	F	T	T	F
T	F	T	T	F	F	T
F	T	T	T	T	F	T
F	F	F	F	T	T	T

   a. If p is T and q is F, what is the truth value for s? For v?
   b. If p is F and q is T, what is the truth value for r? For u?
   c. For which truth values of p and q do r and t have the same truth values?
   d. If q is F, what truth value for p will make r true? What truth value for p will make v true?

6. Form a truth table for
   a. $-p \vee q$
   b. $-p \wedge -q$
   c. $-(p \vee q)$

7. Write each of the following compound statements in symbolic form. Let p represent, "Bill has a job"; and q, "Bill attends college."
   a. Bill either has a job or is attending college.
   b. Bill has a job now, and he is not attending college.
   c. Bill neither has a job, nor is he attending college.
   d. Bill does not have a job, but he is attending college.
   e. Bill cannot possibly have a job and attend college.

8. Under the assumption that p is true and q is false, assign a truth value to each of the statements of Exercise 7.

9. Let p represent "The economy is expanding," and q "Real income is rising." Translate the following symbolic statements into English.
   a. $p \vee q$ b. $p \wedge q$
   c. $-p \vee -q$ d. $p \wedge -q$
   e. $-(p \vee q)$ f. $-(p \wedge q)$
   g. $-p \wedge -q$ h. $-(q \vee -p)$

10. Assume that p has truth value F and q has truth value T. Find the truth values for each of the statements of Exercise 9.

11. Analyze the meaning of the verbal statements formed for Exercise 9. Find as many pairs of statements as possible that have the same meaning.

12. a. Assume that you wish to show that each of the following statements is false—that is, the negation of the statement is true. What evidence would you need in each case?
    i. All chickens are hatched from eggs.
    ii. No man is completely sane.
    iii. Some politicians are dishonest.
    iv. There is only one way to solve that problem.
  b. State the negation of each statement in words.

2.2
CONDITIONAL AND BICONDITIONAL

2.2.1
Conditional

The most frequently used form of statement in science and in deductive reasoning (we could almost say the "natural" form of statement), is the **conditional**. This is the form of statement involving the concept, "**If**..., **then**...," or, "...**implies**...." If we wish to explain what happens under certain circumstances, we use a statement in the form of a conditional. The preceding statement is an illustration of itself. The description of the "circumstances," the "if" part of the statement, is called the **hypothesis**, or **premise**. In the illustrative statement above we hypothesize, or presume, that we wish to explain something. The description of what happens, the "then" part of the statement ("then" may be implied instead of expressed) is called the **conclusion** or **consequence**. In the illustration above, the consequence of the desire to explain is that we use a statement in the form of a conditional.

The following are additional examples of **conditionals**, statements in conditional **form**:

In a criminal trial **if** the jury is not unanimous in the verdict, (**then**) the defendant must be tried again." This statement explains what happens in our judicial system under the circumstances that the jury cannot reach a verdict. The result, or consequence, is that the defendant is tried again.

"The congressman will vote for the bill if he receives a letter from each of his constituents urging him to do so." The **hypothesis** here is that the congressman receives a letter from each of his constituents. The **consequence** is that he will vote for the measure. Note that the **consequence** is stated **before** the **hypothesis** in this statement. The **meaning** indicates which part of the statement is the hypothesis and which is the conclusion. In the symbols of logic the symbolic form identifies the hypothesis and the conclusion. It is important to identify the hypothesis and conclusion correctly. The meaning of a statement is quite different if they are interchanged. We discuss the effect of interchanging the two parts of a conditional in greater detail in a later section.

The symbol used to denote the conditional is an arrow, "$\rightarrow$." $p \rightarrow q$ is the symbolic form for "If p, then q," or "p implies q." The **hypothesis always precedes** the arrow, and the **conclusion follows** it, **regardless of the order used in the verbal form of the statement**.

When using a conditional we are usually most interested in the case in which the hypothesis is true. If the conditional statement is true, the conclusion is expected to be true whenever the hypothesis is. That is, the truth of the conclusion follows as a result of the truth of the hypothesis. However, we must also provide for the other logical possibilities.

Suppose that the hypothesis is true, but the conclusion turns out to be false. That is, what if the things we said would happen under certain circumstances did not happen? The statement in which the "prediction" or "promise," was made must be considered untrue. Thus, for the statement $p \rightarrow q$, when p and q are both true, the statement $p \rightarrow q$ is true; if p is true, but q is false, the statement $p \rightarrow q$ is false.

Let us consider the statement, "If you score above 90% on the final exam, you will get an A in the course." A student who scores above 90% on the final and then receives an A is satisfied that the promise was a valid one. However, a student who scores above 90% on the final and does not receive an A quite justly feels that he has been deceived. The promise was false.

In both of the above situations the hypothesis, p, is true. When the circumstances of the hypothesis are not satisfied (i.e., when p is false), the prediction, or promise, about what will happen is not tested. We have no guide from intuition to suggest a truth value for $p \rightarrow q$. As mentioned earlier, an arbitrary assignment of truth value must be made in this case, and the choice is made so that the final definition is the most useful and consistent possible. Accordingly, the conditional is **defined** to be true in both cases in which the hypothesis is false. This means that the conditional is to be considered true unless there is definite evidence that it is not. As we shall see later, this definition is the most consistent with other variations of the conditional.

Looking again at our example, we see that the student who does not score above 90% on the final exam makes no test of the promise. His grade must, therefore, be based on other considerations. Note that the statement makes no mention of what will happen in the event a student's score is less than 90%. That is, the statement does **not** say that if a student scores less than 90%, he will not get an A. This is an important distinction in the use of the conditional in logic. It states what **will happen** under **certain conditions**. It **makes no mention** of what will happen if these conditions are not satisfied. To specify that you will not receive an A if you do not score above 90% requires a statement different from the one originally made. We shall consider this situation later.

The above considerations are summarized in the truth table definition for the conditional:

Definition 2.2.1

Conditional: "If p, then q"; or "p implies q."

p	q	$p \rightarrow q$
T	T	T
T	F	F
F	T	T
F	F	T

2.2.2
Biconditional

The biconditional embodies the concept expressed by ". . . if, and only if, ..." We have encountered this expression before in definitions of other terms. To say, "p if, and only if, q" is to say "p, if q" **and** "not p, if not q." This is the statement that says p will happen if q does, but p will not happen if q does not. (Note the reversal of the roles of p and q from those in the conditional.) The truth values of the two components are tied together—the truth of one requires the truth of the other, and vice versa. Thus, the biconditional is true whenever the two components have the same truth value, and it is false when they have different truth values. This is summarized in the formal definition below:

Definition 2.2.2

Biconditional: "p if, and only if, q."

p	q	$p \leftrightarrow q$*
T	T	T
T	F	F
F	T	F
F	F	T

Note that, unlike the conditional, the order of the biconditional can be reversed without changing the meaning. $p \leftrightarrow q$ means the same as $q \leftrightarrow p$.

2.2.3
Variations of the Conditional

We have now defined the connectives used in symbolic logic. The most interesting, and most difficult to understand and use correctly, is the conditional. It is the basis of the thought process we call "logical reasoning." It is also the form in which we express our understanding of various situations and phenomena.

*The biconditional is denoted by a two-way arrow, $\leftrightarrow$. $p \leftrightarrow q$ is translated "p, if, and only if, q."

We are interested primarily in conditionals that are true. These are the statements of fact. We shall call such statements **implications**. That is, a statement of fact in the form of a conditional is called an implication. With an implication we know that when the conditions specified in the premise are met, the results stated in the conclusion are surely true. If the conditions in the premise are not met, the conclusion may or may not be true; but the important aspect of an implication is that the truth of the premise **guarantees** the truth of the conclusion.

Often an implication is a statement of cause and effect, particularly in statements of "scientific fact." For example, "If water is heated, it will vaporize" states the effect of heating water. The heat causes the water to evaporate. In other cases an implication is a statement expressing our understanding or knowledge of a situation. "If Mary does not answer the telephone, she is not at home" is a statement based on a knowledge of Mary's reaction to the telephone.

In this section we explore some of the variations in the form of the conditional and some of the many ways a conditional can be stated verbally.

Note that in "pure" mathematics the meanings assigned to the parts of a conditional need have no relationship to each other at all. "If the sun is a star, Tuesday precedes Wednesday" is a conditional having truth value T. It is, therefore, an implication by our definition. Such sophistry is useful, and probably necessary, in working abstractly with the structure of a mathematical system. Our purpose here, however, is to study the application of mathematical concepts to real situations. We shall, therefore, treat only examples in which there is some correlation between hypothesis and conclusion. The formulation and verification of implications is the objective of almost all scientific research. The correlation between the hypothesis and the conclusion is the essential part of the statements. In most branches of mathematics also there is concern about the correlation between hypothesis and conclusion. Implications are usually called **theorems** in mathematics, and they are the means by which mathematical knowledge and understanding are communicated.

One of the most common **misuses** of implications is the assumption that since the truth of the hypothesis establishes the truth of the conclusion, the truth of the conclusion will establish the truth of the hypothesis. In situations that are well understood, this problem does not arise. "If a car runs out of gasoline, its engine stops" is an implication. However, if a car's engine stops, we do not immediately conclude that the fuel supply is exhausted, although this could be the case. The circumstances of the situation influence our judgment. A car drives up to a curb, and the engine stops. Is it out of fuel? Perhaps, but it is more likely that the driver turned off the ignition. But in situations we do not understand or in situations involving our prejudices and emotions, it requires careful thought to identify correctly the hypothesis and to remember that the conclusion cannot be used to verify the hypothesis. The student can undoubtedly think of examples in which other people have made incorrect judgments as a result of using a conclusion to verify a hypothesis.

A word of caution: It is not the case that a conclusion **never** verifies a hypothesis. "If the wax is hot, it melts." The wax melts as the result of its being heated. But we know also that if wax is molten, it is hot. Whenever the statement can be reversed, as in this case, it is actually a **biconditional**, instead of a conditional.

2.2.4 Converse

The statement formed when the hypothesis and the conclusion of a conditional statements are interchanged is called the **converse** of the conditional. In symbols, $q \rightarrow p$ is the converse of $p \rightarrow q$.

Forming the converse of a conditional statement changes the **role** the component parts play in the statement and not necessarily their order in the verbal statement. "If the price is too high, the product will not sell" and "The product will not sell if the price is too high" are the same statement in spite of the change in the order of the verbal components. The converse of this statement is, "If the product will not sell, the price is too high"; or, using the original order of the components, "The price is too high if the product will not sell." Notice that the converse has a different **meaning** than the original statement, not merely a different word order.

2.2.5 Contrapositive

The **contrapositive** form of a conditional statement has the same logical meaning, although the verbal expression sounds quite different. In symbols, $-q \rightarrow -p$ is the contrapositive form of the conditional $p \rightarrow q$. Consider the statement, "If Mary is at home, she will answer the telephone." We used the contrapositive form of this statement earlier as an example, "If Mary does not answer the telephone, she is not home." There are times when the contrapositive form is the more natural form of expression. Both of the above statements have the same logical meaning; but in a situation in which a telephone call is made to Mary's house, the contrapositive form may well be the more natural expression. The statement, "If you are to graduate from college, you must pass a course in English," is not likely to be used in discussing a student's program. The more usual expression is the contrapositive form, "If you don't pass a course in English, you cannot graduate." Both statements have the same meaning—passing a course in English is necessary for graduation.

2.2.6 Inverse

The other variant of a conditional is its inverse, in which the premise and conclusion are each replaced by their respective negations. In symbols, $-p \rightarrow -q$ is the **inverse** of the statement $p \rightarrow q$. The meaning of the inverse, as might be expected, is different from the meaning of the original statement. Consider the statement, "If a person earns $100,000 per year, he can afford a new car each year." The inverse of this statement is, "If a person does not earn $100,000 per year, he cannot afford a new car each year." We are probably inclined to agree with the first statement, and disagree with the second. Clearly, the inverse has quite a different meaning from the original statement.

The truth values of the four variations of the conditional are summarized in the table in Figure 2.2.1.

p	q	$-p$	$-q$	Conditional $p \rightarrow q$	Converse $q \rightarrow p$	Inverse $-p \rightarrow -q$	Contrapositive $-q \rightarrow -p$
T	T	F	F	T	T	T	T
T	F	F	T	F	T	T	F
F	T	T	F	T	F	F	T
F	F	T	T	T	T	T	T

Figure 2.2.1

Note that the conditional and the contrapositive have the same truth values and, thus, the same logical meaning. Note also that the converse and inverse have the same truth values and mean the same in logic. In fact, the inverse is the contrapositive form of the converse.

2.2.7 Verbal Expressions for the Conditional

The use of the conditional is complicated by the variety of ways of expressing it verbally. We do not always use "If..., then..." in stating a conditional. This form often sounds awkward. It is sometimes better to state the conclusion first: "He will run for Congress if he can get enough backing." In the conventional idiom of logic this statement would become, "**If** he can get enough backing, **then** he will run for Congress."

Many of the phrasings carry nuances that are not relevant to the basic logic, but which are important in conveying our feelings about a situation. "I shall go swimming only if the water is warm" is a statement with a contrapositive flavor. The meaning is "If the water is not warm, I shall not go swimming," which could be symbolized, $-w \rightarrow -s$. This is the contrapositive of the statement $s \rightarrow w$, "If I go swimming, the water is warm." My swimming does not cause the water to be warm, of course, but "the water is warm" is a deduction that can be made from the fact that I am swimming in it.

Another form of expression in this same situation is "I shall not go swimming unless the water is warm." The word "unless" can be replaced by "if... not...," and the statement becomes the "standard" contrapositive wording in reverse order, "I shall not go swimming if the water is **not** warm."

The original statement about swimming used the words "only if." Observe that the phrase introduced by "only if" becomes the **conclusion** in the standard phraseology. There is an important difference in the meaning between "if...," which is used to introduce the **premise**, and "only if . . . ," which is used to introduce the **conclusion**. The "if" is used to signify that the conditions in the premise are sufficient to bring about the conclusion. The "only if" is intended to signify **necessity**. These concepts are discussed more fully in the next section.

2.2.8 Necessary and Sufficient Conditions

We have mentioned earlier that the results of scientific experiments or research are frequently stated in the form of conditionals. Much of the effort of scientific research goes into verifying the truth of these conditional statements. In mathematics the statements are called theorems, and proving theorems occupies much of the time of research mathematicians.

The two components of a conditional statement are thought of as "conditions." They are statements whose truth or falsity is a determining factor in what is known about a situation. The truth of the premise, or hypothesis, in the conditional is supposed to bring about, or indicate, the truth of the conclusion. That is, the truth of the premise is **sufficient** to guarantee the truth of the conclusion. If it does not, the conditional itself is false. These conditions are symbolized in the truth-table definition of the conditional, $p \rightarrow q$, by the fact that when p, the hypothesis, has truth value T, and the conditional, $p \rightarrow q$, has truth value T (it is an implication), the conclusion, q, has truth value T also. When p has value T and q has value F, the conditional, $p \rightarrow q$, itself is false.

Consider the statement, "Being born in this country is a sufficient condition for citizenship." This is translated into the standard terminology of logic as, "If a person is born in this country, then he is a citizen." The **sufficient condition** becomes the **hypothesis** in the conditional statement. The statement does **not** say it is **necessary** to be born in this country to be a citizen. In fact, a foreign-born person can be naturalized and attain citizenship also. Thus, although the truth of the hypothesis in an implication is **sufficient** to guarantee the truth of the consequence, it may not be **necessary**.

In the statement, "It is **necessary** to pass English in order to get a college degree," the **necessary condition** becomes the **consequence** when this statement is translated into the standard form. It becomes "If you are to obtain a college degree, then you must pass English." Certainly, we cannot say, "If you pass English, you will receive a college degree." That is, a condition that is necessary does not become a sufficient condition. The usage of terms here may become less confusing if we consider the meaning of "necessary." Something that is necessary is something we cannot get along without. The idea can be illustrated by expressing the example above in contrapositive form, "If you do not pass English, then you cannot obtain a college degree." As we noted earlier, the contrapositive form has the same logical meaning as the standard form although it sounds quite different in words.

Summarizing, a statement specified as a **sufficient** condition becomes the **hypothesis** in a conditional statement. A component statement specified as a **necessary** condition is the **consequence**, or conclusion, in the conditional statement.

If a component statement is specified to be both a necessary and a sufficient condition (a phrase used frequently in mathematics) for the truth of another statement, the resulting combined statement is a **biconditional**. For example, a necessary and sufficient condition for being eligible to vote is to register in the legally prescribed manner. In the standard form of logic, "A person is eligible to vote if, and only if, he registers in the legally prescribed manner."

The frequently used wordings associated with the conditional are summarized in the table below:

Verbal expressions for the conditional, $p \rightarrow q$

1. If p, then q.
2. p implies q.
3. q if p.
4. p only if q.
5. p is a sufficient condition for q.
6. q is a necessary condition for p.

All the above forms are translated into the same symbols: $p \rightarrow q$.

EXERCISE 2.2

1. For each of the following conditionals, identify the hypothesis and the conclusion. **Suggestion**: Rephrase the statements in the form of "if . . . , then . . . ," or ". . . implies . . . ."
   a. You are eligible to vote if you are 18 or over.
   b. If the tree is planted and cultivated properly, it will bear a large quantity of fruit.
   c. A student qualifies for the Dean's list if his grade point average is 3.2 or higher.
   d. Driving a "horseless carriage" can be fun if one has lots of time and money and is handy with tools.
   e. If you have a steady job, you can buy a car on the installment plan, provided you make the monthly payments on time.
   f. A student who neglects his studying will often fail.
   g. Conservationists have become greatly concerned about our natural environment.
   h. Logically equivalent statements have the same truth values.

2. Let p represent "John can pass the aptitude test," and q, "John can handle the job." Translate each of the following into symbols.
   a. If John can pass the aptitude test, he can handle the job.
   b. If John can handle the job, he can pass the aptitude test.
   c. John can handle the job if, and only if, he can pass the aptitude test.
   d. John can pass the aptitude test only if he can handle the job.

3. Let p represent "prices rise"; q, "wages rise"; and r, "there is inflation."
   a. Translate the following into symbols:
      i. If prices and wages rise, there is inflation.
      ii. If there is inflation, prices and wages rise.
      iii. There is inflation if, and only if, prices and wages rise.
      iv. If wages rise and prices do not, there is not inflation.
      v. The fact that there is inflation implies that prices will rise if wages rise.
   b. Translate the following into words:
      i. $(p \wedge r) \rightarrow q$
      ii. $-r \rightarrow (-p \vee -q)$
      iii. $(p \vee q) \rightarrow r$
      iv. $(p \wedge q) \vee -r$
      v. $(p \rightarrow q) \rightarrow r$

4. Express each of the following statements; first, using "only if"; second, using "necessary condition"; and third, using "sufficient condition."
   a. If the wind blows, the lake is rough.
   b. If the gas tank is empty, the car will not start.
   c. If a number is a multiple of 6, it is a multiple of 3.

5. Express each of the following statements using the form, "if . . . , then . . ."; then as a statement using "only if"; and third, as a statement using "necessary condition."
   a. Drinking coffee in the evening keeps me awake.
   b. Every registered student is assigned a permanent file number.
   c. All normal dogs have four legs.
   d. Citizens 21 and over have the right to vote.

6. Write the converse, inverse, and contrapositive of each of the statements of Exercises 4 and 5.

7. For each of the following statements, indicate what evidence would make it false.
   a. It never rains when the sun is shining.
   b. If $A \cap B = \emptyset$, then A is the null set.
   c. If an implication is true, its converse is also.
   d. If that dish falls off the table, it will break.

8. Let p represent "The theory is correct," and q, "The experiment is successful." Translate each of the following into symbols:
   a. The theory is correct only if the experiment is successful.
   b. The experiment's success is a sufficient condition for believing the theory is correct.
   c. It is necessary for the theory to be correct in order that the experiment be successful.
   d. The experiment will be a success if the theory is correct.
   e. The theory is not correct unless the experiment is a success.
   f. The correctness of the theory is a necessary and sufficient condition for the success of the experiment.
   g. The experiment will be a success if, and only if, the theory is correct.

9. Use the conditional "If you own your own home, then you pay property taxes" to
   a. Form the converse.
   b. Write the contrapositive.
   c. Write the inverse.

10. Write the contrapositives of the following: (Advertisers frequently make statements and hope you will act on the contrapositive.)
   a. Good citizens do not litter streets.
   b. Smart shoppers buy at Savabuk's.
   c. If you really care, you'll give her diamonds.
   d. If I am elected, you'll have lower taxes.
   e. The best costs a little more.

11. You wish to prove the **converse** of each of the following statements. Write two different statements for each case, either of which is a statement of what you wish to prove.
   a. If two sides of a triangle are unequal, the opposite angles are unequal.
   b. If a point is on the perpendicular bisector of a line, it is equidistant from the ends of the line.
   c. If alternate interior angles are not equal, the lines are not parallel.
   d. If two chords of a circle are unequal in length, they are unequally distant from the center.

2.3
TRUTH TABLES FOR COMPLEX STATEMENTS

Statements can be combined in an almost endless variety of ways using the connectives that we have studied. Simple statements can be combined to form compound statements; and these can be combined to form more complex compound statements. We wish to study now one way to dissect and analyze such complex statements to obtain their logical meaning. First, we need to see how complex statements are symbolized.

In the same way that it is necessary to indicate groupings of numbers in algebra to make clear how arithmetic operations are to be applied, statement symbols must be grouped to indicate how connectives are to be applied in symbolic logic. We have used simple grouping symbols in the exercises of previous sections in the expectation that the student would understand them, since the symbols used and their meanings are the same as those used in algebra. We shall now formalize the rules for using these symbols.

The basic rule for grouping symbols is that they are used whenever needed to remove ambiguity from a statement. The statement $p \wedge q \vee r$ is ambiguous because it can be interpreted in two ways; as the conjunction of p with $q \vee r$; or as the disjunction of $p \wedge q$ with r. Parentheses are used to make clear which interpretation is intended. The first possibility is written $p \wedge (q \vee r)$; and the second, $(p \wedge q) \vee r$.

Customarily, the negation symbol applies **only** to the **statement immediately to its right**. If it is desired to apply the negation to more than a single symbol, the expression to be negated is enclosed in parentheses or brackets. In the statement $-p \vee q$, for example, the negation applies only to p. If it is intended that the entire disjunction be negated, the statement is written $-(p \vee q)$.

Let us now consider the way in which truth values are assigned to complex statements—statements involving two or more connectives. The assignment of truth values for each of the logical possibilities for a statement determines its logical meaning. It has been noted that the conjunction, disjunction, conditional, and biconditional apply to **two** statements, whereas the negation applies to **one**. Grouping symbols are interpreted to mean "do this first," referring to assigning truth values to the enclosed combination. When multiple grouping symbols are used, the assignment of truth values starts with the innermost group. Each group is considered as a single statement in forming further combinations, and its truth value for each logical possibility is determined by the connective in it. Let us consider again the example, $p \wedge q \vee r$. Let us suppose that p is F, q is T, and r is T. In the interpretation $(p \wedge q) \vee r$, $(p \wedge q)$ is F. Since r is T, however, the disjunction of $(p \wedge q)$ with r has truth value T. In the interpretation $p \wedge (q \vee r)$, the compound statement $(q \vee r)$ is T. The conjunction of this statement with the false statement, p, has a truth value F.

It is necessary to have a truth value for each logical possibility to get the logical meaning of a statement. A truth table is a convenient tabulation of these truth values. When a truth table is constructed for a complex statement, a truth value for each of the logical possibilities is obtained for the **grouped combinations first**. These truth values are used to find the truth values of additional combinations. Thus, the truth table for the entire complex statement is built in a sequence of steps.

Example 2.3.1

Let us construct a truth table for $-(p \vee q)$. The table is started in the standard manner by listing the possible truth values for the component statements, p and q. Next, we consider the disjunction within the parentheses.

p	q	$(p \vee q)$	$-(p \vee q)$
T	T	T	F
T	F	T	F
F	T	T	F
F	F	F	T

Figure 2.3.1

This disjunction is shown in the third column of the table in Figure 2.3.1. The entries in this column are now considered row by row as the truth values for a statement to be negated. In row 1 the statement is T; its negation, therefore, is F. This truth value is entered in row 1 of column 4. In similar fashion we obtain the entries for the remaining rows of column 4. This column gives the truth values for $-(p \vee q)$ for each of the logical possibilities. Columns 1, 2, and 4 are considered to be the truth table for $-(p \vee q)$. Column 3 is used only to facilitate obtaining the truth values in column 4.

Example 2.3.2

Let us consider now a more complex statement, $(p \vee q) \wedge -q$. The basic components are the same as before, p and q. In constructing the truth table (Figure 2.3.2), we list first the four logical possibilities. The statement, $p \vee q$, is enclosed in parentheses. In column 3 we list the truth values for this statement for each logical possibility. In column 4 we list the truth values for $-q$. In each row, of course, the truth value is the opposite of the value listed in column 2 for q. The statements of column 3 and 4 are now considered as **components** for the conjunction, $(p \vee q) \wedge -q$. If we let r represent $(p \vee q)$ and s represent $(-q)$, the statement whose truth table we are constructing is $r \wedge s$. We use the definition of a conjunction to obtain the truth values for this statement. They are entered in column 5 of the table. In row 1, for example, the entry for column 3 (r) is T, and the entry for column 4 (s) is F. Since the conjunction is false when either of its components is false, we enter an F in row 1 of column 5.

(1) p	(2) q	(3) $p \vee q$ (r)	(4) $-q$ (s)	(5) $(p \vee q) \wedge -q$ $(r \wedge s)$
T	T	T	F	F
T	F	T	T	T
F	T	T	F	F
F	F	F	T	F

Figure 2.3.2

The entries in row 2, columns 3 and 4, are both T. The conjunction is true when both its components are true; and the entry for row 2, column 5 is T. The entries for rows 3 and 4 are obtained in a similar manner. The completed truth table for $(p \vee q) \wedge -q$ consists of columns 1, 2, and 5 of Figure 2.3.2. These are the columns that give the truth values for $(p \vee q) \wedge -q$ for each of the logical possibilities. Columns 3 and 4 are auxiliary columns used only to facilitate finding the entries for column 5. (They are very useful auxiliaries, however. Finding the truth value of a complex statement can be very confusing without them.)

Truth tables can be formulated for any compound statement by the same step-by-step process. It can be done more quickly and accurately if the student memorizes the truth tables used to define the basic connectives. Several columns may be needed if there are a number of groupings.

The statement, $[(p \to q) \land (-p \lor q)] \leftrightarrow (q \to p)$, for example, is the biconditional of the conjunction of $p \to q$ and $-p \lor q$ with $q \to p$. In making a truth table for this statement we need a column for each of the compound statements, $p \to q$, $-p \lor q$, and $q \to p$. Then we need a column for the conjunction of $p \to q$ and $-p \lor q$. After we obtain the truth values for these compound statements, we can consider them as components in the definition of the biconditional. The complete table, with the columns of the component statements shown, is given in Figure 2.3.3. Columns 1, 2, and 8 comprise the truth table for the complete statement $[(p \to q) \land (-p \lor q)] \leftrightarrow (q \to p)$.

(1) p	(2) q	(3) $-p$	(4) $p \to q$	(5) $-p \lor q$	(6) (4) $\land$ (5)	(7) $q \to p$	(8) (6) $\leftrightarrow$ (7)
T	T	F	T	T	T	T	T
T	F	F	F	F	F	T	F
F	T	T	T	T	T	F	F
F	F	T	T	T	T	T	T

Figure 2.3.3

2.3.1 Compound Statements with More Than Two Components

The previous paragraphs have discussed complex combinations with just two basic components. A more frequent situation is one in which several different components are combined. We noted that with two basic components the truth table requires four rows to provide for all the logical possibilities. As we add more basic components, the number of possibilities increases. **Each additional basic component doubles the number of rows in the truth table**. If there are three basic components, the truth table has eight rows; if there are four basic components, the truth table has 16 rows; if there are n basic components, the truth table has 2^n rows.

Example 2.3.3

Consider the statement $p \land (q \lor -r)$. The truth values for the basic component statements are listed in the first three columns of the table (Figure 2.3.4).

(1) p	(2) q	(3) r	(4) $-r$	(5) $q \vee -r$	(6) $p \wedge (q \vee -r)$
T	T	T	F	T	T
T	T	F	T	T	T
T	F	T	F	F	F
T	F	F	T	T	T
F	T	T	F	T	F
F	T	F	T	T	F
F	F	T	F	F	F
F	F	F	T	T	F

Figure 2.3.4

All of the possible combinations of truth values for these components must be listed. A pattern for assigning these truth values makes forming the table easier. A frequently used pattern divides the first column into two equal parts. Each row in the upper half has the entry T; and each row of the lower half, F. The second column is divided into four equal sections. Each row in the top section gets a T; each row in the second section, an F; each row in the third section, a T; and each row in the fourth section, an F. The third column is divided into eight equal sections. The entries for each row alternate from section to section, starting at the top with T's. (This is the pattern followed in Figure 2.3.4.)

In the table for this example column 4 is the negation of column 3. Column 5 is the disjunction of the statements in columns 2 and 4. Column 6 represents the statement whose truth values we seek. It is the conjunction of the statements of columns 1 and 5.

As the foregoing examples indicate, truth tables can become quite cumbersome for statements involving a number of basic component statements. However, the value of the truth table is that it provides an **orderly display** of **all** of the logical possibilities in a situation—all of the possible combinations of truth values for the component statements. This is useful in analyzing the results of different circumstances in a described situation (the description takes the form of a combination of statements). It is also useful in formulating a description of a situation to make certain that the description is accurate; that is, that the description correctly states what happens under unusual combinations of circumstances. A description of a situation may sound plausible when considered in terms of expected conditions, but have entirely false implications for conditions that are unusual, but possible.

In a later section we shall use a truth table to check the validity of a deduction made from the facts in a situation. The use of a computer, which can readily be programmed to construct and work with truth tables, provides a way around the cumbersome manipulations. This is one of the many situations in which a computer makes practical a powerful method of mathematical analysis by performing tedious and time-consuming symbol manipulations quickly and accurately.

The other principal asset of the use of truth tables, which should not be overlooked, is that it provides a means for making clear-cut definitions of the concepts (such as the connectives) used in logical analysis.

2.3.2 Logical Equivalents

If we refer to a truth table for a compound statement, the set of T's and F's in the column representing that statement is called its set of truth values, or the **truth values** for that statement. Statements with two basic components have four truth values, one for each logical possibility (each row of the truth table). Although countless compound statements can be formed from two components, there are only two possibilities for each of the four truth values, or a total of $2^4 = 16$ different sets of truth values. Clearly there must be a number of **different** compound statements **formed from the same basic components** that have the **same** set of truth values. Such statements are said to be **logically equivalent**.

Definition 2.3.1

Two statements are **logically equivalent** if, and only if, they are formed from the same basic components **and** they have the same set of truth values.

It is understood, of course, that logically equivalent statements must have the same truth value for each of the possible combinations of truth values for the compound statements, regardless of the order in which they are listed in the truth table. For convenience, we shall always list the tables in the sequence described earlier: The basic components are listed in the first columns at the left of the table, using as many columns as necessary. The truth values for the last of these (the farthest right of the basic components) are alternated TFTF, and so on. The next column to the left is alternated in two's, for example, TTFF. The next column to the left is alternated in four's, and so on. Thus, the first column in the table will have all T's in the top half, and all the entries in the lower half will be F's.

Example 2.3.4

Consider the compound statement $(-p \vee q)$. Its truth table is shown below. (The student should verify that it is correct.)

p	q	$(-p \vee q)$
T	T	T
T	F	F
F	T	T
F	F	T

Note that the truth values here are the same as those for the conditional $p \to q$. Thus, $(-p \vee q)$ is the logical equivalent of $p \to q$.

If we let p represent "John wins the match," and q, "John gets a trophy," $p \to q$ represents, "If John wins, he gets a trophy." $-p \vee q$ represents, "Either John does not win, or he gets a trophy." These two statements certainly do not sound alike, but they have the same **meaning** when interpreted in the strict sense required by logic.

Logical equivalence is analogous to equality for numbers or sets. Any statement can be replaced by its logical equivalent without changing the logical meaning of the statement or argument in which it appears. This makes possible changing statements into more convenient forms, and frequently into simpler forms, to facilitate a logical argument.

The symbol "≡" (and also "↔") is used to denote logical equivalence between two statements. When ↔ is used, it is assumed that the biconditional thus formed has a truth value T.

2.3.3 Tautologies

Definition 2.3.2

A compound statement that has the truth value T for all its logical possibilities is called a **tautology**.

Example 2.3.5

Consider the compound statement $[-a \wedge (a \vee b)] \to b$. We construct the following truth table.

(1) a	(2) b	(3) $-a$	(4) $a \vee b$	(5) $-a \wedge (a \vee b)$	(6) $[-a \wedge (a \vee b)] \to b$
T	T	F	T	F	T
T	F	F	T	F	T
F	T	T	T	T	T
F	F	T	F	F	T

The truth table for $[-a \wedge (a \vee b)] \to b$ consists of columns 1, 2, and 6 of the above table. (These are the columns containing the two basic components and their final combined form.) Since all of the truth values in column 6 are T's, the statement is a tautology.

Any statement that is a tautology is called also a "logical truth." Tautologies play an important role in the process of logical deduction, or proof.

Note: **Logical equivalence** is a **relationship** between **two** compound statements. Statements that are logically equivalent have the **same** truth values (not necessarily T) for the **same** logical possibilities. The term **tautology** applies to **one** statement, a statement that is **always** true.

When two statements are logically equivalent, the compound statement formed when they are connected by a biconditional is a tautology.

Example 2.3.6

(1) p	(2) q	(3) $p \rightarrow q$	(4) $-p$	(5) $-q$	(6) $-q \rightarrow -p$	(7) $(p \rightarrow q) \leftrightarrow (-q \rightarrow -p)$
T	T	T	F	F	T	T
T	F	F	F	T	F	T
F	T	T	T	F	T	T
F	F	T	T	T	T	T

The table shows that column 3 $(p \rightarrow q)$ and column 6 $(-q \rightarrow -p)$ have the same truth values. They are logically equivalent. Column 7 lists the truth values of the biconditional, $(p \rightarrow q) \leftrightarrow (-q \rightarrow -p)$. All of the entries are T's, and the biconditional is a tautology.

2.3.4 Logical Absurdity

A statement whose truth value is false for every logical possibility is called a **logical absurdity** or a **logical falsity**. The simplest form of such a statement is $p \wedge -p$, the conjunction of a statement with its negation. Such statements are used, as we shall see in a later section, in a process of proof called "indirect proof" or "proof by contradiction."

EXERCISE 2.3

1. Construct truth tables for the following:
   a. $-(p \wedge q)$ b. $(p \vee -q) \wedge p$
   c. $-(p \vee q) \vee -(p \wedge q)$ d. $p \vee -(p \wedge q)$
   e. $(p \rightarrow q) \wedge (-q \rightarrow -p)$ f. $(p \rightarrow q) \leftrightarrow -(p \wedge -q)$

2. Let p represent the statement, "The sun is shining," and q, "The sky is blue." Write each of the following in symbolic form, and construct its truth table.
   a. Neither is the sun shining, nor is the sky blue.
   b. It is not the case that the sun is shining and the sky is blue.

3. Construct truth tables for
   a. $(p \wedge q) \vee (p \wedge r)$ b. $(p \wedge -r) \vee (q \wedge -r)$
   c. $(p \rightarrow q) \wedge (q \rightarrow r)$ d. $[p \rightarrow (q \rightarrow r)] \wedge [(p \wedge q) \rightarrow r]$

4. Suppose that the statements a and b are true and the statement c is false. Find the truth value for each of the following:

a. $a \vee (b \vee c)$ b. $-a \vee c$
c. $-(b \wedge c)$ d. $a \wedge (b \vee c)$
e. $-a \vee (-b \wedge c)$ f. $-(a \vee b) \wedge c$

5. Find the truth value for each of the statements of Exercise 4 if a and c are true and b is false.

6. Determine which of the following pairs of statements are logically equivalent.

a. $-(p \wedge q)$ and $-p \wedge -q$
b. $-(p \wedge q)$ and $-p \vee -q$
c. $-p \wedge -q$ and $-(p \vee q)$
d. $p \rightarrow q$ and $(p \wedge -q)$
e. $p \rightarrow q$ and $-p \vee q$
f. $-p \rightarrow (q \wedge -q)$ and p
g. $(p \wedge q) \rightarrow r$ and $p \rightarrow (q \wedge r)$
h. $(p \rightarrow r) \wedge (q \rightarrow r)$ and $(p \vee q) \rightarrow r$

7. Determine which of the following are tautologies:

a. $-(p \vee q) \leftrightarrow (-p \wedge -q)$
b. $[p \wedge (p \rightarrow q)] \rightarrow q$
c. $[-p \wedge (p \vee q)] \rightarrow q$
d. $[(p \wedge q) \rightarrow r] \rightarrow [p \rightarrow (q \rightarrow r)]$
e. $[(p \rightarrow q) \wedge q] \rightarrow p$
f. $[(p \rightarrow q) \wedge (q \rightarrow r)] \rightarrow (p \rightarrow r)$

8. a. Given that $(p \rightarrow q)$, $(q \rightarrow r)$, and p are all true. What are the logically possible truth values for r?
b. Given that $(-p \rightarrow -q)$, $(p \rightarrow r)$, and $-r$ are all true. What are the logically possible truth values for q?

2.4
DEDUCTION AND ARGUMENT

The "scientific method" of study and research was discussed briefly in an earlier section. In bare outline the method involves the formulation of a tentative theory to explain phenomena observed in a particular situation, and the extension of the theory to related situations by "logical deduction." The tentative theory is **assumed** to be correct, and it is used to predict logically what will happen under certain conditions. The theory is tested by performing experiments to determine whether the predicted results actually occur.

An important part of the process is making certain that the predictions follow **logically** from the theory. The theory provides a set of hypotheses, or premises, and the prediction is the consequence. When a consequence does truly follow from a set of hypotheses, it is said to be **valid**. If a valid prediction is not confirmed by experiment, there are mistakes in the hypotheses; that is, the theory needs revision.

We shall discuss briefly the methods by which we determine whether a deduction from a set of hypotheses is **valid**; that is, whether it is a logical consequence of the truth of the hypotheses. One of the main virtues of symbolic logic is that the validity of a deduction can be determined solely from a study of its **logical form**. The meaning of the statements used is not involved.

As we have stated, the process starts with a set of **assumptions**, also called **hypotheses**, or **premises**. These are statements that are assigned a truth value T—they are assumed to be true. (In the scientific method these are the statements that set forth the tentative theory.) The deduction is the **consequence**, or **conclusion**. An **argument** is a statement that puts the hypothesis and the conclusion together as a single statement in the form of a conditional. In symbolic form if we let $p_1, p_2, p_3, \ldots, p_n$ represent the hypotheses (there must be at least two, but there can be as many as desired) and if we let q represent the conclusion, an **argument** is a **statement** of the form

$$p_1 \wedge p_2 \wedge p_3 \wedge \ldots \wedge p_n \rightarrow q$$

We must keep in mind that the premises and the conclusion are all statements in the logical sense. They are formed using logical connectives from simple statements. They have a truth value of T or F for each of the logical possibilities for the simple statements. We are concerned only with those logical possibilities in which all of the premises are true, and the truth value of the conclusion for these possibilities. An argument is **valid** if, and only if, **the conclusion is true for every logical possibility for which the premises are all true.**

In the preceding discussion we focused our attention on the logical possibilities for which the premises are all true. But what if there are none? That is, suppose that for each logical possibility one or more of the premises are false? Such a set of premises is called **logically inconsistent**. **Any argument formed with inconsistent premises is automatically not valid**.

We are now ready to form a definition for a valid argument:

Definition 2.4.1

Valid Argument: An argument is valid if, and only if,

1. It includes at least two premises and a conclusion.

2. There is at least one logical possibility for which **all** of the premises have truth value T.

3. The conclusion has truth value T for **every** logical possibility for which all of the premises have truth value T.

The third part of this definition can also be stated as, "The statement of the argument as a conditional is a tautology."

In a **valid argument** the **conclusion** is said to be a **valid consequence** of the **premises**.

We should keep in mind that a valid argument establishes the **logical truth** of a conclusion, but not necessarily the actual truth. There is no test made of the actual truth of the premises; they are **assumed** to be true. All that is required is that the premises be **consistent** with **each other**. Thus, in the scientific method if an experiment shows a **valid** conclusion to be untrue, one or more of the premises must be untrue. If we let p represent the **conjunction** of the premises and q, the conclusion, the argument in symbolic form is $p \rightarrow q$. A logical equivalent of the argument is its contrapositive form, $-q \rightarrow -p$. Thus, if q is false, p must also be false. p is false whenever one or more of its component statements is false. Thus, one or more of the premises—statements of the theory—have been incorrectly assumed to be true.

2.4.1 Truth-Table Method for Checking Argument Validity

The definition of a valid argument suggests that a truth table can be used to check the validity of an argument. We are concerned with those logical possibilities for which the premises are all true. These can be found readily in a truth table. The following step-by-step method outlines the procedure.

Step 1 The simple statements from which the premises and conclusions are formed are identified and assigned symbols.

Step 2 A truth table is constructed in which a column is provided for each of the simple statements found in Step 1. A column is provided also for each premise and the conclusion.

Step 3 Inspect the truth values in each row in the columns provided for the premises.

a. If **none** of these rows has a truth value T for **each** premise, the premises are **inconsistent**; that is, they cannot all be true at the same time. The **argument** is **invalid** because of inconsistent premises.
b. If there are one or more rows in which each premise has a truth value T, check the truth value of the conclusion in **each** of these rows. If the conclusion is T in **all** of them, the **argument** is **valid**. If there is **any** row in which the premises are **all true**, but the conclusion is **false**, the argument is **not valid**.

Step 4 (Alternate to Step 3)

In the truth table of Step 2, form the conjunction of all of the premises and enter its truth values in another column. Verify that in at least one row this conjunction has truth value T. If it does not, the argument is not valid because the premises are inconsistent.

Let c represent the **conjunction** of the **premises** and q, the **conclusion**. Form the conditional $c \rightarrow q$, and enter the truth values in another column. If, and only if, the conditional $c \rightarrow q$ has truth value T in **every** row, (i.e., if, and only if, $c \rightarrow q$ is a tautology), the argument is valid.

Example 2.4.1

Consider the argument $[(a \vee b) \wedge -b] \rightarrow a$. The premises are $(a \vee b)$ and $-b$. The conclusion is a, which is one of the simple component statements. The argument, if valid, proves that this statement must be true if both premises are true. The truth table for this argument is shown in Figure 2.4.1.

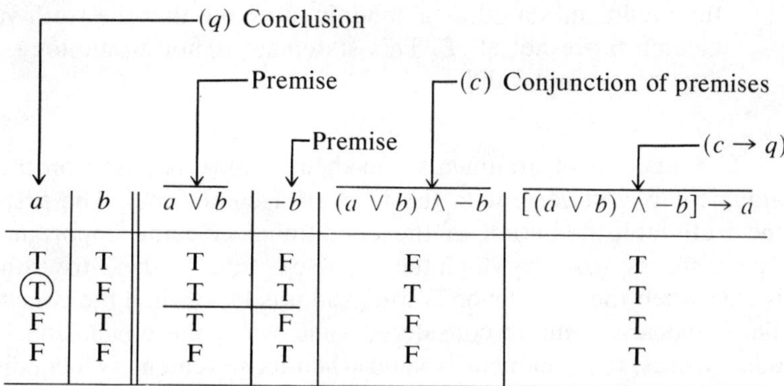

a	b	$a \vee b$	$-b$	$(a \vee b) \wedge -b$	$[(a \vee b) \wedge -b] \rightarrow a$
T	T	T	F	F	T
(T)	F	T	T	T	T
F	T	T	F	F	T
F	F	F	T	F	T

Figure 2.4.1

The truth values for the premises under all the logical possibilities are shown in columns 3 and 4.

Note that the two premises are both true only in the second row. In this row, however, the conclusion, a, (first column) is also true. The argument, therefore, is valid. We can proceed with the truth table and form the conjunction of the premises. The truth values for the conjunction are given in column 5. The last column gives the truth values for the statement of the argument in the form of a conditional. Note that all the truth values are T; that is, this statement is a tautology.

The foregoing example is a widely used type of deduction. In words it says that if either a or b, or both, is true, and b is not, then a must be true. Let us now modify the argument slightly to say, "The disjunction of a and b is true, and a is true. Therefore, b must be false."

Example 2.4.2

The argument above is symbolized $[(a \vee b) \wedge a] \rightarrow -b$. The truth table for this argument is given in Figure 2.4.2.

(1) a	(2) b	(3) $a \vee b$	(4) $-b$	(5) $(a \vee b) \wedge a$	(6) $[(a \vee b) \wedge a] \rightarrow -b$
T	T	T	(F)	T	F
T	F	T	T	T	T
F	T	T	F	F	T
F	F	F	T	F	T

Figure 2.4.2

The truth values for the premises in this argument are listed in columns 1 and 3. We note that both premises are true in row 1 and in row 2. In this case, however, the conclusion, $-b$ (column 4), has truth value T only in row 2. In row 1 the truth value of $-b$ is F. Thus, the argument is not valid. If we continue the truth table to include the entire statement, column 5 gives the truth values for the conjunction of the premises, and column 6, the truth values for the argument stated as a conditional. Note that the truth values in column 6 are not all T. This statement is not a tautology, and the argument is not valid.

The analysis of arguments and their validity is based on the fact that an argument is a statement in the form of a conditional. The first two rows of the truth-table definition of the conditional give the important information. These are the rows in which the premise is true. In these rows the conditional is true when the conclusion is true, and it is false when the conclusion is false. Thus, the argument is considered valid when the conclusion is true; or, in other words, the argument is valid when its statement as a conditional is true.

Note that if we do not require all of the premises to be true simultaneously, we can "prove" anything. Unless the premises are all true for at least one logical possibility, their conjunction is always false. Checking the definition for the conditional, we see that when p is false, the conditional is true whether the conclusion is true or not. Thus, the conditional statement for any argument will always be true if the conjunction of the premises is always false.

Example 2.4.3

Let us consider a simple example of "proving" an obvious fallacy. Let us choose as premises the statements $p \wedge q$, and $-q$. We can show that $[(p \wedge q) \wedge -q] \rightarrow q$ is a tautology. That is, in spite of assuming that q is not true, we can "prove" that it is true. The truth table for the argument is shown in Figure 2.4.3.

p	q	$-q$	$(p \wedge q)$	$(p \wedge q) \wedge -q$	$[(p \wedge q) \wedge -q] \rightarrow q$
T	T	F	T	F	T
T	F	T	F	F	T
F	T	F	F	F	T
F	F	T	F	F	T

Figure 2.4.3

We see that all of the truth values for the conditional statement of that argument are T, and it is a tautology. However, all of the truth values for the conjunction of the premises are F—that is, they form a logical contradiction.

In a simple situation such as that in Example 2.4.3 the fallacy in the argument is obvious. In more complex situations a fallacy may not be quite as apparent. It is often desirable to set up a truth table for the conjunction of the premises. If all the truth values for this conjunction are F, the premises are inconsistent and cannot be used as the basis for a logical argument.

EXERCISE 2.4

1. Construct truth tables to determine the validity of the following arguments:
   a. $[(p \wedge q) \wedge p] \rightarrow q$
   b. $[(p \wedge q) \rightarrow -p] \rightarrow -q$
   c. Premises: $(p \rightarrow q), -q$ Conclusion: $-p$
   d. Premises: $(p \rightarrow q), (q \rightarrow r)$ Conclusion: $(p \rightarrow r)$
   e. Premises: $(p \rightarrow q), (p \rightarrow r)$ Conclusion: $(q \rightarrow r)$
   f. Premises: $p \rightarrow (q \vee r), (p \wedge -q)$ Conclusion: r

2. Determine which of the following sets of premises are consistent and which are inconsistent.
   a. $(p \rightarrow q), -p, -q$
   b. $-(p \vee q), p$
   c. $(p \rightarrow q), (q \rightarrow r), (q \wedge r), -p$
   d. $(p \leftrightarrow r), (q \rightarrow r), r, -q$
   e. $(p \rightarrow r), (q \rightarrow r), -r, (p \vee q)$

3. For each of the sets of consistent premises from Exercise 2, write at least one valid consequence of these premises (other than one of the premises).

Symbolize the following arguments and determine their validity.

4. The applicant is hired only if he passes the aptitude test. The applicant was not hired. Therefore, he did not pass the aptitude test.

5. If prices are raised, sales volume will decrease. If sales volume decreases, there will be layoffs in the shipping department. The management has promised there will be no more layoffs. Therefore, they will not raise prices.

6. If a child is an overachiever, he will be nervous. If a child is nervous, he will not eat properly. Therefore, if a child does not eat properly, he is an overachiever.

7. The displaced farm workers move either to the city or the suburbs. If they move to the suburbs, they must commute to their jobs. They have no access to suitable transportation. Therefore, they move to the city.

8. In an opinion survey, if the interviewees are not selected with great care, the survey results are not representative. On a radio "talk show" the announcer talks with anyone who calls in. Therefore, the opinions expressed on talk shows are not representative of the listening audience.

9. Supply a conclusion that is a valid consequence of the following set of premises. Demonstrate the validity of the conclusion by symbolizing the argument and checking its truth value.
   a. The company will move its factory unless the city council lowers the tax rate.
   b. If the company moves the factory, many of the city's residents will lose their jobs.
   c. The city cannot afford to have more unemployed citizens.

10. The general manager of a company held a staff meeting in which the following statements were made:
    Production manager: "We must have a higher sales volume."
    Advertising manager: "We can increase sales if we do more advertising."
    Sales manager: "We can increase sales if we lower prices, but not otherwise."
    Controller: "We do not have the money needed to increase advertising."

    Is there a course of action the general manager can take that will be consistent with the statements of each of the staff members? If so, what can he do?

2.5 ARGUMENT FORMS

A truth table provides the complete story about a set of premises. Using the information in the truth table we can develop all of the valid consequences that can be drawn from the set of premises. Any compound statement formed from the simple component statements that is true for every logical possibility for which the conjunction of the premises is true is a valid deduction. When there are more than four or five basic components, however, the truth table becomes so large that it is unmanageable unless a computer is available.

An argument involving several premises can be tested without recourse to a large truth table if we break it up into a number of subarguments. These subarguments can be tested with truth tables of manageable size. In fact, because the subarguments usually have one of a few standard forms, it is often not necessary to use truth tables at all. The process is that which we regularly use in reasoning through a situation to obtain "logical conclusions."

Let us consider the following "argument." "If it is raining, Joe cannot play tennis. Joe told me that if he did not play tennis this afternoon, he would go to the library to study. It has been raining hard all day. Therefore, I conclude that Joe is in the library studying." The step-by-step reasoning follows a common pattern.

"It is raining, and if it is raining, Joe cannot play tennis." First conclusion: "Joe is not playing tennis." We continue: "If Joe does not play tennis, he is going to the library to study." Thus, we conclude that Joe is studying in the library. The complete argument was broken into two parts according to the following symbolic pattern:

Let p represent: "It is raining";
q represent: "Joe plays tennis";
and r represent: "Joe goes to the library."

The premises are:
(1) $p \to -q$
(2) $-q \to r$
(3) p

The conclusion is r.

The first part of the argument combines the first and third premises: $[p \land (p \to -q)] \to -q$. Since this is a valid argument, $-q$ is true whenever both p and $(p \to -q)$ are true. It can, therefore, be used as a **substitute in truth value for these two premises** in the rest of the argument.

The second step in the argument combines the first conclusion, $-q$, with the second premise:

$$[-q \land (-q \to r)] \to r$$

This is a valid argument: this statement is a tautology. Thus, r is a valid conclusion of these premises.

The argument above used the same basic form twice to reach the conclusion. There are four other basic argument forms used to break up long arguments into small steps. These are listed in Figure 2.5.1.

Argument Forms

Form Number	Premises	Conclusion
1	p and $(p \to q)$	q
2	$-q$ and $(p \to q)$	$-p$
3	$(p \to q)$ and $(q \to r)$	$p \to r$
4	$(p \lor q)$ and $-p$	q
	or	
	$(p \lor q)$ and $-q$	p
5	$p \to (q \land -q)$	$-p$

Figure 2.5.1

Argument Form 1 is the straightforward use of the conditional. The conditional states the consequence that will ensue if the premise is true. The premise is stated to be true. Therefore, the consequence is established as true.

Argument Form 2 involves the conditional in contrapositive form. If we change the premise $p \to q$ to its contrapositive $-q \to -p$, Argument Form 2 has the same pattern as Argument Form 1.

Argument Form 3 is a statement of the transitive property of the conditional. Establishing its validity is left as an exercise.

Argument Form 4 is a statement of proof by the "process of elimination." The disjunction $(p \lor q)$ states that either p or q must be true. The conclusion follows from the fact that if one of the components is not true, the other one must be.

Argument Form 5 is a statement of proof by "reduction to absurdity."

Since $(q \wedge -q)$ is a contradiction, it is always false. If p leads to a contradiction, p must be false also.

The structure of an argument, as a conditional whose hypothesis is the conjunction of the premises of the argument, makes it possible to make the following modifications to premises:

1. The component statements of any premise stated as a **conjunction** can be used individually when it is convenient. That is, a premise stated in the form $p \wedge q$ can be separated into the separate premises p and q. Similarly, two separate premises can be combined to form a conjunction. p and q listed as separate premises can be combined to form $p \wedge q$.
2. A statement given as a premise in an argument can have **any other** statement joined to it by a **disjunction**. Thus, if p is a premise, it can be replaced by $p \vee q$, where q is **any** statement. $p \vee q$ is true whenever p is true. Since p is a premise, it is always true in the argument; and, thus, we are not concerned with the truth values of $p \vee q$ when p is false. **Warning**: A disjunction given as a premise **cannot** be separated into its components. That is, $p \vee q$ cannot be replaced by p or by q as a premise. Neither of these statements is true for all the logical possibilities in which $p \vee q$ is true.

2.5.1 Direct Proof

The process of direct proof of an argument consists of finding a sequence of ways to combine the premises of the argument in subarguments, by using the argument forms shown in Figure 2.5.1, until the conclusion is exhibited as a valid consequence. The premises can be used in any order. Any premise can be replaced by its logical equivalent whenever it is convenient to do so.

The justification for the process comes from the fact that in **any** valid argument, the conclusion is true in all logical possibilities for which the premises are all true. It can, therefore, be used as a simple substitute for the premises on which it is based. Note that the conclusion is not the logical equivalent of the premises, but it is true whenever all the premises are—the only consideration of importance in an argument.

Example 2.5.1

Let us use the method of direct proof first in a simple argument, given in symbolic form:

Premises: $(p \rightarrow q)$, $(q \rightarrow r)$, and p

Conclusion: r

Proof: Using Argument Form 3 we can combine the premises $(p \rightarrow q)$ and $(q \rightarrow r)$ to form $(p \rightarrow r)$.
Using Argument Form 1 we can combine $(p \rightarrow r)$ with the other premise, p, to obtain r as the conclusion.
Thus, r has been established, or "proved" to be a valid consequence of the premises.

We note that the sequence used in the proof is not the only one possible. In all but the simplest arguments there are many sequences of steps leading to the same conclusion. As an alternate sequence in this argument we have:

First, using Argument Form 1 we can combine $(p \rightarrow q)$ with p to establish q as true.

Second, using the same argument form we can combine $(q \rightarrow r)$ with q to establish r as true.

The argument of Example 2.5.1 was presented in symbolic form. It represents a pattern of reasoning we should recognize. Let p represent "Production costs increase"; q, "Prices increase"; and r, "Sales volume decreases." The argument in verbal form sounds like this: "Our production costs are increasing (p). Increased production costs will force us to raise prices $(p \rightarrow q)$. If we raise prices, we can expect a decrease in sales $(q \rightarrow r)$. Therefore, we are projecting a lower sales volume (r)."

In the verbalization of this argument, the "laws" of business economics were the source of the second and third premises stated. The only premise that came out of the particular business involved was the fact that production costs were increasing. This is the case with much of the reasoning we do – the situation supplies many of the premises. Often such premises are left implied instead of stated. We should be aware of their existence, however; and we should be careful not to use any that are not valid.

2.5.2 Use of Logical Equivalence

The statements of the premises in an argument are not always in a form that fits directly into one of the argument forms. We can change the form of any statement into another form that is **logically equivalent** at any time without affecting the logical meaning of the statement. Figure 2.5.2 gives a list of logically equivalent forms that are useful in simplifying compound statements or in making them more convenient to use.

Logical Equivalents

Form 1	Form 2
1. $(p \rightarrow q) \wedge (r \rightarrow q)$	$(p \vee r) \rightarrow q$
2. $(p \rightarrow q) \wedge (p \rightarrow r)$	$p \rightarrow (q \wedge r)$
3. $p \rightarrow q$	$-q \rightarrow -p$
4. $p \rightarrow q$	$-p \vee q$
5. $p \vee q$	$-p \rightarrow q$
6. $-(p \vee q)$	$-p \wedge -q$
7. $-(p \wedge q)$	$-p \vee -q$
8. $(p \leftrightarrow q)$	$(p \rightarrow q) \wedge (q \rightarrow p)$
9. $p \vee (q \wedge r)$	$(p \vee q) \wedge (p \vee r)$
10. $p \wedge (q \vee r)$	$(p \wedge q) \vee (p \wedge r)$

Figure 2.5.2

The statements listed under Form 1 can be used to replace those under Form 2 or vice versa. It is true also that the negation of a statement can be replaced by the negation of its logical equivalent. Additional useful replacements can be formed from the statements in Figure 2.5.2 when the negations are used. It is useful to keep in mind that the order of any conjunction, disjunction, or biconditional (but **not** conditional) can be reversed without affecting its meaning.

Let us consider a simple illustration of the use of logical equivalents in establishing the validity of an argument.

Example 2.5.2

Premises: p, $(p \rightarrow q)$, and $(r \rightarrow -q)$

Conclusion: $-r$

Proof: Using Argument Form 1, the premises p and $(p \rightarrow q)$ can be combined to establish q as true.
With Logical Equivalent 3 we can replace the premise $r \rightarrow -q$ with $-(-q) \rightarrow -r$, or $q \rightarrow -r$.
Then, using Argument Form 1, we can combine q with $q \rightarrow -r$ to establish $-r$ as a valid consequence of the premises.

2.5.3 Formal Proof

It is sometimes necessary to present a proof in a formal manner to convince others that your deductions are valid. In the usual format of a formal proof each successive step is listed and numbered. Each step is in the form of a statement accompanied by a reference to the justification for the step. The justification can be that the statement is one of the premises, a logical equivalent for one of the previous steps, or the result of combining two of the previous steps with one of the argument forms. The process is the same as discussed in the previous sections. The premises or their logical equivalents are put together in a series of subarguments. The goal is to find a sequence of steps that leads to the conclusion as a valid consequence.

Example 2.5.3

Premises: $p \rightarrow q$, $-(q \vee r)$

Conclusion: $-p$

Statement	Justification*
1. $p \rightarrow q$	*pr*
2. $-(q \vee r)$	*pr*
3. $-q \wedge -r$	(2), LE6
4. $-q$	(3), simplification†
5. $-p$	(1), (4), AF2

*The symbol "*pr*" is used to mean "premise." "AF . . ." is used to designate which argument form from Figure 2.5.1 is being used. "LE . . ." is used to designate one of the logical equivalents from Figure 2.5.2. The numbers in parentheses refer to the number of the statements in the proof being used to obtain the result stated.

†"Simplification" means that we have taken one of the components of a **conjunction** and used it by itself. This was established as a valid procedure earlier in the discussion of the argument forms.

In statement 5 only the conclusion is stated. As noted in the justification column, statement 1, $(p \rightarrow q)$, is combined with statement 4, $(-q)$, to form $(p \rightarrow q) \wedge (-q)$. (Remember that all of the premises or their logical equivalents are connected by conjunctions in an argument.) Argument Form 2 says that the valid consequence of this combination of statements is $-p$.

It may be necessary to use trial and error to find a sequence of subarguments that leads to the conclusion to be proved. A few procedural tips may help to shorten the task.

1. Find the symbolic statement of the conclusion, or its logical equivalent in the premises. (It must be there somewhere or it cannot be proved.)
2. From the form of the statement involving the conclusion, check the argument forms to see which one can be used to establish the conclusion.
3. Determine what other statement or statements need to be true to use the argument form selected. Find one of these statements in another premise.
4. Repeat the process until the necessary statement is found to be true directly in one of the premises.
5. Work your way back through the premises you have listed, using the appropriate argument forms until the conclusion is reached.

Example 2.5.4

Let us use the above process on the following argument:

Premises: (1) $-p \rightarrow q$
(2) $-r \vee -s$
(3) $q \rightarrow s$
(4) r

Conclusion: p

The only premise involving the conclusion, p, is the first one. We note that in this premise the negation of p is the hypothesis of a conditional. However, we can replace this premise with its logical equivalent, $-q \rightarrow p$. We note that, using Argument Form 1 if we can establish $-q$ as true, then our conclusion p will be true.

We now seek a way to establish $-q$. We note that premise 3 contains the simple statement q as the hypothesis of a conditional. Argument Form 2 can be used with this statement to establish $-q$ as true provided $-s$ is true.

We find $-s$ as one component of a disjunction with $-r$. Argument Form 4 shows $-s$ to be true provided $-(-r)$ is true. We can replace $-(-r)$ with its logical equivalent r. We can now construct the proof, starting with r.

	Statement	Justification
1.	r	*pr*
2.	$-r \vee -s$	*pr*
3.	$-s$	(1), (2), AF 4
4.	$q \rightarrow s$	*pr*
5.	$-q$	(3), (4), AF 2
6.	$-p \rightarrow q$	*pr*
7.	$-q \rightarrow p$	(6), LE 3
8.	p	(5), (7), AF 1

Example 2.5.5

Let us consider an argument in verbal form and present a formal proof for it.

"Inflation will continue if either wages or prices continue to rise. Continued inflation will bring about either strict controls on the economy or the need for subsidy payments to people on fixed incomes. If there are subsidy payments, the budget will be unbalanced. The business community will not tolerate controls, and the administration cannot survive another unbalanced budget. Therefore, we expect to see some strong pressure by the administration to keep wages down."

The basic statements for this argument are:

w: Wages rise.
p: Prices rise.
i: Inflation continues.
c: Controls are imposed.
s: Subsidy payments are made.
u: The budget is unbalanced.

Using these basic statements, we can symbolize the argument as follows:

First sentence: $(w \vee p) \rightarrow i$
Second sentence: $i \rightarrow (c \vee s)$
Third sentence: $s \rightarrow u$
Fourth sentence: $-c \wedge -u$

Conclusion: $-w$

A formal proof of the argument can be made as follows:

	Statement	Justification
1.	$(w \vee p) \rightarrow i$	*pr*
2.	$i \rightarrow (c \vee s)$	*pr*
3.	$(w \vee p) \rightarrow (c \vee s)$	(1), (2), AF3
4.	$-c \wedge -u$	*pr*
5.	$-u$	(4) simplification
6.	$s \rightarrow u$	*pr*
7.	$-s$	(5), (6), AF2
8.	$-c$	(4) simplification

9.	$-c \wedge -s$	(7), (8), conjunction
10.	$-(c \vee s)$	(9) LE6
11.	$-(w \vee p)$	(3), (10), AF2
12.	$-w \wedge -p$	(11) LE6
13.	$-w$	(12) simplification

We have now established $-w$ as a valid consequence of the four premises.

Not all arguments have valid conclusions. Sometimes the conclusion, although it may be true, cannot be proved from the premises.

Example 2.5.6

Consider the argument:

Premises: $(p \rightarrow q)$ and $(q \vee r)$

Conclusion: $r \rightarrow -p$

Proof:	**Statement**	**Justification**
1.	$p \rightarrow q$	*pr*
2.	$-q \rightarrow -p$	(1), LE3
3.	$q \vee r$	*pr*
4.	$-q \rightarrow r$	(3), LE 5

Statements 2 and 4 cannot properly be combined to form the conclusion, and the argument is not valid. That is, the conclusion does not follow as a valid consequence of the premises.

Note that if q is not true, both r and $-p$ will be true. In this case the conclusion $r \rightarrow -p$ is a true statement.

However, if q is true, r could be either true or false, and p could be either true or false. In the case that r and p are both true, $r \rightarrow -p$ is not a true statement.

Since the premises do not specify a truth value for q, we cannot establish whether the statement $r \rightarrow -p$ is true or false.

2.5.4 Indirect Proof

In some arguments a direct proof is very difficult, if not impossible, to establish. In these cases it is often possible to use an indirect proof—a proof by contradiction. Recall that any statement in logic must be either true or false. Thus, the truth of a statement can be established by showing that it cannot be false. Recall further that it is impossible for any statement to be both true and false. These two properties of statements form the basis for an indirect proof.

To establish the truth of a conclusion by indirect proof, the conclusion is assumed to be false; and this assumption is shown to lead to an impossibility—a statement that is both true and false. Since this assumption means that the **negation** of the conclusion is **assumed** to be true, it becomes, thereby, one of the premises. The new conclusion is the conjunction of any statement with its negation—a logical contradiction and always false. The indirect proof consists of showing the logical contradiction is a valid consequence of the original premises plus the negation of the conclusion.

In symbols, if

$$(p_1 \wedge p_2 \wedge p_3 \wedge \ldots \wedge p_n) \rightarrow q$$

is the argument we are trying to prove valid, it is changed to

$$(p_1 \wedge p_2 \wedge p_3 \wedge \ \ldots \wedge p_n \wedge -q) \rightarrow (r \wedge -r)$$

The statement, r, can be **any statement formed from the basic components in the argument.**

Recall that in a direct proof, the conclusion must be true whenever all of the premises are true. Thus, the negation of the conclusion will be false whenever all the premises are true. The conjunction of the negation of the conclusion and the premises will always be false. An indirect proof shows this to be the case; and it is, therefore, the logical equivalent of a direct proof.

Example 2.5.7

Let us establish the argument of Example 2.5.3 by an indirect proof.

In that example the premises were $(p \rightarrow q)$ and $-(q \vee r)$. The conclusion was $-p$. In an indirect proof we change $-p$ to p and add it to the list of premises. We now have

Premises: $(p \rightarrow q)$, $-(q \vee r)$, p

Conclusion: any contradiction

1.	$p \rightarrow q$	*pr*
2.	p	*pr*
3.	q	(1), (2), AF1
4.	$-(q \vee r)$	*pr*
5.	$-q \wedge -r$	(4) LE6
6.	$-q$	(5) simplification
7.	$q \wedge -q$	(3) and (6)

Thus, we have shown p to be inconsistent with the premises in the argument and, therefore, $-p$ is a valid consequence of these premises.

Let us consider now an example in which it is very difficult to establish a direct proof, but in which an indirect proof is relatively simple.

Example 2.5.8

Consider the argument:

Premises: $(a \leftrightarrow b)$, $(b \rightarrow c)$, $(-c \vee d)$, $(-a \rightarrow d)$

Conclusion: d

In this argument a single statement is the conclusion, whereas each of the premises is a compound statement. It is virtually impossible to "separate out" a single statement from compound statements unless the compound statement is a conjunction. In this argument the conclusion, d, appears in two of the premises, $(-c \vee d)$ and $(-a \rightarrow d)$. In the first of these, $(-c \vee d)$, we need to establish c as true to establish d. In the second, $(-a \rightarrow d)$, we need $-a$ to establish d. There is no way to establish either of these statements from the other two premises. Thus, we try an indirect proof.

The new argument becomes

Premises: $(a \leftrightarrow b), (b \rightarrow c), (-c \vee d), (-a \rightarrow d), -d$

Conclusion: Any contradiction

1.	$-c \vee d$	*pr*
2.	$-d$	*pr*
3.	$-c$	(1), (2), AF4
4.	$a \leftrightarrow b$	*pr*
5.	$(a \rightarrow b) \wedge (b \rightarrow a)$	(4) and LE8
6.	$a \rightarrow b$	(5) simplification
7.	$b \rightarrow c$	*pr*
8.	$a \rightarrow c$	(6), (7), AF3
9.	$-a \rightarrow d$	*pr*
10.	a	(2), (9), AF2
11.	c	(8), (10), AF1
12.	$c \wedge -c$	(3) and (11)

Since we have established a contradiction as a valid consequence of the premises and the negation of the conclusion, the conclusion, d, is a valid consequence of those premises.

The method used in an indirect proof, that of proving a contradiction, can also be used to test a set of premises for consistency. For example, if the original set of premises of Example 2.5.8 had included the statement, $-d$, they would have been inconsistent. This could be shown by steps 1–12 of the proof.

EXERCISE 2.5

1. Verify Argument Form 3, using a truth table.

Check the following arguments for validity, using the argument forms and logical equivalents from Figures 2.5.1 and 2.5.2.

2. **Premises:** $p \rightarrow (r \rightarrow s), (-r \rightarrow -p), p$
   Conclusion: s

3. **Premises:** $(p \rightarrow q), -(q \vee r)$
Conclusion: $-p$

4. **Premises:** $(c \wedge d) \rightarrow (-a \vee -b), (a \wedge b)$
Conclusion: $(c \wedge d)$

5. **Premises:** $(-t \rightarrow -r), (t \rightarrow w), -(r \vee s)$
Conclusion: s

6. **Premises:** $(w \vee v) \wedge -p, (w \rightarrow s), (s \rightarrow p)$
Conclusion: v

7. **Premises:** $(p \rightarrow q), (-r \rightarrow -q), (r \rightarrow t), p$
Conclusion: t

8. **Premises:** $(p \rightarrow q), (q \rightarrow -s), (-t \rightarrow s), (m \rightarrow -t)$
Conclusion: $p \rightarrow m$

9. **Premises:** $(p \rightarrow q), (s \rightarrow -q), (s \vee t), (p \rightarrow -t)$
Conclusion: $-p$ (**Hint:** Use indirect proof.)

What conclusion can validly be drawn from the following?

10. **Premises:** $(p \rightarrow q), (-r \rightarrow -q)$
Conclusion: ?

11. **Premises:** $(-p \vee q), -q$
Conclusion: ?

12. **Premises:** $-(p \wedge -q), -q$
Conclusion: ?

13. **Premises:** $p \rightarrow (s \vee t), (-s \wedge -t), s \rightarrow p$
Conclusion: ?

Symbolize the following arguments and check their validity.

14. If I do not pass this course, I cannot continue college. If I cannot continue college, I must either go into the service or find a job. I cannot find a job, and I do not want to go into the service. Therefore, I will study and pass the course.

15. To be elected in that state it is necessary to get most of the farm vote. A candidate cannot get the farm vote unless he has an acceptable position on farm subsidies. Candidate Smith has endorsed the farm-subsidy program sponsored by the two largest farmers' organizations. (**Hint:** What does this say about Smith's position on farm subsidies?) Therefore, Smith will win the election.

16. If I get a promotion, I shall have to work hard in my new duties. If I get a promotion, I shall buy a new car. However, if I buy a new car, I cannot afford a vacation. If I do not have a vacation, I will not be able to work hard. Either I get a promotion or I will quit. Therefore, I will quit. (**Hint:** Use indirect proof.)

2.6
QUANTIFIED PROPOSITIONS

In the previous sections we have presented the foundations for a system of logical reasoning and analysis. The discussions have included the structure of arguments made with declarative statements—statements made up of simple facts. Much of the work done in business and the behavioral sciences focuses on studying the characteristics and properties of sets of people or subjects, and the implications of these characteristics for individual members of these sets. As was mentioned earlier, statements made about the characteristics or properties of the members of sets are called propositions. These contain indefinite terms that are intended to refer to the members of specific sets. The truth value of the proposition is dependent on whether or not a particular replacement for the indefinite term is a member of the appropriate set. This aspect of propositions makes it necessary to use special methods in working with arguments involving them.

Whenever we use sets in any way, there is a universal set made up of all the elements relevant to the situation. Propositions are separated into two basic classifications with respect to their universal sets: those pertaining to the entire universal set and those pertaining only to **some** of the elements, but not to others.

2.6.1
Universal Propositions

Propositions that pertain to **every** member of a set are called **universal** propositions. Such propositions as, "Every corporation has a board of directors"; or, "All applicants must take an aptitude test"; are examples of universal propositions. The first applies to all business establishments; the second, to all people. Depending on the context, the universal set for the second proposition probably has a more restricted universal set than the set of all people.

In mathematical discussions the unidentified term in a proposition is usually assigned a letter to represent it. Letters at the end of the alphabet, such as "x," "y," and "z," are commonly used for this purpose.

The example propositions above, when expressed in mathematical language, become:

"For all x, if x is a corporation, x has a board of directors."

"For all x, if x is an applicant, x must take an aptitude test."

The universal nature of these propositions is expressed by the phrase "for all x." It shows that the accompanying proposition applies to every element of the appropriate universal set. All business establishments are not corporations. The proposition states that every business that **is** a corporation has a board of directors. There is no corporation without a board of directors.

When symbols are assigned to propositions, the symbol representing the unidentified term is specified in parentheses along with the symbol for the statement. We can let $p(x)$ represent "x is a corporation" and $q(x)$, "x has a board of directors." The example proposition becomes, "For all x, $p(x) \rightarrow q(x)$."

The symbol "$\forall$," which looks like an upside down "A," is used to denote "for all," or "for every." Thus, when completely symbolized, the proposition becomes

$$\forall x,\ p(x) \rightarrow q(x)$$

In words this is translated, "For all x, p of x implies q of x"; or "For all x, if p of x, then q of x."

Universal propositions are applied to particular elements of the universal set by replacing the x, called the **variable** in the proposition, with the name or identity of that element. This process is used regularly in algebra when numbers are substituted for the letters in algebraic expressions.

For example, if we let a represent the ABC Company, the example proposition from above becomes

$$p(a) \rightarrow q(a)$$

Translating this statement into words we have, "If the ABC Company is a corporation, it has a board of directors."

We use particularizations of universal statements frequently. In many cases we do not use the full logical form. The following example illustrates the complete logical process for applying a universal statement to a particular case.

Example 2.6.1

"All doctors make lots of money. John is a doctor. Therefore, John makes lots of money."

As we can see, the application of a universal statement takes the form of a logical argument. We check the validity of the application by determining whether the argument is valid.

The first statement, the "generalization," can be symbolized as

$$\forall x,\ d(x) \rightarrow m(x)$$

which, when translated literally, becomes, "For all x, if x is a doctor, x makes lots of money."

The second statement, the particularization, can be symbolized, letting a represent John:

$$d(a)$$

that is, "John is a doctor."

The conclusion becomes

$$m(a)$$

"John makes lots of money."

Since the first statement is a **universal** statement, applicable to any replacement for x, we can substitute a for x to obtain

$$d(a) \rightarrow m(a)$$

The complete argument in symbols is

$$[(d(a) \to m(a)) \land d(a)] \to m(a)$$

We recognize this as having the same pattern as Argument Form 1 of Figure 2.5.1. It is, therefore, valid, and $m(a)$ is a valid conclusion.

This pattern of reasoning should be familiar to the student, although we often leave one, or both, of the premises implied, instead of expressed. We also frequently state the conclusion first with the justifying premises afterwards. "Oh, John makes lots of money. He is a doctor, you know." The premise, "All doctors make lots of money," is left implied. In another variation, "John makes lots of money; all doctors do," the fact that John is a doctor is assumed to be known to the listener.

When universal statements are used, the following requirements must be kept in mind:

First, when two or more universal statements are used in an argument, the **same universal set** must be used for each of them.

Second, when a variable is replaced by a particular element of the universal set, the **same element** must be used to replace **that** variable **each time it occurs**.

2.6.2 Existential Propositions

The other basic classification of quantified propositions, the **existential proposition**, relates to one or more elements of a set, but not to all of them. A word frequently used as the quantifier is the word "some." We use the proposition, "Some men are happy in their work," to mean that there is at least one man who is happy in his work. That is, the set of men who are happy in their work is not empty. In logic such words as "some," "few," "many," and "at least one" are all synonymous as far as their use as quantifiers is concerned. They all specify that the set of whatever it is they are modifying is not empty. They are all symbolized with the existential quantifier "$\exists$." If h is used to represent "is happy in his work," our example statement is symbolized, $\exists x \in M, h(x)$. "There exists an x that is an element of the set of men, such that x is happy in his work." An alternate expression, which does not specify the applicable universal set, is

$$\exists x, m(x) \land h(x)$$

This is translated, "There exists an x such that x is a man, and x is happy in his work."

Note the difference in the symbolization of an existential and universal proposition. The universal proposition is symbolized as a conditional, the existential proposition as a conjunction. A universal proposition **applies to all** of the elements of the universal set **although it may not be true of all** of them. An existential statement says that there is at least one element of the universal set to which the proposition applies; that is, something exists that is an element of a specified set and that also has the property described by the proposition. If we make the proposition that "**All** businesses are profitable," as soon as something is identified as a business, it follows that it is profitable. That is, $\forall(x), b(x) \rightarrow p(x)$. However, the proposition, "Some businesses are profitable," merely states that there is at least one thing that is both a business and profitable. That is, $\exists x, b(x) \wedge p(x)$.

2.6.3 Negations of Quantified Propositions

The negation of a quantified proposition, as with other statements in logic, is a proposition that has the opposite truth value to the original proposition. That is, if a proposition is true, its negation is false, and vice versa. Consider the statement, "All men are honest." This is a universal proposition that can be symbolized,

$$\forall x, m(x) \rightarrow h(x)$$

The negation of this statement is formed symbolically by placing the negation symbol in front of the entire statement:

$$-[\forall x, m(x) \rightarrow h(x)]$$

The question is, what does this negated statement mean? We need a more convenient form, a logical equivalent, that we can combine with other statements in arguments.

Let us consider the circumstances under which the proposition, "All men are honest," is false. These are the circumstances under which its negation is true. We note that if there is just one dishonest man, the proposition, "**All** men are honest," is not true. The statement that there is at least one dishonest man is symbolized, as we have seen,

$$\exists x, m(x) \wedge -h(x)$$

This is an existential proposition. **In all cases the negation of a universal proposition is an existential proposition.**

Let us note here that the negation of our example is **not** "**All** men are not honest." That is, we did not merely insert a negation symbol into the proposition. To make the meaning clear, it is necessary to change the negation of a universal proposition to an existential proposition.

We have the first rule for the negation of quantified propositions: the negation of a universal proposition is an existential proposition.

Let us examine the form of this existential proposition. We have seen that in symbols

$$-[\forall x, m(x) \rightarrow h(x)]$$

becomes

$$\exists x, -[m(x) \rightarrow h(x)]$$

From the table of logical equivalents in Figure 2.5.2, 4 permits us to change this to

$$\exists x, -[-m(x) \vee h(x)]$$

Logical Equivalent 6 replaces this with

$$\exists x,\ [-(-m(x))\ \wedge -h(x)]$$

Removing the double negative, we have

$$\exists x,\ m(x) \wedge -h(x)$$

A literal translation of this is "There exists an x, such that x is a man and x is not honest." A more normal statement is "Some men are not honest." This agrees with our earlier appraisal of the circumstances under which "All men are honest" is considered false.

As with the universal proposition, the negation of an existential statement must be formed so that the meaning is correct. Consider the proposition, "Some highways are heavily traveled." There is a temptation to form the negation by inserting a "not" into the proposition in a plausible location, such as "Some highways are not heavily traveled." This proposition, however, is not true when the other is false, as is required for a negation.

If we examine the meaning of the proposition, we find that the negation is that "No highways are heavily traveled." An alternative phrasing for "Some highways are heavily traveled" is "At least one highway is heavily traveled." When we deny this statement, we are saying, "No, not even one highway is heavily traveled." We find that **the negation of an existential proposition is a universal proposition.**

We can symbolize our example:

$$\exists x,\ h(x) \wedge t(x)$$

where $h(x)$ represents "x is a highway," and $t(x)$, "x is heavily traveled." The negation

$$-[\exists x,\ h(x) \wedge t(x)]$$

becomes, as we have seen,

$$\forall x,\ -[h(x) \wedge t(x)]$$

By a process similar to that used for the universal proposition, we find that this is equivalent to,

$$\forall x,\ h(x) \rightarrow -t(x)$$

Literally, "For all x, if x is a highway, x is not heavily traveled." In ordinary language, "No highway is heavily traveled."

Our verbal expressions for quantitified propositions are often ambiguous. One of the uses of logic is to provide a test for our verbal expressions (by symbolizing them) to make certain they state what we mean.

For example, consider the statement, "All the applicants today are not going to be hired." What does it mean? One possibility is

$$\forall x,\ a(x) \rightarrow -h(x)$$

This statement could also be translated, "None of today's applicants will be hired." The choice of phrasing depends on the context.

On the other hand, the speaker could be emphasizing the "all" to mean

$$-[\forall x, a(x) \rightarrow h(x)]$$

or

$$\exists x, a(x) \wedge -h(x)$$

It is likely that the voice inflections when the statement is spoken could convey the meaning intended. But written statements have no intonations, and they are often misunderstood. The results of scientific research are more often written than spoken. Clarity of expression is extremely important. The methods of symbolic logic provide a handy check on this clarity.

Let us summarize the concepts of the negations of quantified propositions into a set of logical equivalents, as shown in Figure 2.6.1. Note that in these equivalences $p(x)$ can be any proposition, simple or compound. If it is a compound statement, further logical equivalents are probably needed, as we have seen earlier, to put the negation into usable form.

	Statement	**Logical Equivalent**
1.	$-[\forall x, p(x)]$	$\exists x, -p(x)$
2.	$\forall x, p(x)$	$-[\exists x, -p(x)]$
3.	$-[\exists x, p(x)]$	$\forall x, -p(x)$
4.	$\exists x, p(x)$	$-[\forall x, -p(x)]$

Figure 2.6.1

EXERCISE 2.6

1. Write each of the following as a quantified proposition (a statement using one of these quantifiers: all, every, no, none, some, a few, many, etc.) and symbolize:

a. If you earn more than $600 in a year, you must file an income tax return.
b. If you are a good citizen, you will not litter the streets.
c. If you are a college graduate, you are considered literate for the purpose of voter eligibility.
d. He is a gentleman and a scholar.
e. If a person cannot use mathematics, he cannot be an engineer.

2. Write the negations of each of the propositions of Exercise 1 in ordinary language. Write these negations as quantified propositions, and symbolize each.

3. Write the following propositions as conditionals (i.e., in the form, if . . . , then . . .).

a. All musicians are temperamental.
b. No liberal could ever be elected in that district.

c. Barking dogs never bite.
d. No student has ever failed that course.
e. It is not true that some automobiles produce no smog.

4. Write the negations of each of the propositions of Exercise 3 in ordinary language. Symbolize each and check the meaning.

5. Consider the old adage, "All that glitters is not gold." Write two possible different meanings and symbolize each. Which meaning is more likely intended?

Determine the validity of the following arguments:

6. All successful corporations have capable presidents. ABC, Inc. has a very capable president. Therefore, ABC, Inc. is successful.

7. People who live in cities must be able to endure noise and confusion. No sensitive person can stand noise and confusion. Therefore, people who live in cities are insensitive.

8. A well-organized experiment requires careful planning. A well-organized experiment produces usable data. Therefore, (some? all?) carefully planned experiments produce usable data.

9. Consider a situation in which a psychologist publishes a paper giving the results of a study that he has made of 1000 tone-deaf men. He observed that each of these men is left-handed. In the paper he proposes a theory that says, "All tone-deaf men are left-handed."
a. What relationship does his theory state concerning the set of left-handed men and the set of tone-deaf men?
b. Suppose that you disagree with his theory? What evidence do you need to disprove it?
c. You know a left-handed musician who is definitely not tone-deaf. Can you use him as an example in your attack on the psychologist's theory? Explain.
d. State the theory in symbols. What is the statement, in symbols, that refutes the theory?

10. A management consultant firm studied the organizational structure of a small corporation. Some of the observations by one member of the consultant team were written in set symbols.

U = Employees of the corporation
D = Board of Directors
E = Executives
M = Male employees
W = Female employees
P = Production-department employees
F = Field sales personnel
T = Employees making $20,000 per year or more

The following are the observations:

i. $D \subset E$
ii. $F \subset M$

iii. $E \subset T$
iv. $W \cap F = \emptyset$
v. $P \cap T \neq \emptyset$

a. Rewrite each of the observations as a quantified proposition (using all, none, some, etc.) and do **not** use any set terminology ("is an element of," "is a subset of," etc.)
b. Symbolize each of the above statements, using the symbols of logic.
c. Formulate five other different statements, expressing relationships that could exist among the sets of employees. Express these statements in both set symbols and the symbols of logic.

3

relations and functions

3.1 GENERAL DESCRIPTION

In almost all of our activities and fields of interest, we try to find or establish **relationships** among various ideas, things, and phenomena. These relationships are of many different types. Some are mere associations of one thing or person with another. Others are closer relationships in which one component has a controlling, or determining, influence over the other. An illustration of the former is the association we form between people and their possessions. "Joe has a red convertible." "Susie has a darling yellow dress." We sometimes use these associations as a means of identifying people. "Joe is here. I see his car parked over there." "That's Susie over by the table. She's the one in the darling yellow dress."

As an example of a relationship in which one component determines the other, consider a person planning a trip in an automobile. As part of his planning he wishes to know how much gasoline he will need. He determines this by dividing the distance to be traveled by the "mileage" (the number of miles he can travel on one gallon of gasoline) obtained with the automobile. The mileage is a property of his particular automobile and his driving habits. Thus, the distance to be traveled is the quantity that **determines** the amount of gasoline needed.

Associations, or relationships, such as the foregoing are called **relations** in mathematics. The concept of ownership, or possession, is a relation between people and things. One aspect of travel by automobile is a relation between the distance traveled and the amount of gasoline needed. Studying the nature and properties of relations, and of a particular type of relation called a **function**, constitutes an important branch of mathematics. In this chapter we shall discuss some of the basic structure of relations and functions and study in greater detail an important and useful function: the linear function. At first we shall discuss relations between two items. These are called **binary relations**. In the next chapter we shall extend our discussion to relations involving more than two items.

Although mathematics is usually concerned with relations involving numbers, the properties of relations apply equally well to other types of things. Let us consider further the illustration used above, the concept of ownership. We use many different words and phrases to indicate the existence of this relationship between a person and a thing. The essence of all these expressions, however, can be stated simply as "y owns x," where y represents the owner, and x, the object owned. Note that this sentence is similar in form to the propositions studied in the previous chapter, except that now there are two unspecified terms instead of one. If the x is replaced by a specific object, and y, by the name of a person, we have a statement that is either true or false.

The proposition, "y owns x," states a relationship between the elements of two sets: the set of objects and the set of people. We can form a set of pairs—each pair consisting of an object and its owner—that can be used with the proposition to make it a true statement. The set of all such pairs is called the **truth set** for the proposition. Note that each element of the truth set is a **pair** of items: an object, and its owner. From our illustrations above we would have (red convertible, Joe) and (darling yellow dress, Susie). The symbol for a "typical" element of the truth set would be the pair, (x, y).

The role of mathematics in dealing with such concepts as ownership is to study the **structure** of the relationship, and to determine the properties relationships have in common and the characteristics that distinguish one type of relationship from another. For example, a businessman is interested in companies and their products or services. A relationship between them can be expressed in the proposition, "y sells x." This relation has many of the same properties as the algebraic relation, "y is a number less than x." A sociologist studies people and their occupations, a relation characterized by the proposition, "y is the occupation of x." This relation has many of the same properties as the algebraic relation, "y is the largest integer smaller than x." In the following sections we name and define the fundamental terms and entities in the system of relations.

3.1.1 Variables—Independent and Dependent

We noted earlier that a binary relation could be described as a proposition in which there are two unspecified terms. In our examples we used x and y to represent them. The truth set for the proposition consists of pairs of these terms: one, a replacement for the x; and the other, a replacement for the y. The form of a typical element of a truth set is (x, y). The foregoing discussion is often shortened in mathematics texts to state simply that a relation* consists of a set of ordered pairs (x,y).

One of the first requirements in working with a relation is that the pairs making up the truth set be **ordered**; that is, the components are always listed in the same order to keep clear the role each plays in the proposition. The **components** of a **typical pair**, the x and the y, are called **variables**. Their replacements **vary** from one pair to another. The **first component** of the pair is the **independent variable**; and the **second**, the **dependent** variable. Although x is usually used as the first component and y as the second, this is not always the case. However, regardless of the symbol used to represent it, the **first** component is **always** the **independent variable**. Thus, if the pairs of a relation were denoted by (y,x), y would be the independent variable and x the dependent variable.

The terms "independent variable" and "dependent variable" suggest their relative status. When working with the elements comprising a relation, the first component (variable) is designated first. Its designation is done "independently." Then the proposition characterizing the relation specifies the appropriate second component to be paired with it. The choice of replacement for the dependent variable **depends** on the identity of the choice for the independent variable.

The set of elements that can be used as replacements for the independent variable is determined by the nature of the proposition defining the relation. This set is called the **domain** of the relation. A relation is said to be "defined over the domain." While the nature of the relation determines the complete set comprising its domain, a person using the relation can specify a subset of this complete set as the domain for any particular discussion. Once the domain has been established, the set of replacements for the dependent variable needed to "match up with" all of the elements of the domain designated are determined by the proposition defining the relation. This set is called the **range** of the relation.

A proposition defining a relation may be such that there is more than one choice possible for the dependent variable for a given choice for the independent variable. In this case a different ordered pair is formed using each of the possible second elements.

Example 3.1.1

Consider the relation defined by the proposition, "y is a counting number less than x." This is a relation in which x could be any real number greater than one. Let us arbitrarily restrict the domain so that there are a reasonable number of elements in the truth set. Let

$$\text{Domain} = \{x \mid x \text{ is a counting number between 1 and 100}\}$$

From the proposition we can now determine that the range for the restricted relation is

$$\text{Range} = \{y \mid y \text{ is a counting number less than 99}\}$$

* In mathematics the defining proposition for a relation does not have to "make sense." It can state merely that y is the element associated with x in a specified set of ordered pairs. Thus, **any** set of ordered pairs comprises a relation.

If we choose five as the replacement for x, the counting numbers 1, 2, 3, and 4 will each make the proposition a true statement. Thus, the ordered pairs, (5,1), (5,2), (5,3), and (5,4) are all elements of the truth set.

The entire truth set of pairs using each element of the domain constitutes the "restricted" relation. If we designate this relation as R_1,

$$R_1 = \{(2,1), (3,1), (3,2), \ldots, (99,98)\}$$

There is a total of 4851 pairs in R_1.

In the symbology of mathematics this relation would be described as

$$R_1 = \{(x,y) \mid x,y \in N,^{*}\ 1 < x < 100, \text{ and } y < x\}$$

In words, "The relation, R_1, is the set of ordered pairs (x,y) such that x and y are counting numbers (elements of the set of counting numbers), x is greater than one and less than 100, and y is less than x."

Note that the specification for the restricted domain is included in the description of the set.

In practice, as much of the description of a relation as possible is left to context and understanding. If the context indicates that the domain and range include only counting numbers, the relation of Example 3.1.1 is defined merely as $y < x$, in the expectation that the reader understands that a relation is a set of ordered pairs. The restriction to the domain, however, must be stated. If no restriction is specified, it is assumed that the domain is the largest set for which the specified relationship "makes sense." That is, the largest set for which the relationship can be defined.

Let us summarize the concepts covered this far:

Binary relation: The truth set for a proposition with two unspecified terms. One of these (the independent variable) is the **primary** term. Its replacement **determines** the identity of the second term.

Ordered pair: An element of a binary relation. The replacement for the independent variable is listed first.

Independent variable: The **primary**, or determining, term in the proposition defining a binary relation. The **first** component in an ordered pair of the relation. To form a particular ordered pair, a replacement for this variable is first chosen arbitrarily from the domain. (x is frequently used as a symbol for this variable, but **any other** symbol could be used instead.)

Dependent variable: The **second** component in an ordered pair of a binary relation. In a particular ordered pair its identity is determined by the choice of the first component (independent variable) and the defining

*N is widely used to designate the set of counting numbers, which are also called the **natural** numbers.

proposition for the relation. (y is usually used as a symbol for this variable, but any other symbol could be used instead.)

Domain: The set of all the possible replacements for the independent variable of a relation.

Range: The set of all the replacements for the dependent variable of a relation. The elements of this set are specified by the defining proposition following the designation of the domain.

3.1.2 Functions

The term "binary relation" is the general expression denoting any set of ordered pairs. A given element of the domain of a relation can be associated with any number of different elements for the range. In many relationships the identity of the replacement for the **dependent variable** is **determined without ambiguity** once a choice has been made for the independent variable. A relation with this property is called a **function**. The expression, "is a function of," is used in everyday language to denote a situation in which one thing is determined by another. A salesman's earnings in commissions **is a function of** how much he sells. He can determine his earnings exactly by multiplying the value of his sales by the rate of commission he is paid. The time required to move a specified distance **is a function of** the speed of travel. The time of travel is determined by dividing the given speed into the distance to be traveled. The name of the capital city **is a function of** the name of the state. Specifying one of the states determines the name of the associated capital without ambiguity.

The study of functions comprises a large part of mathematics. A function is defined formally as follows:

Definition 3.1.1

> **Function:** A **function** is a **relation** such that the choice for the first element of any ordered pair in the relation **determines uniquely** the second element.

This definition is often given in mathematics texts in an indirect, although perhaps more elegant manner:

Definition (Alternate) 3.1.1A

> **Function:** A **function** is a **relation** such that, for any two ordered pairs of the relation (x_1, y_1) and (x_2, y_2), if $y_1 \neq y_2$, then $x_1 \neq x_2$.

The definition says that whenever the **second elements** of **any two pairs** of a relation are **different**, if it follows from this fact that **corresponding first elements** are **also different**, the relation is a **function**. That is, for a relation to be a function, the same first element cannot be paired with two different second elements.

Notice that neither form of the definition says anything about a situation in which the same second element is associated with different first elements. By implication, then, it is possible for this to happen in a function. Examples of functions in which the same second element can be associated with different first elements of the ordered pairs in the function:

The price list in a supermarket. In the list of prices for the different items, the name of the commodity is the independent variable; and the price, the dependent variable. No items will be listed with more than one price, and the price list is, therefore, a function. Several items, however, may have the same price.

Test scores in an examination. The set of people taking the examination is the domain. The set of scores is the range. Several people may get the same score, but no one will get more than one score. The set of pairings, each person with his score, constitutes a function.

In mathematics the defining propositions for functions are usually written in the form of equations. The most general of these is $y = f(x)$, which is read, "y is a function of x"; or often, "y equals f of x." The particular function intended is the formula, or mathematical expression, represented by f. The (x) part is used to show that x is the symbol representing the independent variable. Individual ordered pairs belonging to the function are obtained by selecting a value for x from the domain, substituting it into the formula represented by f, and performing the indicated computations. The result of the computations is the value for y, the dependent variable in the ordered pair. The function is the set of all the ordered pairs obtainable in this way. The symbol $f(x)$ is **not** used to indicate multiplication of f by x, although parentheses are often used for this purpose. The context should make clear which meaning is intended.

Example 3.1.2

Let us consider a specific formula that defines a function. If we let the "f" of $y = f(x)$ represent the formula "$3x + 6$," we have the equation for a function:

$$y = 3x + 6$$

To obtain ordered pairs belonging to the function, we select values for x and perform the computation. If $x = 1$, the corresponding value for y is obtained by replacing the x in the formula with 1. This substitution is symbolized as $f(1)$.

$$f(1) = 3(1) + 6 = 9$$

"f of 1" ⟶ ↑ (points to $f(1)$) ↖ Multiplication of 3 times 1

Thus, nine is the value for y when one is the value for x, and the ordered pair, (1,9), is an element of this function.

If $x = 2$, y is given by $f(2)$:

$$\begin{aligned} f(2) &= 3(2) + 6 \\ &= 12 \end{aligned}$$

The ordered pair, (2,12), is an element of this function. Other ordered pairs in the function are found in a similar manner. In a later section we shall study functions of this type in greater detail.

The complete description of this function, designated by f, is

$$f = \{(x,y) \mid x,y \in R,^* \text{ and } y = 3x + 6\}$$

In words, "f is the set of ordered pairs, (x,y) such that x and y are real numbers and y is the sum of three times x added to six."

A few verbal descriptions of this type give one an appreciation for the simplicity and clarity of mathematical symbols.

EXERCISE 3.1

1. Write defining propositions for four binary relationships that exist among the members of a family. Write the proposition so that (x,y) is the typical element of the truth set. Choose the four relationships so that two of them are functions and two are not.

2. Write defining propositions for three binary relationships that exist among the students at a college. Write the propositions so that (x,y) is the typical element of the relationship. State whether each relation is a function or not, and give examples to show why.

3. Use the elements of the set $\{1,2,3,4\}$ as the possible values for x and y; that is, let the set $\{1,2,3,4\}$ be both domain and range. List the elements for these relations:
- a. y is equal to x.
- b. y is greater than x.
- c. y is one less than x.
- d. y is a multiple of x.

4. Which of the relations of Exercise 3 are functions? List examples to show why the others are not functions.

5. Form two relations, one a function, one not, using the alphabet as domain and range.

6. Form two functions in which the domain is a set of people and the range is a set of numbers.

7. Specify suitable domains and the corresponding ranges for the following:
- a. y is a sibling of x.
- b. y is the president of x.
- c. y is x's age.

*R is the symbol usually used to designate the set of real numbers.

d. $y = 2x + 3$.
e. $y < 3x - 2$.

8. Specify suitable domains and ranges for the following relations. Indicate which relations are also functions.
a. y is the length of x.
b. y is the sum of seven and four times x.
c. y is the weight of x.
d. y is the altitude of x.
e. y is x's sales volume for 1971.

9. For which of the relations in Exercises 7 and 8 is it possible for the domain and range to be the same set?

10. List three elements of each of the relations of Exercise 8, using the domains and ranges you have specified.

11. In the function defined by $y = f(x)$, let $f(\ )$ represent $2(\ ) - 3$.
a. Find the numbers associated by this function with the following values for x:

5, 3, 1, 9

b. Compute $f(4)$, $f(6)$, $f(2)$, $f(0)$.

c. Supply second components so that the following number pairs belong to this function:

(7,), (1,), (−2,), ($1\frac{1}{2}$,)

12. Let $3x + 2y = 30$ be the defining proposition for a function, whose typical element is (x,y).
a. Find the numbers associated by this function with each of the following values for x:

2, 4, 0, −2, 10

b. Supply second elements so the following pairs belong to this function:

(6,), (−10,), (1,) (5,), (30,)

c. If the domain of this function is specified to be all the numbers from 0 to 10 inclusive, what is the corresponding range?

13. A relation, R, consists of the following set:

$$\{(u,v) \mid u,v \in \text{alphabet} \quad \text{and } v \text{ precedes } u\}$$

List five elements of R.

14. Referring to the set $\{a, b, c\}$ and its subsets and elements:
a. Define the domain and range for the relation, y is an element of x.
b. List four elements of the relation, y is a subset of x.

3.2
GRAPHS

Relations and functions, as the truth sets for propositions, are essentially compilations of information. The relation, "y sells x," mentioned in the previous section is an important one in our economy. Portions of its truth sets are widely publicized in various forms. The usual format is tabular. Every purchasing agent has a collection of catalogs in which the suppliers for various items are listed. The ordinary citizen has access to a tabulation of a part of the truth set in the "yellow pages" of his telephone directory. Such tabulations, while not the ultimate in convenience, are the best way we have found thus far to do a necessary job.

When a relation involves **quantities**, it is possible to portray it visually in a picture called a **graph**. Pictures convey much more information than can conveniently be put into words and other abstract symbols. For this reason graphs are used extensively in the study and application of mathematics. The student has undoubtedly seen, used, and made graphs already. However, in this section we shall review briefly some of the important considerations in the making of graphs. In particular, the section on Cartesian coordinates should be studied carefully. It contains material about the reasons behind graphing systems the student may not remember.

Quantities are expressed by numbers. The set of numbers has the unique property of being **completely ordered**. That is, if we draw a line and select an arbitrary reference point on it, every real number can be associated with a unique point on the line; and every point on the line represents a unique number. The "size" of the number determines the **distance** its associated point on the line will be from the reference point. Further, the "standard" numerals used to express the numbers tell us immediately the **relative location** of any two numbers on the line. This latter information comes from the trichotomy property of real numbers. With any two real numbers, represented by a and b, one of the following must be true: a and b can represent the same number, in which case $a = b$; a is less than b ($a < b$); or, a is greater than b ($a > b$).

The student is probably aware of these properties of numbers, but not in the formalized language of algebra. It is not our purpose here to delve into the theory of numbers, but only to make the student aware of the implications of concepts that he regards as commonplace and natural.

3.2.1
Line Graphs

One type of graph used in the behavioral sciences is a "preference scale" or "opinion scale." These graphs are used to "quantify" what are normally regarded as subjective concepts. A common form of such a scale uses a line with scale markings from 0 to 10, as illustrated:

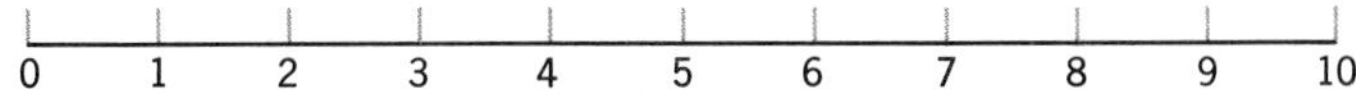

The graph enables a person to express his feelings about something without being required to verbalize them. For example, suppose a personnel consultant is studying employee attitudes in a company. He asks each employee to "rate" his supervisor in several different areas on scales such as those shown in Figure 3.2.1. There are many varieties of this general type. They yield numerical values for feelings and opinions. These can be averaged, or combined to obtain a "consensus" much more easily than can words such as "indifferent" and "reasonably capable."

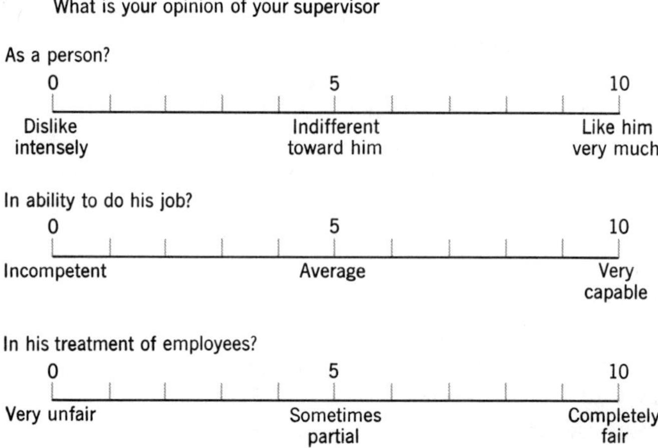

Figure 3.2.1

A second type of line graph is used to display numerical values for measurable quantities in place of a tabulation. The visual impact of the graph conveys the relative magnitudes of these quantities much more vividly than do columns of numbers. Figure 3.2.2 is a typical line graph, showing comparative population figures. The message can be read at a glance; the population increase in the 1950–60 decade occurred entirely in urban areas. The relative amount of the population change can also be judged from the length of the lines. The

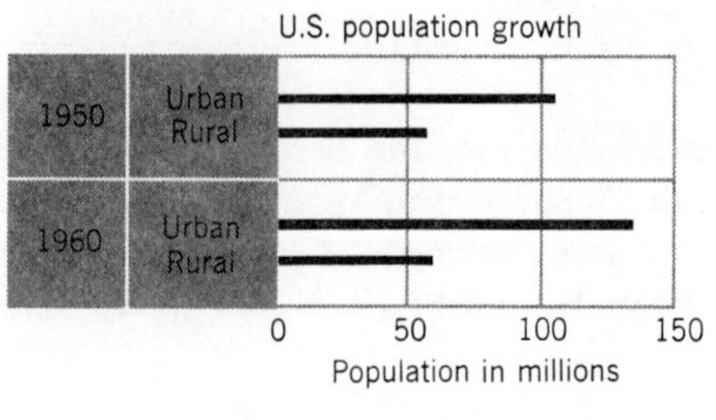

Figure 3.2.2

same information can be obtained from a tabulation of numbers, of course, but most people can read a graph more easily.

Note that the graph is used to portray a function. A line graph of the type shown in Figure 3.2.2 is used when the domain of the function is not basically numerical, but the range is. In Figure 3.2.2 the domain is the type of community—either urban or rural—and the range is the population, a number. Separate lines, whose length varies as the size of the numbers, are drawn for each element of the domain. There are many variations of this general type of graph used to display numerical information. The aspect of interest in the graphs in this discussion is that they make use of **visual images whose size is in direct proportion to the "size" of numbers being represented**.

3.2.2 Cartesian Plane

In most of the binary relations and functions studied in mathematics both the independent and dependent variables represent numbers. It turns out that a plane surface, a sheet of paper for example, is beautifully suited for displaying such relations graphically. In making a graph we correlate **the size of the number** to a **distance**. Distances can be measured in two distinct directions on a sheet of paper—distances **across** the page, and distances **up and down** the page. A key consideration is that we can lay out a distance across a page while staying at the same place up and down on the page, and vice versa. This permits setting up two completely independent scales for numbers. Thus, distances across a page can be used to represent one of the numbers of a number pair; and distances up and down can be used to represent the other. The distances across a page are called **horizontal** distances; and the distances up and down are called **vertical**.

A graph is formed by establishing first a starting place for measuring horizontal distances. This starting place takes the form of a **vertical line**, which provides a reference point for horizontal measurements at any place up and down the page. Similarly, the starting place for vertical measurements is a **horizontal line**. Customarily, the horizontal distances are used to represent the independent variable; and vertical distances, the dependent variable. Next, the correlation between distances and numbers, the **scale**, is chosen. A scale is established by choosing arbitrarily a convenient distance to represent the number, one. The distances representing all other numbers are suitable multiples of that distance. **For convenience**, the scale is usually marked off along the reference lines, starting with zero at their intersection. The point of intersection of the two reference lines, denoted by the point (0,0), is called the **origin**, and is symbolized simply as "O". Positive numbers are customarily measured either to the right or up; negative numbers, to the left or down. The resulting framework for graphing is called a Cartesian plane, or Cartesian coordinate system, in honor of the French mathematician, Rene Descartes (1596–1650), who first developed the correlation between points on a plane and pairs of numbers, which underlies this graphing procedure.

A typical Cartesian coordinate system is shown in Figure 3.2.3. Notice that the two reference lines are perpendicular to each other. The horizontal reference line is called the **horizontal axis, or x-axis;** and the vertical reference line, the **vertical axis** or **y-axis**.

An important fundamental aspect of a Cartesian coordinate system is that each individual number pair is associated with a unique point in the plane area of the coordinate system, and each point in this plane represents a unique number pair. (This is not true in all coordinate systems—e.g., polar coordinates.) This fact makes it possible to transfer back and forth, without ambiguity, between the number pairs in a relation and their corresponding points in the coordinate system whenever it is convenient. In fact, the correlation between number pairs and points is so close that the number pairs are frequently called "points" even when no graph is involved.

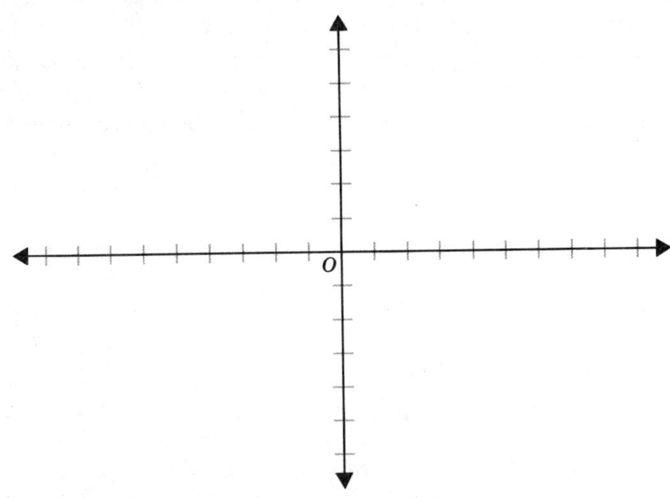

Figure 3.2.3

To locate the point on a graph associated with a particular number pair (x,y) we proceed as follows.

Starting at **any point** on the **vertical** axis, measure a distance of x units in a horizontal direction (to the right for a positive number; to the left for a negative number). Draw a **vertical** line at this distance from the vertical axis. Any number pair whose first component has this particular value will lie somewhere on this line.

Next, starting at **any** point on the **horizontal** axis, measure a distance of y units in a vertical direction (up for a positive number, down for a negative). Draw a **horizontal** line at this distance from the horizontal axis. Any number pair whose second component has this particular value will lie somewhere on this line.

The point of intersection of these two lines is the point representing the number pair.

Figure 3.2.4 shows the result of locating the point representing (3,2).

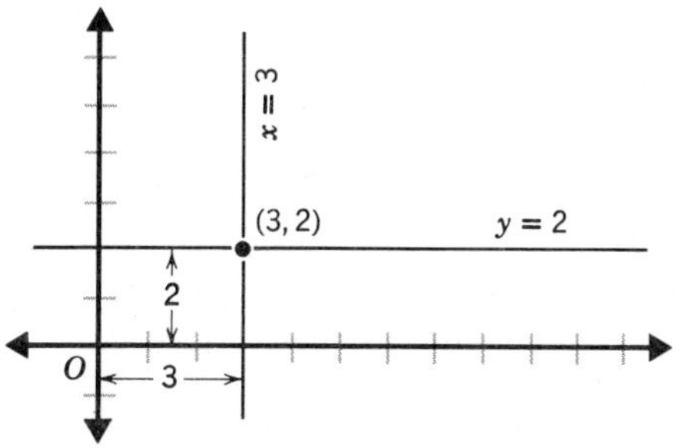

Figure 3.2.4

In this discussion the use of the scales marked on the two axes has been purposely avoided. We wish to emphasize that the value for x is **not** associated with a **point** on the horizontal axis—in particular the point whose scale marking is the value for x. The value for x is represented by a **distance from the vertical axis**. Thus, in discussion about number pairs if we specify that $x = 3$ and give no value for y, **any** point on the vertical line three units to the right of the vertical axis will meet the specification. In this context, therefore, the statement $x = 3$ describes a **vertical line** on the graph. Similarly, if we specify that $y = 2$, but give no value for x, the graphical representation for this specification is a horizontal line two units above the horizontal axis.

The point on the horizontal axis associated with the **scale mark**, 3, is (3,0). Any point on the horizontal axis has a y-value of zero. Similarly, the point on the vertical axis with the scale mark, 2, is (0,2). Any point on the vertical axis has an x-value of zero.

The scale marks on the axes are used for convenience in measuring the distances representing the numbers. For example, in locating the point (1,2) we can use the scale on the horizontal axis to locate a point at a distance of one unit to the right of the vertical axis. We then draw a vertical line through this point to locate any point whose "x-value" is one. For this reason the horizontal axis is commonly called the x-axis—it is used in locating the x-value of a point. The axis itself, however, is described by the specification, $y = 0$! That is, the x-axis is the graph of the set of all points whose y-value is zero. Similarly, the vertical axis is called the y-axis; and it is used in locating the y-value of a point. As a line, it is described, or specified, by the statement, $x = 0$.

Let us digress for a moment to discuss nomenclature. There are many different terms that strictly apply to specific contexts, but that are used more or less interchangeably because of the close correlation between relations and their graphs.

> The symbol (x,y) is commonly used to represent a typical **number pair** in a relation, or a **point** on a graph. The individual numbers in a **number pair** are **components**. In a **point** they are called **coordinates**. The "x" in (x,y) is the **independent variable**. Any one of the following terms refers to specific numbers used to replace the x.
>
> x-component
> x-coordinate

x-value
First component
First coordinate
Value for independent variable
Abscissa*

The y in (x,y) is the **dependent variable**. Any one of the following terms refers to a specific replacement for y:

y-component
y-coordinate
y-value
Second component
Second coordinate
Value of dependent variable
Ordinate*

3.2.3 Graphs of Relations

The regarding of a numerical relation as a set of number pairs and the correlation between number pairs and the points on a graph combine to provide a method for "drawing a picture" of a relation. We may, also, start with a collection of points on a graph and construct a relation to fit them. (This is often done in the analysis of data. The process required, however, involves mathematical considerations beyond the scope of this text.)

In our discussions here we shall consider only relations described by algebraic expressions which are also functions. They are the most useful for the applications we shall consider in later sections. They are by no means, however, the only type of relations whose graphs are useful and important.

We shall illustrate the procedure for drawing a graph of a function with some simple algebraic equations, making use of a property of algebraic equations—their graphs are all "smooth" curves.

The graphing process consists of three steps:

1. Find, by computations with the algebraic equations, a set of several number pairs belonging to the function. Select convenient values for x over a reasonable interval for use in the computations. **Note**: This selection requires a decision on the part of the grapher. There is no universal "standard" set of values to use. In a particular application use the values that are of most interest.

2. Locate on the graph the points associated with the number pairs found in Step 1.

3. Draw a smooth curve through the points, progressing through the points in the order of their x-values. Start with the point farthest to the left and move to the right, passing through each point in sequence. If there is some

*The terms "abscissa" and "ordinate" are used almost exclusively in discussing graphs. Modern usage favors the more descriptive terms, "x-coordinate" and "y-coordinate."

doubt about the shape of the curve in some region, choose more values for x in that region and find additional points. Keep in mind that if the graph is representing an algebraic equation, it will not have any "angles" in it. All changes in direction are made with smooth curves.

Example 3.2.1

Let us find the graph for the function described by the equation, $y = 2x + 4$.

We start by selecting some convenient values for x, say 0, 1, 2, 3, and 4. We compute the corresponding values for y and form the resulting number pairs or points:

For $x = 0$, $y = 2(0) + 4 = 4$ Point: $(0, 4)$
For $x = 1$, $y = 2(1) + 4 = 6$ Point: $(1, 6)$
For $x = 2$, $y = 2(2) + 4 = 8$ Point: $(2, 8)$
For $x = 3$, $y = 2(3) + 4 = 10$ Point: $(3, 10)$
For $x = 4$, $y = 2(4) + 4 = 12$ Point: $(4, 12)$

We set up a coordinate system and plot these points, labeling each point with its coordinates, as shown in Figure 3.2.5.

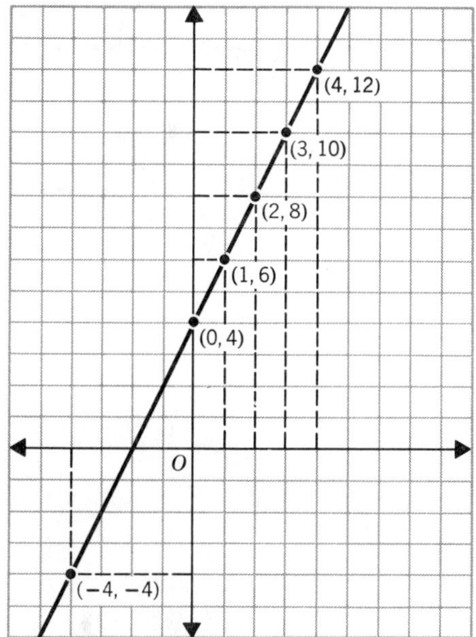

Figure 3.2.5

The "curve" drawn through the points looks very much like a straight line. We can verify that it is by extending the line down to the left and computing the coordinates for a point in that region.

Let $x = -4$. The corresponding value for y is given by

$y = 2(-4) + 4 = -4$ Point: $(-4, -4)$

When we place this point on the graph, we find that it "falls on" the line we have drawn.

Checking the shape of the graph by plotting another number pair from the function illustrates an important and fundamental property of graphs: Any number pair belonging to the relation, that is, which "satisfies" the equation describing the relation, will lie somewhere on the graph of the relation. The coordinates of any point on the graph of the relation will "satisfy" the equation describing the relation.

Example 3.2.2

Let us now consider a more complex equation:

$$y = x^2 - 4x + 3$$

We start by selecting values for x, say $-2, 0, 2, 4$; and compute the corresponding values for y:

When $x = -2$,

$$y = (-2)^2 - 4(-2) + 3 = 4 + 8 + 3 = 15 \qquad \text{Point: } (-2, 15)$$

By similar computations we obtain the points:

$$(0, 3), \quad (2, -1), \quad (4, 3)$$

Figure 3.2.6 shows these points plotted on a graph.

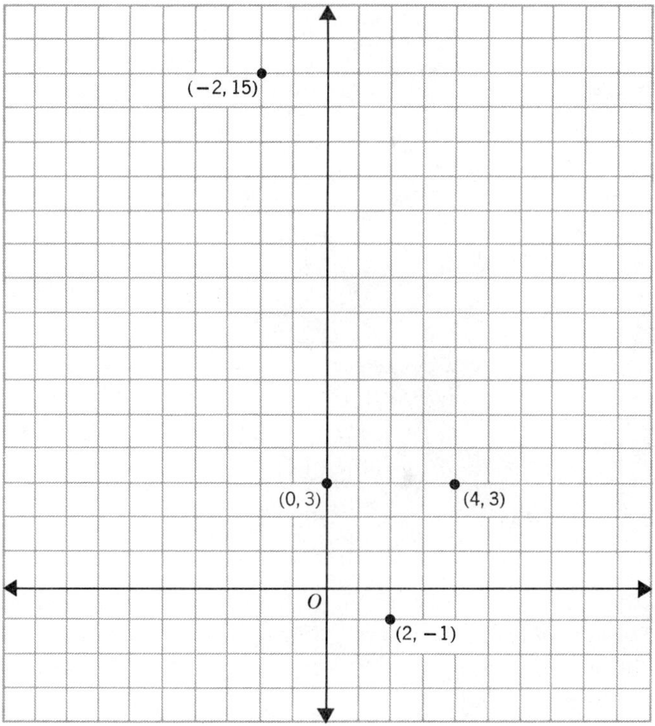

Figure 3.2.6

Clearly, these points do not lie on a straight line. We need more points between 0 and 4 to find the shape of the curve in that region, and another point or two beyond $x=4$ to determine where the curve is going as x takes on larger values.

Let us try $x = 1$, 3, and 5. We obtain the points, (1,0), (3,0), and (5,8). When we add these to the graph we can draw a smooth curve, as shown in Figure 3.2.7.

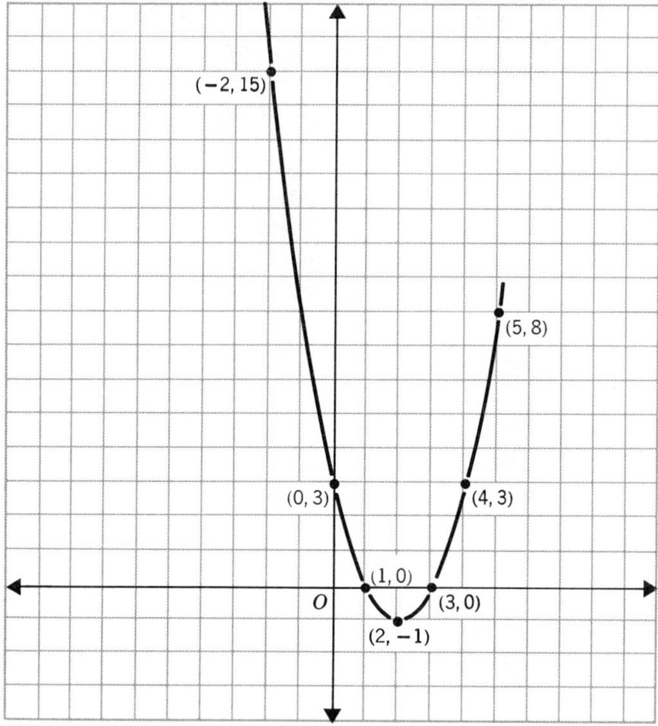

Figure 3.2.7

Remember that all of the number pairs belonging to a relation appear on its graph, and the coordinates of any point on the graph form a number pair belonging to the relation. A study of the "behavior" of the graph, its peculiarities and areas of interest, is in effect a study of the relation itself. The curve for the relation of Example 3.2.2 shows that there are no values for y less than -1. This is a characteristic of this particular function: it has a minimum value of -1, which is associated with the x-value of 2. That is, the dependent variable takes on its minimum value when $x=+2$.

The graphs in Figures 3.2.5 and 3.2.7 show only a small portion of the complete functions. A graph can be drawn for any region of the domain containing information of interest; but since these functions have the entire set of real numbers for a domain, they cannot be graphed in their entirety. The pattern of computations for the points show the "trends" of the graphs outside the region graphed.

Students are sometimes confused by the process of choosing values for the independent variable. They do not know which ones to choose. In mathematics, where the structure and properties of relations are studied, the answer is to choose arbitrarily as many different values as necessary to determine the peculiarities of the particular relation under consideration. In other situations, where mathematics is **applied**, the situation itself supplies the choices for the variables.

At this point we should mention how the graph of a function differs from the graph of a relation that is not a function. Recall that the distinguishing characteristic of a function is that any value of the independent variable is associated with only one value for the dependent variable. If we draw a vertical line on a graph, it will contain all of the points whose x-coordinate has the value determined by the distance of the line from the y-axis. Since a function can contain no more than one number pair with that x-value, its graph will cross the vertical line no more than once. Thus, if **any** vertical line on the graph of a relation crosses the curve at **no more than one point**, the relation is a function. If there is even one vertical line that intersects the curve at more than one point, the relation is not a function.

Example 3.2.3

Let us consider a simple equation, which describes a relation that is not a function. Consider the equation, $y^2 = x$.

To plot the graph of this relation, we choose some values for x, say 0, 1, and 4.

When $x = 0$.
$y = 0$ — Point: (0,0)

When $x = 1$,
$y^2 = 1$, and y can be either $+1$ or -1 — Points: (1,1), (1,–1)

When $x = 4$,
$y^2 = 4$, and y can be either $+2$ or -2 — Points: (4,2), (4,–2)

These points are shown on the coordinate system in Figure 3.2.8.

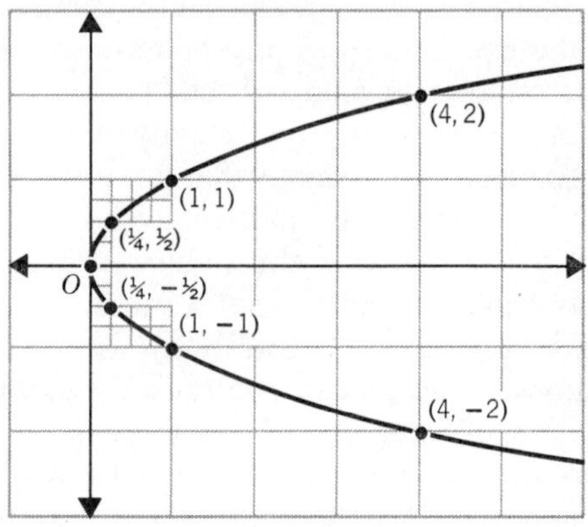

Figure 3.2.8

The domain of the relation contains no negative numbers. There are no values for y, whose squares will be negative. The shape of the curve near $x = 0$ is seen more clearly if we note that when $x = 1/4$, y is either $+1/2$ or $-1/2$.

Note that any vertical line drawn to the right of the origin intersects the curve at two points. Thus, the relation is not a function.

EXERCISE 3.2

Note: Use of quadrille ruled or graph paper will make working the problems requiring graphs simpler and more accurate.

1. Draw a **line graph** showing the sales volume for a company in its six years of operation, as shown in the table:

	Volume
1st year	$ 200,000
2nd year	300,000
3rd year	700,000
4th year	1,000,000
5th year	1,200,000
6th year	1,100,000

2. Draw a **line graph** showing the total sales of the 10 largest U.S. Corporations (1970).

10 Largest Companies (1970)	*Sales (in billions)*
General Motors	$18.8
Standard Oil	$16.6
Ford Motor Company	$15.0
General Electric	$ 8.7
International Business Machines	$ 7.5
Mobil Oil Corp.	$ 7.3
Chrysler Corp.	$ 7.0
International Telephone and Telegraph	$ 6.4
Texaco	$ 6.3
Western Electric	$ 5.9

3. The numbers below are the first test scores for the students in a math class. Count the number in each 10-point interval (0-10, 11-20, 21-30, etc.) and draw a **line graph** showing the distribution of scores (the number in each interval).

37	88	28	41	59	56	92	59
48	68	76	88	51	45	26	61
58	79	97	23	36	82	39	57
27	37	97	45	83	45	48	52
19	51	77	57	46	71	55	45

4. Locate the following points on a Cartesian coordinate system, and label each one.

(1,3), (−3,1), (−2,−3), (4,−2), (−2,0), (0,3), (2,2), (−2,2)

5. Referring to Figure 3.2.9, find the coordinates of the points A, B, C, D, E, and F.

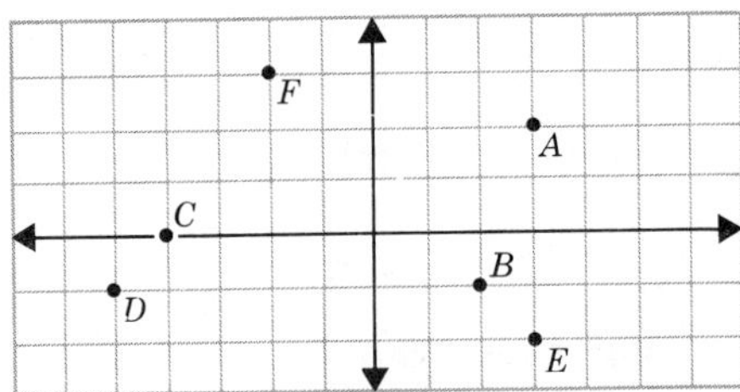

Figure 3.2.9

6. Consider the following projection for population growth over a 100-year period:

Year	*Estimated population (billions)*
1960	3.0
1970	3.7
1980	4.5
2000	6.7
2020	10.0
2040	14.5
2060	22.1

a. Use these data to form a line graph showing the projected population in each of the years listed.

b. Choose an appropriate interval and scale on the horizontal axis of a Cartesian coordinate system, using time (year) as the independent variable, and plot the points given. Draw a smooth curve through the points. (The algebraic expression producing a curve of this shape is called an **exponential**. The data shown were obtained using an annual growth rate of 2%—approximately the current rate.)

7. For each of the following functions, for which we assume that $y = f(x)$, find $f(0)$, $f(1)$, $f(-1)$, and $f(2)$.
   a. $y = 3x - 2$
   b. $2y = 5x + 6$
   c. $2x + 3y = 10$
   d. $y = 3x^2 - 2x + 5$
   e. $3y = x^2 - 6x + 2$
   f. $x - 2y + 6 = 0$

8. Plot each of the following functions on a set of Cartesian coordinates:
   a. $y = x - 3$
   b. $3y = x + 6$

c. $x + 2y = 8$
d. $3x - 4y = 12$
e. $y = x^2 + 4x - 5$
f. $y = 4 - x^2$

9. On a set of Cartesian coordinates plot the points (2,2) and (8,6). Draw a straight line through the two points. Assume that this line is the graph of the function, $y = f(x)$.

Using the graph,

a. Find $f(5)$; $f(-4)$.
b. Determine for what value of x, $y = 0$.

10. A man is renting chairs for a meeting. He is told there is a \$10 delivery charge no matter how many chairs he rents. The rental charge is \$1 each.

a. Let y represent the total cost; and x, the number of chairs rented. Write an equation that will give the total cost for renting the chairs.
b. Let y represent the cost of renting **each** chair after the delivery charge is added to the rental.
Write an equation for y as a function of x, the number of chairs rented. (Note that your equation does not make sense for fewer chairs than one.)
c. Plot each of these functions on the same set of coordinate axes. Extend the graphs to show the function values up to and including $x = 20$.

11. Graph the relation, $x^2 + y^2 = 16$. What is the shape of the curve in the graph?

12. Consider the two functions, $y = x + 1$ and $y = x^2 - 4x + 5$.

a. Draw the graphs for these functions on the same set of coordinates.
b. Use the graphs to find pairs of values for x and y that belong to both functions.

3.3 LINEAR FUNCTIONS

In the previous section there were several functions in the examples and exercises whose graphs were straight lines. Such functions are called, appropriately enough, **linear functions**. They are so frequently used that further study of their characteristics is merited. Many quantities are related naturally by linear functions, and the relationships between many others can be approximated by linear functions with acceptable accuracy. One of the attractive characteristics contributing to their wide use is their relative simplicity.

When we work with related quantities, one of the questions of interest is, "When one of the quantities is **changed** by a given amount, how much **change** will there be in the second quantity?" A person who owns stock looks at the "net change" column in the stock-market report. The number listed there, when multiplied by the number of shares he owns, tells him how much his stock has changed in value that day. A property owner follows the deliberations of a local government when they are setting the annual property tax rate. Discussion focuses on the **change** in tax rate. This number, multiplied by the assessed value of his property, tells him how much his taxes are going up (or down) for that year.

In each of these situations a key quantity is the **rate of change** of the dependent variable with respect to the independent variable. The amount of the change in the dependent variable is found in each case by multiplying the change in the independent variable by the **rate of change**.

The **rate of change** of the dependent variable with respect to changes in the independent variable—that is, the **ratio** of a **change** in the dependent variable to the **corresponding change** in the independent variable—is called the **slope** of a function. Figure 3.3.1 shows portions of the graphs of three

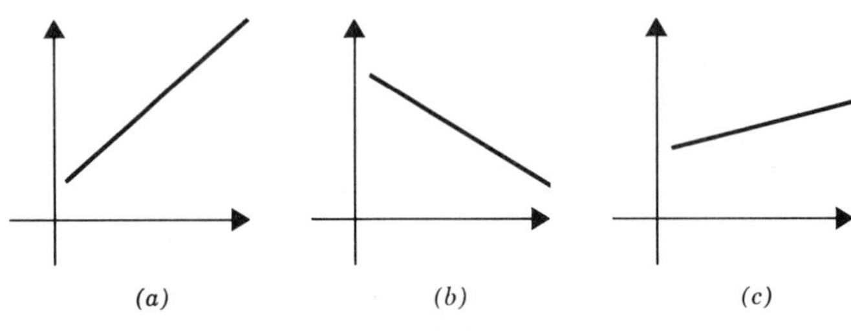

Figure 3.3.1

functions. In (a) the line "slopes" **upward** from left to right across the graph. It is said to have a **positive slope**. In (b) the line slopes **downward**. It is said to have a **negative slope**. The magnitude of the rate of change—the "size" of the slope—is a measure of the "steepness" of the graph: the steeper the curve, the larger the magnitude of the slope. In (c) the line slopes upward, and the slope is positive; but is smaller than the slope in (a). When the graph is horizontal, the slope is zero; there is no change in the dependent variable when the independent variable changes.

A **linear function** is a function whose slope is **constant**—it has the same value everywhere on the graph. It is the only type of function for which this is true. We shall limit our discussions here to linear functions. The methods for working with functions whose slope changes as the independent variable changes form the basis for calculus. They are beyond the scope of this text.

In the first of the examples mentioned earlier in this section, both of which are linear functions, the slope is the number of shares of stock the man owns. The independent variable is the market price of the stock, and the dependent variable is the total value of the stock. The man can, of course, change the slope of the "stock-value function" by buying or selling shares of stock; but for any particular period of time, the number of shares remains constant. The total value of his stock is given by the product of the number of shares he owns and the market price. If we let x represent the market price of the stock,

y the total value of the stock, and m the number of shares of stock, the function in symbols becomes $y = mx$. Graphs of this function for $m = 10$ and for $m = 20$ are shown in Figure 3.3.2. (Note the difference in the scales on the two axes

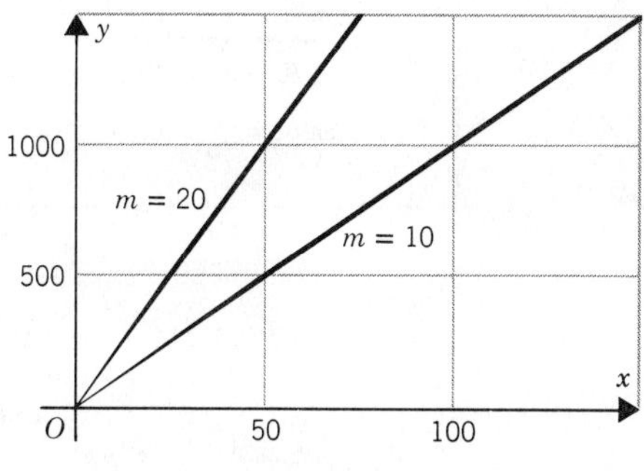

Figure 3.3.2

in the graph.) In Figure 3.3.3 the determination of the slope from the graph is illustrated for two different portions of the line for $y = 10x$.

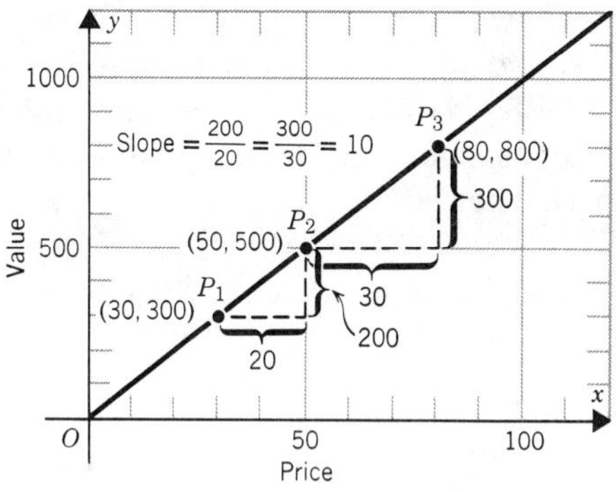

Figure 3.3.3

The linear function of Figure 3.3.3 has a special property—it passes through the origin. That is, when x has the value of zero, y is also zero. Let us consider a linear function for which this is not the case.

Example 3.3.1

Automobiles can be rented by the day for a fixed fee plus a charge for each mile the car is driven. Suppose a man wishes to rent a car for a day. The fixed fee is $5.00, and the charge for driving the car is 10¢ per mile. If y represents the total rental cost and x, the number of miles driven, the function giving the total cost is $y = 5 + (.10)x$. A graph of this function is shown in Figure 3.3.4.

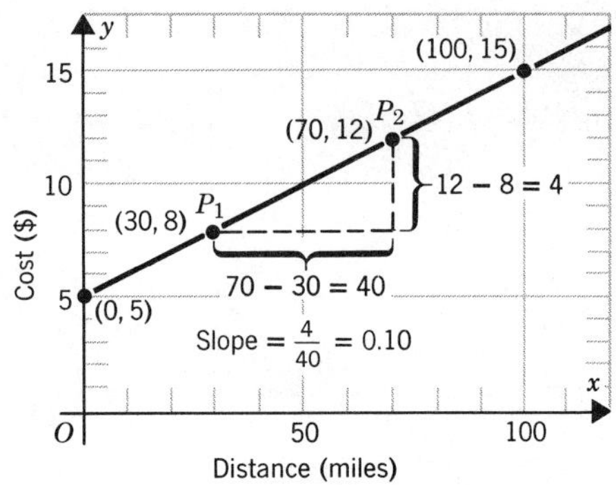

Figure 3.3.4

Since the graph is a straight line, it can be obtained from two points. When $x = 0$, $y = 5$ and when $x = 100$, $y = 5 + (.10)100 = 15$. We have the points (0,5), (100,15). We can obtain the slope of the line in the following manner: two different points on the line, P_1 and P_2, are chosen and the coordinates for these points determined. Any two points can be used. Call the coordinates of P_1, (x_1, y_1); and the coordinates of P_2, (x_2, y_2). Starting with either point, find the differences between the respective coordinates, $x_1 - x_2$ and $y_1 - y_2$. (Or, $x_2 - x_1$ and $y_2 - y_1$.) In our example $x_2 - x_1 = 70 - 30 = 40$; $y_2 - y_1 = 12 - 8 = 4$. The slope is the ratio of the difference in y-coordinates (the change in y) to the difference in x-coordinates (the change in x). That is,

$$\text{Slope} = \frac{y_2 - y_1}{x_2 - x_1} \qquad (3.3.1)$$

In this example the slope is 4/40, or .10. Note that in the function for the cost, (.10) is the cost per mile—the multiplier for x.

We can generalize the function of the example by letting b represent the daily rental ($5); and m, the mileage cost (.10). The function becomes

$$y = b + mx \qquad (3.3.2)$$

Equation 3.3.2 is one of the general forms for a linear function. Any linear function can be described in this form by an appropriate choice of values for b and m. The number, b, is called the **y-intercept** (the line crosses the y-axis at $y = b$). The number, m, is the **slope**. b and m can be any real numbers, either positive or negative.

The graphs for the linear functions in which $b = 5$, $m = -3/2$ and for which $b = -3$, $m = 2/3$, are shown in Figure 3.3.5.

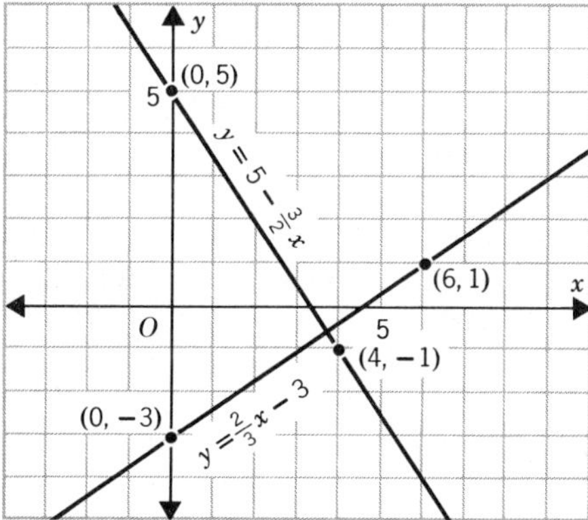

Figure 3.3.5

3.3.1 Alternate Forms for Linear Functions

In many cases translating a situation into a mathematical model produces a linear function with a form different from that of Equation 3.3.2. For example, suppose a factory requires 100 gallons of a special oil each day for one of its production processes. The factory produces varying amounts of this oil as a by-product of one of its other processes, but not enough to supply all of its needs. The additional oil needed is purchased from another company. If x represents the amount of oil produced by the plant itself on a given day, and y the amount purchased, the relation between x and y is given by the function,

$$x + y = 100 \tag{3.3.3}$$

In one version of a guaranteed income plan, the government guarantees an income of at least \$250 per month to a family. That is, the government supplements the income of the family to make it at least \$250. As an incentive for the family to earn as much of this as possible, only 60% of the family's earnings are counted in computing the amount that the government supplies. If x represents the family's monthly earnings and y the government's contribution, the mathematical model for the payments under the plan is

$$.60x + y = 250 \tag{3.3.4}$$

In each of the above examples the equations can be changed into the form of Equation 3.3.2 by the algebraic manipulation of subtracting the x term from each side of the equation. Equation 3.3.3 becomes $y = 100 - x$. The y-intercept is 100, and the slope is -1. Equation 3.3.4 becomes $y = 250 - .60x$. The y-intercept is 250, and the slope is $-.60$.

The computation of the y-intercept and the slope in each of these cases is simplified by the fact that the coefficient for y is one. A two-step computation is required if the coefficient for y is some number other than one. Consider a case in which a person needs a nutritional supplement in his diet. He needs 12 units of this supplement daily. There are two foods, A and B, that are recommended as sources for the nutrient. A contains two units per ounce, and B three units per ounce. If x represents the number of ounces of A he eats in a day and y the number of ounces of B, the mathematical model for his diet supplement is

$$2x + 3y = 12 \tag{3.3.5}$$

To find the slope and y-intercept, we first subtract the x-term from each side of the equation to obtain

$$3y = 12 - 2x$$

Next, we divide each side of the equation by 3, the coefficient of y, to obtain

$$y = 4 - \tfrac{2}{3}x$$

The y-intercept is now seen to be 4, and the slope is $^{-}\frac{2}{3}$. The interpretation of these quantities is that four represents the amount of B he would need if he ate no A at all; the slope tells him that for each ounce of A that he eats, he can reduce the amount of B by $\frac{2}{3}$ ounce.

Equations 3.3.3 and 3.3.4 can be generalized into an equation of the form,

$$Ax + By = C \tag{3.3.6}$$

Any equation of this form represents a linear function.

The symbols A and C can be replaced by *any* real numbers. The symbol B can be replaced by any number except zero. (Recall that in finding the slope we divided by the coefficient of y. We cannot divide by zero.)

3.3.2 Equation of Line from Two Points

If a function is known to be linear, its equation can be obtained if the coordinates of any two points in the function are known. Let the points be designated as (x_1, y_1) and (x_2, y_2). As we have seen earlier in Equation 3.3.1, the slope is given by

$$m = \frac{y_2 - y_1}{x_2 - x_1}$$

After the slope has been computed, it can be used together with the coordinates of either point in an equation of the form of Equation 3.3.2 to find the y-intercept.

Example 3.3.2

Let us find the equation for the linear function that includes the points (3,4) and (2,6).

The slope is given by

$$m = \frac{6 - 4}{2 - 3} = \frac{2}{(-1)} = -2$$

Note that in setting up the computation for the slope, the coordinates for the **same** point are used in the first position. It does not matter which point is used first. In this computation if we use the point (3,4) first instead of (2,6) we have

$$m = \frac{4-6}{3-2} = \frac{-2}{1} = -2$$

First point (3, 4); Second point (2, 6) [arrows point to 3 and 2 in the denominator]

If we use the value computed for m, Equation 3.3.2 becomes

$$y = b - 2x \tag{3.3.7}$$

Equation 3.3.7 represents a "family" of lines, each with the same slope, but at different "locations," depending on the value for b. Figure 3.3.6 shows three members of this family.

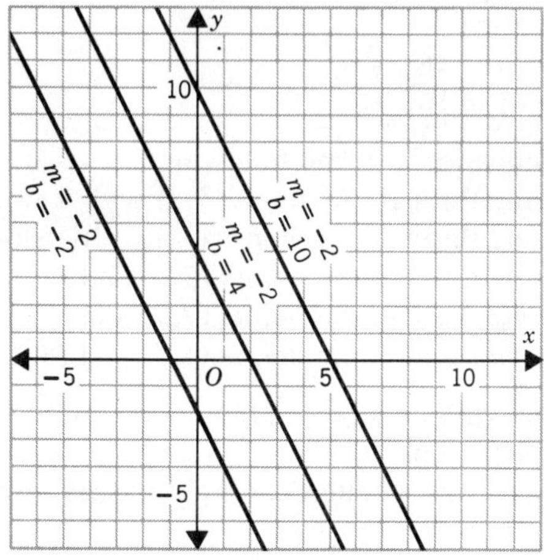

Figure 3.3.6

Note that the lines are parallel.

We now must find the particular line in this family that passes through (3,4) and (2,6). The equation for a function specifies the relationship between the coordinates of any point on its graph. Therefore, if we substitute the coordinates of either point into Equation 3.3.7, we have an expression from which the value for b can be obtained.

If we use the point (3,4), we have

$$4 = b - 2(3)$$

The value for b in the equation we are seeking must make this expression a true statement. Thus, $b = 10$. If we had used the point (2,6).

$$6 = b - 2(2) \qquad \text{and} \qquad b = 10$$

The equation for the line through the points (3,4) and (2,6) is $y = 10 - 2x$.

EXERCISE 3.3

1. Plot the graphs of the following linear functions:
   a. $y = 2x - 6$
   b. $y = 3 - \frac{3}{4}x$
   c. $y = 4 + \frac{2}{3}x$
   d. $3x - 2y = 12$
   e. $5x + 2y = 10$

2. Use the graphs plotted for Problem 1 and find the value of x for which $y = 0$.

3. Find the slope and the y-intercept for the following:
   a. $4x - 2y = 9$
   b. $x + 2y = 3$
   c. $3x - y = 6$
   d. $5x + 5y = 25$
   e. $4x - 3y = 16$

4. Find the equations for the following lines:
   a. Passing through the point (5,3) with a slope of −2.
   b. Passing through the point (1,1) with a slope of 3/2.
   c. Passing through the points (5,4) and (2,3).
   d. Passing through the points (2,7) and (5,3).
   e. Passing through the points (5,4) and (8,4).

5. a. A line with a slope of 2/3 passes through the point (−3,2). What is its y-intercept?
   b. A line whose y-intercept is −3 passes through the point (8,6). What is its slope?

6. The x-coordinate of the point at which a line crosses the x-axis is called its x-intercept. What is the equation of a line whose x-intercept is 5 and whose y-intercept is 3?

7. a. A salesman rented a car for a day. When he made up his expense account, he remembered the mileage fee was 10¢ per mile, but he could not remember the daily rental charge for the automobile. He drove 70 miles, and the total bill was $15. What was the rental charge?
   b. In another city he remembered the rental charge was $7 per day, but he could not remember the mileage fee. This time he drove 60 miles and the bill was $14.20. What would his bill have been if he had driven 80 miles?

8. A jet airplane uses 1500 gallons of fuel per hour when flying at a steady speed. Write an equation for the amount of fuel remaining in the tanks as a function of the number of hours flown. The tanks hold 12,000 gallons.

9. There are two temperature scales in regular use, Centigrade and Fahrenheit. The correlation between the two is a linear function that can be obtained from this information: The temperature of freezing water is 0°

Centigrade and 32° Fahrenheit; and the temperature of boiling water is 100° Centigrade and 212° Fahrenheit.

a. Write an equation for the correlation between temperature scales using degrees Fahrenheit as the independent variable and degrees Centigrade as the dependent variable.
b. What specific information about these temperature scales is contained in the slope of this function?

10. The amount of water in a reservoir at noon of a given day was 10,000,000 gallons. At noon of the following day it was 7,600,000 gallons. Assuming that the flow rate of water is constant, write an equation giving the amount of water in the reservoir at any time of that day. Define your symbols clearly.

11. An income tax table shows the tax owed as a function of net income. The table shows several different "domains" for the income, and the formula for computing the tax. In each domain the tax is a linear function of the income. If x represents the amount of net income:

Domain	*Formula for Tax*
$0 < x \leq 1000$	14% of x
$1000 < x \leq 2000$	\$140 + 15% of amount of income over \$1000
$2000 < x \leq 3000$	\$290 + 16% of amount of income over \$2000
$3000 < x \leq 4000$	\$450 + 17% of amount of income over \$3000
$4000 < x \leq 8000$	\$620 + 19% of amount of income over \$4000

For each of the domains shown above:

a. Write a linear function in algebraic form showing the amount of tax owed, y, as a function of net income, x.
b. Draw a graph showing the amount of tax owed as a function of net income from 0 to \$8000.

3.4 APPLICATIONS

3.4.1 Straight-Line Depreciation

In business accounting it is necessary to provide in the records for the decreasing value of "capital equipment" such as machinery and buildings. When an item of capital equipment is purchased by a business concern, its cost is recorded as one of the concern's assets. It is assigned a period of time over which it is expected to be useful. At the end of this period it is assumed to have depreciated in value to zero, or to its scrap value. As the equipment decreases in value with time, an accounting of this depreciation must be made in the concern's records. One accounting procedure for accomplishing this is called "straight-line depreciation."

When straight line depreciation is used, the recorded value of the equipment is reduced at **constant rate** over its assigned period of usefulness. (The actual useful life of the equipment may or may not match its assigned period. For accounting purposes this is irrelevant.)

The fact that the value of the equipment decreases at a **constant** rate means that the function giving the value at any given time is a linear function.

Twenty years is a commonly used period of usefulness for machinery. Let us suppose that a company purchases a machine that costs \$20,000. It is expected that at the end of 20 years the machine will be worth nothing. Let us find a function that will provide the "current value" of machine at any time after the purchase.

We symbolize the time since purchase as t, and the current value of the machine as V. When the machine is purchased, t is equal to zero, and the value is \$20,000. Thus, the point (0,20,000) belongs to the current value function. When t is 20 years, V is zero; and we have the point (20,0) as an element of the function. The slope of the linear function that includes these two points is given by

$$m = \frac{20{,}000 - 0}{0 - 20} = -1000$$

That is, the value of the machine will decrease by \$1000 per year. The fact that the slope is negative means that the value is decreasing.

Since we know that $V = 20{,}000$ when the independent variable, t, is zero, we can write the equation for the value as

$$V = 20{,}000 - 1000t$$

This function gives the value of the machine at any number of years, up to 20, after the purchase.

3.4.2
Simple Interest and Loan Discounting

Interest is a rental fee paid, or charged, for the use of money. With simple interest the amount of this fee is determined by multiplying the length of time the money is kept times a fixed rate. The interest is, therefore, a linear function of time. (With compound interest, interest is paid on the interest, and the resulting function has a different form.)

When a loan is made on simple interest, the amount of the interest is often added to the principal of the loan and the entire sum paid at one time. Under such an arrangement the amount owed increases at a constant rate and is, therefore, a linear function of time.

For example, suppose a person borrows \$100 at simple interest of 12% per year (1% per month). The amount he must repay increases by 1% of \$100, or \$1, each month. At the end of a year he would owe \$112 instead of the \$100 he borrowed.

If we let A represent the amount to be repaid, P the amount borrowed (usually called the "principal"), I the interest rate, and t the time, the equation giving the amount owed is

$$A = P + (P \cdot I)\, t \tag{3.4.1}$$

Since in any transaction the principal and interest rate are fixed, Equation 3.4.1 is a linear function between two variables: A (dependent) and t (independent).

If we compare Equation 3.4.1 with the standard form for a linear function, $y = b + mx$, we see that A has replaced y as the dependent variable; P plays the role of b, the y-intercept; the slope, m, has become $P \cdot I$, the dollar amount of the interest for a unit time period; and t has replaced x as the independent variable. The use of different symbols in no way changes the properties of the various quantities in the function.

An alternate procedure for paying the interest charged on a fixed-term loan is called "discounting." A fixed-term loan is one that is repaid in a lump sum at the end of a specified period of time. When such a loan is "discounted," the amount of interest on the loan is subtracted from the principal at the time the loan is made. The borrower receives the remainder. If a person borrows $100 for a period of a year and the interest charged is $12, he receives $88 when the loan is made; but he repays the entire $100.

The amount the borrower receives at the time the loan is made is a linear function of the length of the period of the loan. Loan offices usually have values for this function tabulated for different loan periods and interest rates.

Let us write the equation of the function giving the amount that the borrower receives on a "standardized" loan of $1. The actual amount he would receive is obtained by multiplying the principal value of his loan by the value given by this function.

Let A represent the amount the borrower receives and t, the period of the loan in months. If we write the equation in the form of Equation 3.3.2, the value for b is 1, and the slope is the rate of interest per month. Since the value, A, is decreased at t grows larger, the slope will have a negative sign. For an interest rate of 1% per month, we have

$$A = 1 - .01t$$

Business firms often obtain discounted loans when they need money for a short period of time for some special purpose. The reason for keeping the period of such loans short is revealed if we compute the equivalent simple interest rates of a discounted loan as a function of the period of the loan. This is not a linear function, and its algebraic expression means little to a non-mathematician. We can obtain quite a bit of information, however, from a graph.

To construct the graph, we need to compute the values for several points. Let us consider a discounted loan with an interest rate of 1% per month. This rate is equivalent to a simple interest rate of 12% per year. If the period of a discounted loan is six months, the borrower would receive 94¢ for every dollar in the "face value," or principal of the loan. The amount of interest is 6¢. 6¢ simple interest on 94¢ (the amount the borrower actually receives) is .06/.94 or 6.4% for six months, which is equivalent to 12.8% per year.

Similar computations for other loan periods give the following data:

Loan Period (t)	Total Interest Paid	Equivalent Simple Interest (per year)
12 months	13.6%	13.6%
18 months	22.0%	14.6%
24 months	31.6%	15.8%
36 months	56.3%	18.8%
60 months	150%	30%

These data can be plotted to obtain the graph in Figure 3.4.1.

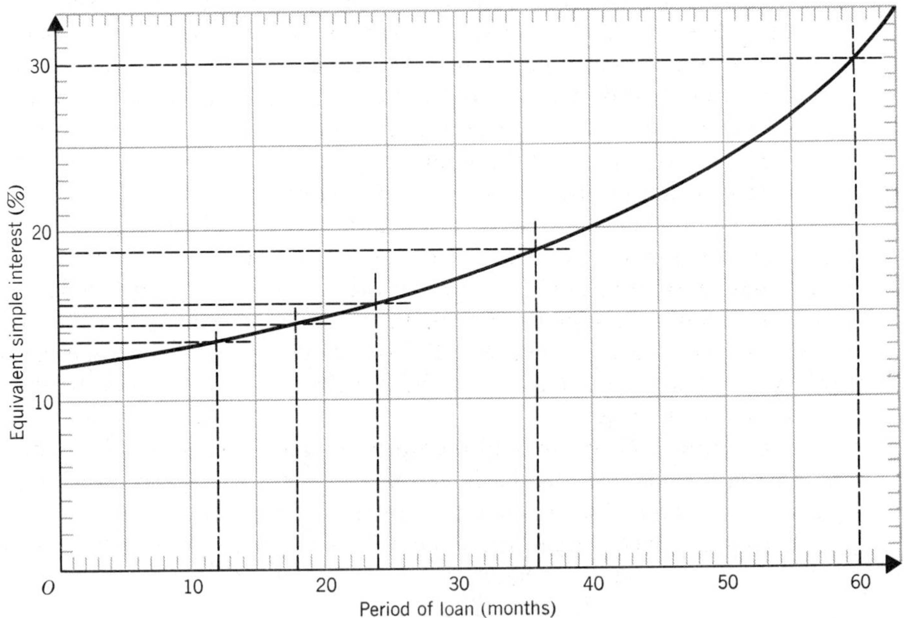

Figure 3.4.1

The curve of Figure 3.4.1 rises at an ever-increasing rate. When the period of the loan reaches $8\frac{1}{3}$ years, the borrower receives no money at all, but is committed to repay the amount of the loan—an infinitely large interest rate. Obviously, no one would make such a commitment. Discounted loans are rarely made for more than a year.

3.4.3 Break-Even Analysis

An important question that faces almost every business from time to time is the decision on whether to invest money in a new product, in an advertising campaign, or in plant improvement, and so on. Such a decision is often made on the basis of how long it is expected to take before the business gets the money back through increased profits. Although there are usually complicating factors in each case, much information can be obtained from a relatively simple analysis.

Suppose that a company is considering a new product. It is estimated that the development costs to prepare for production are \$10,000. The material and direct labor required to make one of the items is estimated at \$1.20. They expect to be able to sell the items for \$2.50 each. How many must they sell to "break even"; that is, to recover all of their costs?

On a graph on which both the costs and the income are plotted together, the break-even point is the point at which the costs and the income are the same—the intersection of the cost curve and the income curve. In our hypothetical situation the costs can be described as a function of the number produced. If C represents the cost and n, the number of items produced and sold, we have

$$C = 10{,}000 + 1.20n$$

This equation states that the cost of the program is the sum of the \$10,000 development cost and \$1.20 times the number of items produced.

The gross income from the product is \$2.50 times n, assuming that they could sell all they produced. If S represents the income,

$$S = 2.50n$$

Since the dependent variable in each of these functions is the number of dollars involved, and the independent variable in each case is the number of items, we can plot both functions on the same graph, as shown in Figure 3.4.2.

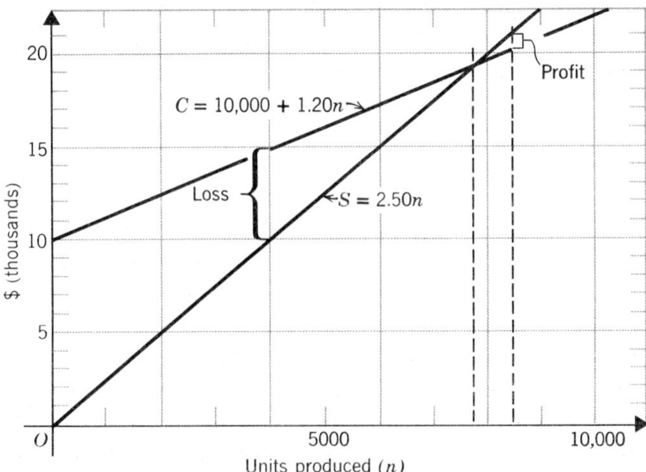

Figure 3.4.2

The graph shows that the cost will be recovered when approximately 7700 items are sold. If the total number sold turns out to be less than this number, the project loses money. Or, if it is estimated that it will take too long a time to sell so many of these items, the project would not be undertaken.

If, however, the company executives expect to sell more than 7700 items within a reasonable time, the project would look attractive to them. The amount of the loss or profit for a given number sold is the difference between the y-value of the cost line and the y-value of the income line at that value for n. If the cost line is above the income line at that value of n, there is a loss; if it is below the income line, there is a profit.

3.4.4 Linear Correlation and Projections

In this chapter we have been discussing various ways that one quantity is related to another, and the uses to be made of these relationships. The primary utility of functions and relations, particularly functions, is to provide a means by which numerical values for one quantity, which are available directly, can be used to determine numerical values for another quantity, which are not available. For example, we saw how to use a specified period of time (known) to determine the value of a piece of machinery (not known); and the number of miles a rented car was driven (known) was used to determine the rental charge (not known). In each of these cases we have used a knowledge of the situation to construct the equations relating one quantity to another.

In a great many situations there are no direct methods, such as we have used, to determine the values of the parameters* in the functions. Functions relating quantities are none the less useful in such situations. For example, many colleges require that students "pass" a qualifying examination before being permitted to enroll for certain courses. The qualification requirement constitutes a function. The domain is the set of possible scores on the exam; and the range consists of two elements: "qualified" and "not qualified." The function assigns one or the other of these elements to each possible score. The assignment is usually made, of course, by partitioning the domain into two subsets, those below a certain score that are associated with "not qualified," and those above a certain score that are associated with "qualified." There is no analytical way to determine the "cut-off" score. It is established as the result of experiment. Before adopting the test as a criterion for qualification, we let a large number of students take the examination, and they all take the course regardless of their score. Their success or failure in the course is observed. Statistical methods are then used to determine the "best" cut-off score to use.

Often it is desired to have a relationship that will give numerical values for the dependent variable, instead of mere yes-or-no results. An equation is needed to accomplish this, and the equation for a linear function is the simplest form available. In the absence of knowledge that would yield the function as a result of analysis, statistical methods are used with experimental data to develop the functions. The details of these methods are outside the scope

* Parameters are the numerical quantities in a function **other than** the variables. In a linear function the slope and the y-intercept are parameters. The assignment of values to the parameters is completely arbitrary in the mathematical use of parameters, whereas the values possible for the variables is determined by the form of the function.

of this text—the student will learn them in future study—but we can get an idea of how the methods are used by considering the following example.

Let us suppose a man operates a lunch counter catering to the employees of nearby business offices. He observes that during the summer his sales of iced tea go up on the warmer days and go down when the weather is cooler. This suggests a correlation between iced-tea sales and the temperature. He loses sales on a hot day if he runs out of iced tea (customers drink only iced water), but his profit margins are so small that he cannot afford to make a large quantity of tea unless he can sell it. By the time he can tell how hot it is going to be, he is so busy with other duties that he cannot take time out to make tea. He decides to use the temperature forecast in the morning weather report as a guide in deciding how much tea to make.

There is no analytical way to translate a temperature into the number of gallons of tea to make. He must experiment to determine the form of the correlation. He decides to make a large quantity of tea each day for a week, and note carefully both the temperature and the amount of iced tea he sells. He obtains the following data:

Day	Temperature	Tea Sold (gallons)
Monday	74°	3
Tuesday	80°	5
Wednesday	90°	7
Thursday	85°	6
Friday	78°	$4\frac{1}{2}$

He plots the data on a graph, as shown in Figure 3.4.3.

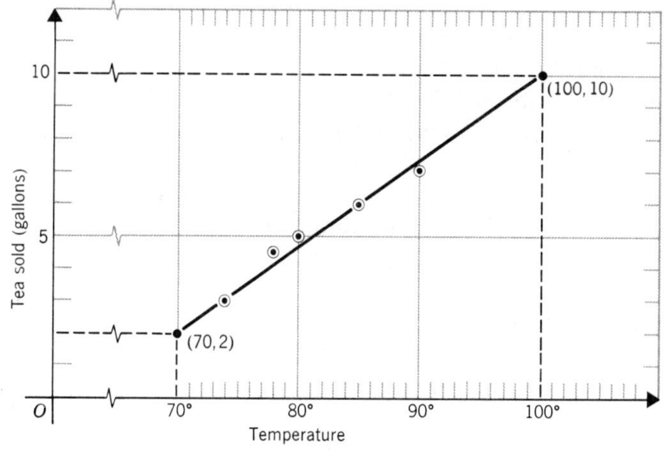

Figure 3.4.3

As might be expected, the points do not lie in a straight line. However, a line drawn as shown in Figure 3.4.3 is close to each of the points. We can write a linear function for the line shown over the domain from 70° to 100°. A point is chosen near each end of the line. The points (70, 2) and (100, 10) are suitable choices. The slope of the function is

$$m = \frac{10 - 2}{100 - 70} = \frac{8}{30} = \frac{4}{15}$$

Using the point (100, 10) in the general form of a linear function, we have

$$10 = b + \tfrac{4}{15}(100)$$

and

$$b = -17$$

The complete equation is

$$y_{(\text{gallons})} = \tfrac{4}{15}T - 17 \tag{3.4.2}$$

where T is the predicted temperature for the day.

In this example we drew the straight line to "fit" the points by visual inspection. There are statistical methods that give the equation for the line giving the "best fit" for the points plotted.

The function described by Equation 3.4.2 is called a **linear correlation**. It describes a relationship between two variables using a linear function. Such correlations obtained from observations in a manner similar to the example are used frequently in both business and the behavioral sciences. In business they are often the basis for forecasts of expected future business conditions. In the behavioral sciences the nature of the phenomena studied almost never permits the establishment of correlations by direct analysis. It is necessary to use statistical methods on experimental data almost exclusively.

In the iced-tea example, note that the "formula" gives a negative number for the amount of tea to be made when the temperature is forecast at 63° or lower. It is not possible to make a negative amount of tea. The formula tells how much tea to make for impossibly high values for the temperature. The formula is probably reasonably accurate for temperatures from 65° to 100°. This is the **applicable domain** of the function. Almost all linear functions used to approximate observed results will be applicable over a **limited domain**. It is important to specify this domain when discussing the function.

Let us consider a business forecast made using a linear correlation. These are usually made by recording values of the quantity of interest over a period of time—time is the independent variable in making forecasts. These values are plotted in a graph and a line drawn coming as close as possible to all of the points. A linear function can be formed **using points on the line** (not the actual observations!) that provides **approximate values** for the quantity of interest over a limited domain—for a limited time into the future. The slope of this function is often called a "trend." It tells the rate at which the quantity is changing with time. Forecasts **under the assumption that the observed trend will continue** are made by extending the line on the graph. Values that the quantity of interest are expected to attain in the future can be obtained either from the graph or from the linear function describing it.

Let us suppose that the annual sales volume of a business has been increasing over a period of several years. The figures are shown in the following table:

Year	*Sales*
1965	$350,000
1966	375,000
1967	425,000
1968	450,000
1969	500,000
1970	600,000

The company is interested in the sales volume that it can expect in 1975, assuming the rate of growth will remain the same. The data in the table are plotted in a graph in Figure 3.4.4. By extending the line as shown, the forecast for 1975 is seen to be approximately $725,000. Similar forecasts for other years can be obtained from the graph also.

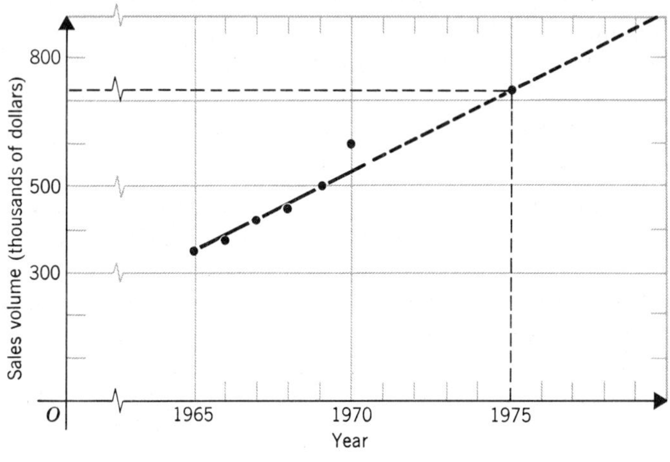

Figure 3.4.4

EXERCISE 3.4

1. A company buys a machine costing $250,000, and they plan to depreciate it completely over a period of 25 years.
   a. Write an equation that will give the current value of the machine at any time during that period, and draw the graph for the function.
   b. If the machine was purchased in 1965, what will the company records show its value to be in 1975?

2. A company buys a machine for $225,000. They plan to depreciate it over a period of 20 years. Their plant engineer estimates the machine will have a scrap value of $5000 at the end of that time.

a. Write an equation giving the current value of the machine at any time after its purchase.
b. After they have had the machine 12 years, technological development forces them to replace the machine with a completely new one. They are able to sell the machine for $50,000. What is the amount of loss they sustain in the transaction?

3. A man obtains a loan of $300 at simple interest of 8% per year. How much must he repay if he keeps the money for two years?

4. A man obtains a loan at 9% simple interest per year. At the end of 18 months he pays off the loan with a payment of $454. How much did he borrow?

5. A man obtains a loan for $500, which is discounted at the rate of $1\frac{1}{2}$% per month. If he agrees to repay the money in six months, how much does he receive?

6. A man is negotiating for a loan of $1000 to be discounted at $1\frac{1}{2}$% per month.
a. What is the maximum period for the loan if he is to receive at least $500?
b. What is the simple interest rate on the amount he receives if he keeps it for the period found in part **a**?

7. A man is contemplating setting up a booth at a county fair to sell a "revolutionary new can opener." The rental charge for the booth is $500. The can openers cost him 10¢ each, and he plans to sell them for 50¢. Draw a graph showing his costs and his receipts as a function of the number of can openers he sells.
a. How many must he sell to break even?
b. How much will he make (or lose) if he sells 1000? 2000?

8. A student organization is planning a dance to raise funds to finance its activities. It will cost them $250 for the combo, $40 to rent a hall, and $35 for advertising and tickets. A catering firm agrees to supply refreshments for 40¢ per person. The goal is to clear $500 for the organization. They discuss the price to charge for tickets. The suggested prices are $1, $1.50, and $2.
a. Write a linear function showing the total cost as a function of number of persons attending.
b. Write a linear function giving their receipts as a function of number of tickets sold for each of the ticket prices suggested.
c. How many tickets must they sell at each price to "break even"?
d. How many tickets must they sell at each price to reach their profit goal?

9. A behaviorist was hired as a consultant by a sporting goods manufacturer to study the correlation between monthly income and the amount spent each year on a certain kind of sports equipment.
In a preliminary survey of 10 people with different incomes, he obtained the following data:

Monthly Income	*Amount Spent Per Year*
450	20
500	15
600	30
650	50
750	70
900	80
1050	130
1100	110
1300	150
1450	180

It is assumed that the correlation is linear.

a. Plot the data on a graph and draw a straight line to approximate the function. Use points on the line drawn to write the equation that gives approximately the amount spent per year as a function of monthly income.

b. Assume the same function applies to people of higher income, and find the approximate amount a person with an income of \$2000 per month spends.

10. A company is concerned about the price that it must pay for some of the materials that it uses in a manufacturing process. Its records show the following prices:

Year	*Average Price Per Pound*
1965	\$1.02
1966	1.05
1967	1.15
1968	1.15
1969	1.21
1970	1.25

Assume that the price trend can be approximated satisfactorily by a linear function.

Plot the data on a graph and draw a straight line as nearly as possible through the points.

a. What is the average increase in price per year as shown by the line drawn? In the equation for the line, what role does this quantity play?

b. Assuming that the price trend continues, what is the expected price in 1980?

4

matrices and vectors

4.0 INTRODUCTION

In the first chapter we studied one of the mathematical concepts used in organizing and working with information, the set. We turn our attention now to another means for dealing with data, the matrix. Matrices, unlike sets, are composed exclusively of numbers, and the numbers are arranged in a particular order.

As an introduction to the nature of a matrix, consider the following situation: You are a new employee in a company, and you are assigned to a department in which your fellow workers regularly eat lunch at their desks. The usual fare is sandwiches; and, as the newest employee, you have the privilege of going to a nearby lunch counter to get them. The lunch counter sells four kinds of sandwiches—ham, cheese, tuna, and egg—on a choice of white, wheat, or rye bread.

There are 38 employees in your department, and they have a variety of tastes. On the first day you make a list of the orders something like this:

3 Ham on white
2 Ham on wheat
4 Ham on rye
1 Cheese on white
2 Cheese on wheat
5 Cheese on rye
3 Tuna on white
4 Tuna on wheat
1 Tuna on rye
2 Egg on white
5 Egg on wheat
1 Egg on rye

After a day or two of such tedious list making, you devise a shorter way of recording the order. You arrange a chart, such as

	Ham	Cheese	Tuna	Egg
White	3	1	3	2
Wheat	2	2	4	5
Rye	4	5	1	1

The columns in the chart represent the different kinds of sandwiches; and the rows, the different kinds of bread. Before long you have the headings memorized, and you record only the quantities ordered:

$$\begin{matrix} 3 & 1 & 3 & 2 \\ 2 & 2 & 4 & 5 \\ 4 & 5 & 1 & 1 \end{matrix}$$

This rectangular array of numbers, each in a particular column and row, is called a **matrix.**

Matrices are used to tabulate numerical information in a manner similar to the above. The great variety of uses for such tabulations, together with the convenience of performing some kinds of operations with the individual entries in a matrix, have made the study of matrices an important branch of mathematics. We shall study only the elementary properties of matrices along with a few examples of their use in the solution of problems.

4.1 BASIC OPERATIONS WITH MATRICES

A matrix is a rectangular array of numbers. The numbers are arranged in **rows** and **columns.** The **rows** are sets of numbers in **horizontal lines**; the **columns** are sets of numbers in **vertical lines.** The individual numbers in a matrix are called **elements.** Each of the rows in a matrix has the same number of elements, and each of the columns has the same number of elements. The number of elements in the rows, however, may be the same or different from the number of elements in the columns.

The array of numbers constituting a matrix are enclosed by parentheses or brackets. Capital letters are usually used as symbols to represent matrices, and lower-case letters are often used to represent the numbers which are the elements of a matrix. Thus, the array

$$B = \begin{bmatrix} 3 & -1 & 2 & 4 \\ 7 & 0 & -3 & 2 \\ 1 & 0 & 7 & -6 \end{bmatrix}$$

is a specific matrix, here designated as B. The **generalized** matrix with the same number of rows and columns as B, which we shall designate here as A, is

$$A = \begin{bmatrix} a_{11} & a_{12} & a_{13} & a_{14} \\ a_{21} & a_{22} & a_{23} & a_{24} \\ a_{31} & a_{32} & a_{33} & a_{34} \end{bmatrix}$$

The representations for the individual elements, the a's, have **subscripts** to locate them in the matrix. The first element, a_{11}, (read, "a, sub one-one") is the element in the first row, first column. The first subscript designates the row; the second, the column. Thus, a_{32}, ("a, sub three-two"—**not** "thirty two") is the element in the third row, second column. These symbols, although rather bewildering at first, are quite convenient and informative once you are used to them.

The **dimension** of a matrix is determined by the number of rows and columns. An $m \times n$ matrix (read "m by n") has m **rows** and n **columns.** The **number of rows is always given first** in specifying the dimension of a matrix.

A matrix may have any number of rows and columns. A matrix with only one row **or** one column is often called a **vector.*** A matrix having **one row only** is called a **row vector**. A vector with **one column only** is called a **column vector**.

$[a_{11}, a_{12}, a_{13}, a_{14}]$ is a four-element row vector.

$\begin{bmatrix} a_{11} \\ a_{21} \\ a_{31} \end{bmatrix}$ is a three-element column vector.

4.1.1 Equality of Matrices

Two matrices are equal when each element of one is the same as the corresponding element of the other. This implies the two matrices cannot be equal unless they have the same dimension.

Definition 4.1.1

Equal Matrices: Two matrices are equal if, and only if,

a. They have the same dimension; **and**

b. Each element of one is equal to the **corresponding** element of the other.

For example,

$$\begin{bmatrix} 3 & 1 & 2 \\ 4 & -1 & 7 \end{bmatrix} = \begin{bmatrix} 3 & 1 & 2 \\ 4 & -1 & 7 \end{bmatrix}$$

However,

$$\begin{bmatrix} 3 & 1 & 2 \\ 4 & -1 & 7 \end{bmatrix} \neq \begin{bmatrix} 3 & 4 \\ 1 & -1 \\ 2 & 7 \end{bmatrix}$$

*Vectors are widely used in mathematics in many different kinds of contexts. Depending on the context, they are defined in different ways. Two other definitions the student may have encountered:

"A vector is a **directed** line segment."

"A vector is an **ordered set** of real numbers—a set of numbers in which the order of listing is important."

Allowing for the differences in the context in which they are used, we can show all of these definitions to be equivalent.

These latter two matrices are not equal, even though they contain the same elements, because they have different numbers of rows and columns. That is, they do not have the same dimension. The matrix on the left is 2×3 (2 rows, three columns), whereas the matrix on the right is 3×2 (3 rows, 2 columns.)

$$\begin{bmatrix} 1 & 2 & 3 \\ 1 & -1 & 2 \\ 3 & 1 & 4 \end{bmatrix} \neq \begin{bmatrix} 1 & -1 & 2 \\ 1 & 2 & 3 \\ 3 & 1 & 4 \end{bmatrix}$$

These matrices are not equal, even though they have the same dimension (3×3) and the same elements, because the elements in **corresponding** rows and columns are not the same.

We designate the "general" element of an $m \times n$ matrix, A, with the symbol a_{ij} (read "a sub i, j"). This means the element in the "ith" row and "jth" column of the matrix. i can represent any counting number from 1 to m, the number of rows in the matrix. j can represent any counting number from 1 to n, the number of columns in the matrix. Thus, we say, in symbols

$$A = B, \text{ iff } a_{ij} = b_{ij}, \text{ for all } i, j$$

A and B represent two matrices. a_{ij} is the element in the ith row, jth column of A. b_{ij} is the element in the ith row, jth column of B. If we say that $a_{ij} = b_{ij}$ for all i, j, we mean that for every possible choice for the i's and j's in the two matrices, the specified elements are equal. Thus, if A and B were 3×4 matrices, i could be 1, 2, or 3; and j could be 1, 2, 3, or 4. There would be $3 \cdot 4 = 12$ choices for different values for i–j, each representing a different one of the 12 elements of the matrices.

4.1.2 Sum of Two Matrices

The sum of two matrices, A and B, is a matrix, $A + B$, each of whose elements is obtained by adding the corresponding elements of A and B. This statement implies that A and B must be of the same dimension in order for their sum to exist. If we designate the sum of A and B as C; that is, if $C = A + B$, then each element of C is the sum of the corresponding elements of A and B. In the symbols of the previous section, $C = A + B$ if, and only if, $c_{ij} = a_{ij} + b_{ij}$ for all i, j.

Example 4.1.1

Let $A = \begin{bmatrix} 1 & 3 & 2 \\ 1 & 0 & -1 \end{bmatrix}$ and $B = \begin{bmatrix} 3 & -1 & 2 \\ -4 & 6 & 8 \end{bmatrix}$

Then $C = A + B = \begin{bmatrix} 1+3 & 3-1 & 2+2 \\ 1-4 & 0+6 & -1+8 \end{bmatrix} = \begin{bmatrix} 4 & 2 & 4 \\ -3 & 6 & 7 \end{bmatrix}$

Note that, if $A = \begin{bmatrix} 1 & 3 & 2 \\ 1 & 0 & -1 \end{bmatrix}$ and $B = \begin{bmatrix} 3 & -4 \\ -1 & 6 \\ 2 & 8 \end{bmatrix}$

no sum exists for these two matrices.

4.1.3 Multiplication of a Matrix by a Scalar

The term **scalar** is used in matrix algebra to designate an ordinary number.

A matrix can be multiplied by a scalar by multiplying each element of the matrix by that number. That is, if A is any matrix, and k is any scalar, (any real number), kA is the matrix obtained by multiplying **each element** of A by k.

Example 4.1.2

$$\text{Let } A = \begin{bmatrix} 1 & 3 & 2 \\ 1 & 0 & -1 \end{bmatrix} \quad \text{and } k = 3$$

$$\text{Then } 3A = \begin{bmatrix} 3(1) & 3(3) & 3(2) \\ 3(1) & 3(0) & 3(-1) \end{bmatrix} = \begin{bmatrix} 3 & 9 & 6 \\ 3 & 0 & -3 \end{bmatrix}$$

In symbols, if A is a matrix whose elements are represented by a_{ij}, then kA is a matrix whose elements are ka_{ij}. An even more compact notation is sometimes used in discussions of the properties of matrices. The matrix, A, is specified as $[a_{ij}]$, meaning the matrix with i rows and j columns whose elements are the $i \cdot j$ values for a_{ij}. In this notation

$$kA = [ka_{ij}]$$

Note that multiplication of a matrix by a scalar includes multiplication by -1. Thus, for every matrix, A, there exists another matrix, $(-1)A$, or $-A$; and $A + (-A) = [0]$. That is, when A is added to $-A$, a matrix is obtained in which each element is zero. Such a matrix is called a **zero matrix**, and is designated $[0]$. Zero matrices exist for every possible dimension. When the symbol for the zero matrix, $[0]$, is used, it is implied that it represents the zero matrix whose dimension is appropriate for the particular use.

4.1.4 Associative Property of Matrix Addition

The addition of matrices is an operation performed with exactly two matrices. When more than two matrices are to be combined by addition, the matrices can be paired in any order and the sum will be the same. This property is symbolized in the statement,

$$(A + B) + C = A + (B + C)$$

The parentheses are used to indicate the order in which the additions are to be performed.

The statement says the sum of matrices A and B when added to matrix C gives the same result as when the matrix A is added to the sum of matrices B and C. This is called the **associative property of matrix addition**. It is directly analogous to the associative property for the addition of numbers.

Commutative Property of Matrix Addition

The order of the matrices in a sum can be reversed and the sum remains the same. That is, for two matrices, A and B,

$$A + B = B + A$$

This statement is the **commutative property of matrix addition**. It is directly analogous to the commutative property for addition of numbers.

EXERCISE 4.1

1. a. Given:

$$\begin{bmatrix} 3 & a & 2 & -1 \\ 4 & 6 & 1 & b \\ -1 & c & d & 2 \end{bmatrix} = \begin{bmatrix} 3 & 3 & 2 & -1 \\ 4 & 6 & 1 & -2 \\ -1 & 2 & -2 & 2 \end{bmatrix}$$

Find the values for a, b, c, and d.

b. If

$$\begin{bmatrix} x & y & 3 \\ 1 & 2 & -2 \\ w & -1 & z \end{bmatrix} = \begin{bmatrix} 2 & -1 & 3 \\ 1 & 2 & -2 \\ -4 & -1 & 1 \end{bmatrix}$$

What are the values of w, x, y, and z?

2. Let $A = \begin{bmatrix} 3 & 2 & 4 \\ -1 & 7 & 6 \end{bmatrix}$ $\quad B = \begin{bmatrix} 1 & -1 & 0 \\ 3 & -2 & 4 \end{bmatrix}$

$$C = \begin{bmatrix} 4 & 7 \\ 3 & -1 \\ 2 & 0 \end{bmatrix} \qquad D = \begin{bmatrix} 1 & -3 & 6 \\ 2 & 0 & 1 \\ 4 & -1 & 2 \end{bmatrix}$$

$$E = \begin{bmatrix} 2 & -1 \\ 1 & 3 \\ 0 & 4 \end{bmatrix} \qquad F = \begin{bmatrix} 2 & 1 & 1 \\ -1 & 3 & -2 \\ 4 & 0 & -3 \end{bmatrix}$$

a. Form all the sums of matrices possible.
b. Find the following, where possible:

$3A - B =$ $\qquad C - 2E =$ $\qquad 2C + D =$
$A + D =$ $\qquad D - F =$

3. Using the matrices A, B, C, D, E, and F as defined in Problem 2
a. What is the dimension of each matrix?
b. $a_{13} =$ $\qquad d_{21} =$ $\qquad c_{22} =$
$b_{23} =$ $\qquad f_{33} =$ $\qquad e_{31} =$

4. If

$$\begin{bmatrix} 3 & 1 & x & -4 \\ 2 & y & 1 & z \\ 4 & -1 & w & 2 \end{bmatrix} = \begin{bmatrix} a & 1 & 2 & -4 \\ 2 & 2 & b & -1 \\ 4 & c & -5 & d \end{bmatrix}$$

Find the values for a, b, c, d, w, x, y, and z.

5. Let

$$N = \begin{bmatrix} 2 & -1 \\ 3 & 1 \end{bmatrix} \qquad P = \begin{bmatrix} 1 & 3 & -2 \\ 4 & -1 & 0 \end{bmatrix} \qquad R = \begin{bmatrix} 1 & 9 & -6 \\ -3 & 5 & -4 \end{bmatrix}$$

$$Q = \begin{bmatrix} 4 & 4 & -1 \\ -3 & 2 & 6 \\ 7 & 9 & -8 \end{bmatrix} \qquad S = \begin{bmatrix} 1 & 3 & -2 \\ 9 & -7 & 6 \\ 4 & 2 & -5 \end{bmatrix} \qquad T = \begin{bmatrix} 8 & 8 \\ -6 & 5 \\ 2 & -1 \end{bmatrix}$$

a. Form all the sums of matrices possible.
b. Where possible, find the following:

$2N + R \qquad 3N + T \qquad Q + \frac{3}{2}T$
$3P - 2R \qquad Q - 2S \qquad P + R - T$

6. For the matrices, N, P, R, Q, S, and T of Problem 5
a. What is the dimension of each matrix?
b. $n_{22} =$ $\qquad q_{22} =$ $\qquad s_{32} =$
$p_{22} =$ $\qquad t_{22} =$ $\qquad q_{23} =$

7. Given that

$$\begin{bmatrix} 2 & 3 & -1 & w \\ x & 1 & -3 & 2 \\ 4 & y & -1 & z \end{bmatrix} + \begin{bmatrix} a & -1 & 2 & 2 \\ 3 & b & 2 & -1 \\ -4 & 1 & c & 2 \end{bmatrix} = \begin{bmatrix} 4 & d & 1 & 6 \\ 1 & 3 & e & 1 \\ f & 2 & -4 & 0 \end{bmatrix}$$

Find the values for a, b, c, d, e, f, w, x, y, and z.

8. Let

$$A = \begin{bmatrix} 7 & 2 & -1 \\ -4 & 6 & 3 \\ 5 & -2 & 1 \end{bmatrix} \quad \text{and } B = \begin{bmatrix} 8 & -1 & 2 \\ 1 & -3 & 4 \\ 2 & 3 & -3 \end{bmatrix}$$

If $A + C = B$, find C.

9. Let

$$A = \begin{bmatrix} 4 & -7 & 2 \\ 3 & 0 & -5 \\ 6 & 6 & -2 \end{bmatrix} \qquad B = \begin{bmatrix} 3 & -9 & 8 \\ 4 & 2 & 6 \\ -3 & -5 & 2 \end{bmatrix}$$

If $A - C = B$, find C.

10. At the start of each week an automobile dealer checks his inventory of new cars. He records the results in a matrix of the following form:

	2-door Hardtop	*4-door Sedan*	*2-door Sedan*	*Station Wagon*
Economy	4	1	5	3
Standard	2	3	4	1
Deluxe	2	1	2	3

Suppose that this matrix represents his inventory at the start of a week.

On that day he receives a shipment of the following:

Standard line: One 2-door hardtop and two station wagons

Economy line: One 2-door hardtop, five 4-door sedans, and one station wagon.

Deluxe line: Three 4-door sedans and three 2-door sedans.

a. Assuming there are no cars sold between the time of his inventory and the receipt of the shipment, write down the new matrix of cars on hand.

b. During that week the matrix for cars sold is

$$\begin{bmatrix} 2 & 1 & 1 & 2 \\ 1 & 2 & 3 & 1 \\ 1 & 0 & 1 & 0 \end{bmatrix}$$

What is the inventory matrix for the start of the next week?

11. The dealer of Problem 10 has found that the following matrix represents the best "balance" for his stock:

$$\begin{bmatrix} 5 & 6 & 6 & 4 \\ 5 & 6 & 6 & 5 \\ 4 & 5 & 4 & 4 \end{bmatrix}$$

Using the inventory obtained in problem 10*b*, what cars should he order to bring his stock into balance?

12. Students take aptitude tests when they enter college. The results of these tests are given as three scores: verbal aptitude, quantitative aptitude, and a composite aptitude. Three students obtained the following scores: On the verbal aptitude test the first student scored 56; the second, 31; and the third, 78. On the quantitative test the first student scored 27; the second, 62; and the third, 48. The composite scores were for the first student, 46; the second, 67; and the third, 19. Write the scores for **each** student as a row vector.

13. What conclusions might you draw about a student described by the vector [36, 24, 35]?

14. An appliance dealer stocks television sets in three models: standard, deluxe, and regal. He has both black and white and color sets with 21-inch and 25-inch screens. Make up a matrix with appropriate numbers to represent his stock on hand.

15. The dealer of Problem 14 receives a shipment of the following:

Color sets: Three 21-inch standard; one 21-inch regal; four 25-inch deluxe; one 25-inch regal.

Black and White: Four 21-inch standard; three 21 inch deluxe; two 25-inch standard; three 25-inch regal.

a. Make a matrix of the same form as for Problem 14, showing the sets received in the shipment.

b. Add this matrix to the stock-on-hand matrix.

16. Verify the commutative property for matrix addition, $A + B = B + A$.

4.2
MULTIPLICATION OF MATRICES

We saw in the previous sections how the concept of a matrix evolved as a usable entity. When such entities are developed, it is only natural that we try to make as full use of them as possible. One of the roles played by mathematicians in our society is to explore such concepts, learn as much about their nature as possible, and then develop ways to use them. This process involves the "invention" of operations to be performed with and on the new concepts being studied.

In the previous section the operation of addition for matrices was defined. We are interested now in a procedure for multiplying matrices that will yield the most useful type of result. We recall that in the algebra of real numbers multiplication was developed originally as a shortcut for adding a number to itself many times. This particular purpose is served in matrix algebra by the multiplication of a matrix by a scalar. One obvious possibility for a multiplication procedure is to multiply corresponding elements of matrices together in a manner analogous to the addition of matrices. Such a procedure could be used, for example, in finding the total value of each type of item in an inventory matrix. If matrices were used only for inventories and similar purposes, this procedure would probably have been used as the definition of matrix multiplication. It turns out, however, that there are many other applications for matrices in which another, less obvious procedure is much more useful. Let us look first at this procedure itself. In later sections we shall see examples of its usefulness.

4.2.1
Product of Two Vectors

We start by considering the product of two vectors and, in particular, the product of a row vector and a column vector. We specify one of each type, not because it is necessary for this particular situation, but because the pattern we shall use in obtaining the product fits more closely the pattern we need for matrices of other dimensions. To illustrate how the operation works, let us use an example. Suppose that the components of the row vector are the quantities of several items we are buying, and the components of the column vector are their respective unit prices. The product of these two vectors will give the **total cost** of **all** the items.

Suppose we are buying one pound of cheese, two loaves of bread, and four cans of soup. The cheese costs 75¢ per pound, the bread is 35¢ per loaf, and the soup is 20¢ per can. The item vector is $[1, 2, 4]$. The price vector is

$$\begin{bmatrix} .75 \\ .35 \\ .20 \end{bmatrix}$$

The total cost of our purchase is given by

$$(1)(.75) + (2)(.35) + (4)(.20) = \$2.25.$$

2.25 is defined to be the **product** of the two vectors.

There are several aspects of this procedure to note carefully.

First, the **row vector** is the **multiplier**; the **column vector** is the **multiplicand.** The operation of multiplication is symbolized by placing the vectors side by side, as in number algebra. For our illustration we have

$$[1, 2, 4]\begin{bmatrix}.75\\.35\\.20\end{bmatrix} = 2.25$$

Second, the number of components in the row vector must equal the number of components in the column vector.

Third, no matter how many components there are in the two vectors, their product is a **single number.**

A generalized definition for the product of two vectors follows:

Definition 4.2.1

Product of Vectors: If A represents an n-component **row vector,** whose components are $a_1, a_2, a_3, \ldots, a_n$; and B represents an n-component **column vector** with components $b_1, b_2, b_3, \ldots, b_n$; the product AB is the **sum** of the products of the respective components, $a_1b_1 + a_2b_2 + a_3b_3 + \ldots + a_nb_n$.

At this point we shall not define the product BA, the product of a column vector times a row vector. It is, however, quite different from the product AB.

4.2.2 Product of Two Matrices

We are ready now to define the product of two matrices. The matrix that is the **multiplier** is considered to be a **set** of **row vectors**; and the matrix that is the multiplicand is regarded as a **set** of **column vectors.** In the product AB, where A and B are matrices. A would be considered a set of row vectors; and B, a set of column vectors.

The **product** AB is a **matrix** whose **elements** are the respective **vector products** of the **rows** of A times the **columns** of B. In the product matrix these elements are placed in the same row as the row vector from A, and in the same column as the column vector from B. This means the product matrix will have the same number of rows as A and the same number of columns as B.

Let us examine this procedure more carefully. First, we note that the row vectors from A must have the same number of elements as the column vectors from B. This requires that **the number of columns of A must equal the number of rows of B.** This requirement follows from the fact that the number of elements in a row of A is determined by the number of columns of A—each element is in a different column;—and the number of elements in a column of B is determined by the number of rows in B. If this requirement is not met, the product matrix does not exist.

There is no limitation placed on the number of rows in A, nor on the number of columns in B. Considering the dimensions of A and B, if A has the dimension $m \times n$, B must have a dimension $n \times p$ in order for the product AB to exist. There are no limitations on m or p.

Example 4.2.1

Let $A = \begin{bmatrix} 2 & -1 & 3 \\ 1 & 2 & 4 \\ -1 & 3 & -2 \end{bmatrix}$ and $B = \begin{bmatrix} 1 & 0 & 2 \\ 3 & -1 & 4 \\ -2 & 1 & 3 \end{bmatrix}$

Since the number of columns in A equals the number of rows in B, the product AB exists. We obtain the product as follows:

$$\underset{A}{\begin{bmatrix} 2 & -1 & 3 \\ 1 & 2 & 4 \\ -1 & 3 & -2 \end{bmatrix}} \underset{B}{\begin{bmatrix} 1 & 0 & 2 \\ 3 & -1 & 4 \\ -2 & 1 & 3 \end{bmatrix}} = \underset{AB}{\begin{bmatrix} \textcircled{-7} & \textcircled{4} & \textcircled{9} \\ \cdot & \cdot & \cdot \\ \cdot & \cdot & \textcircled{4} \end{bmatrix}}$$

Row vector Column vector Element

(1st row of A) (1st column of B) = (1st row, 1st column of AB)

$$\begin{bmatrix} 2 & -1 & 3 \end{bmatrix} \begin{bmatrix} 1 \\ 3 \\ -2 \end{bmatrix} = (2)(1) + (-1)(3) + (3)(-2) = \textcircled{-7}$$

(1st row of A) (2nd column of B) = (1st row, 2nd column of AB)

$$\begin{bmatrix} 2 & -1 & 3 \end{bmatrix} \begin{bmatrix} 0 \\ -1 \\ 1 \end{bmatrix} = (2)(0) + (-1)(-1) + (3)(1) = \textcircled{4}$$

(1st row of A) (3rd column of B) = (1st row, 3rd column of AB)

$$\begin{bmatrix} 2 & -1 & 3 \end{bmatrix} \begin{bmatrix} 2 \\ 4 \\ 3 \end{bmatrix} = (2)(2) + (-1)(4) + (3)(3) = \textcircled{9}$$

.
.
.

(3rd row of A) (3rd column of B) = (3rd row, 3rd column of AB)

$$\begin{bmatrix} -1 & 3 & -2 \end{bmatrix} \begin{bmatrix} 2 \\ 4 \\ 3 \end{bmatrix} = (-1)(2) + (3)(4) + (-2)(3) = \textcircled{4}$$

In the complete process each row of A multiplies each column of B. The final result is

$$AB = \begin{bmatrix} -7 & 4 & 9 \\ -1 & 2 & 22 \\ 12 & -5 & 4 \end{bmatrix}$$

In Example 4.2.1 both matrices have the same number of rows and columns. A product can also be obtained with nonsquare matrices if the number of columns of A is equal to the number of rows of B.

Example 4.2.2

Let $A = \begin{bmatrix} 1 & 2 & 1 \\ -1 & 0 & 3 \\ 2 & 1 & 0 \end{bmatrix}$ and $B = \begin{bmatrix} 3 & 1 \\ 2 & -1 \\ 0 & 2 \end{bmatrix}$

The product exists: A has three columns, and B has three rows. Each row of A is multiplied times each column of B, using the procedure of vector multiplication.

$$AB = \begin{bmatrix} \overbrace{(1)(3) + (2)(2) + (1)(0)}^{\text{1st row of } A}_{\text{1st column of } B} & \overbrace{(1)(1) + (2)(-1) + (1)(2)}^{\text{1st row of } A}_{\text{2nd column of } B} \\ \overbrace{(-1)(3) + (0)(2) + (3)(0)}^{\text{2nd row of } A}_{\text{1st column of } B} & \overbrace{(-1)(1) + (0)(-1) + (3)(2)}^{\text{2nd row of } A}_{\text{2nd column of } B} \\ \overbrace{(2)(3) + (1)(2) + (0)(0)}^{\text{3rd row of } A}_{\text{1st column of } B} & \overbrace{(2)(1) + (1)(-1) + (0)(2)}^{\text{3rd row of } A}_{\text{2nd column of } B} \end{bmatrix}$$

$$= \begin{bmatrix} 3+4+0 & 1-2+2 \\ -3+0+0 & -1+0+6 \\ 6+2+0 & 2-1+0 \end{bmatrix} = \begin{bmatrix} 7 & 1 \\ -3 & 5 \\ 8 & 1 \end{bmatrix}$$

Note that the product matrix in Example 4.2.2 has the same number of rows as A (3) and the same number of columns as B (2). This is true in general. If A has the dimension $m \times n$ and B has the dimension $n \times p$, the product, AB, will have the dimension $m \times p$.

If the number of columns in the multiplier matrix does not equal the number of rows in the multiplicand, the operation cannot be performed.

Example 4.2.3

If we use the matrices from Example 4.2.2 and try to find the product BA, we have, for the element in the first row, first column:

$$(ba)_{11} = [3 \quad 1] \begin{bmatrix} 1 \\ -1 \\ 2 \end{bmatrix} = (3)(1) + (1)(-1) + (?)(2)$$

There is no element in the first row of B to combine with the third element in the first column in A. Thus, the multiplication operation cannot be performed.

Notice that with matrices, unlike the multiplication of ordinary numbers, the result of multiplying AB is not the same as BA, except in very special cases. With the matrices of Example 4.2.2 the product can be found in one order, but not in the other. Even when the product can be found in both directions, it is not generally the same. With nonsquare matrices the products in reverse order, when they both exist, will have different dimensions. If A is 2×3 and B is 3×2, for example, AB will have the dimension 2×2, whereas BA will be 3×3. Examples of this fact are provided in the exercises.

With square matrices, whenever the product exists in one order, it exists in the other. The two product matrices, however, are the same only in special cases.

Example 4.2.4

Consider again the matrices of Example 4.2.1

$$BA = \begin{bmatrix} 1 & 0 & 2 \\ 3 & -1 & 4 \\ -2 & 1 & 3 \end{bmatrix} \begin{bmatrix} 2 & -1 & 3 \\ 1 & 2 & 4 \\ -1 & 3 & -2 \end{bmatrix} = \begin{bmatrix} 0 & 5 & -1 \\ 1 & 7 & -3 \\ -6 & 13 & -8 \end{bmatrix}$$

This result is substantially different from the product AB. (The student should verify the result of multiplying BA above.)

4.2.3 Associative Property of Matrix Multiplication

Examples 4.2.3 and 4.2.4 show that matrix multiplication does not have the commutative property. What now about the associative property? Since matrix multiplication is a binary operation (it involves **exactly** two matrices), how do we handle the product of three matrices? It can be shown (although it is a rather tedious task in the general case) that matrix multiplication is associative. That is, if we wish to find the product ABC, AB can be formed first and then used to multiply C; or BC can be found first and then multiplied by A. This fact is expressed in the usual form for the associative property:

$$(AB)C = A(BC)$$

It is assumed, of course, that the dimensions of A, B, and C are such that the products are defined.

The dimensional requirements for ABC to exist are that the number of columns of A equal the number of rows of B, and that the number of columns of B equal the number of rows in C. The resulting product, ABC, will have the same number of rows as A, and the same number of columns as C. Using symbols, if A has dimension $m \times n$, then B must be $n \times p$ and C must be $p \times q$. The product ABC has the dimension $m \times q$.

Note that

$A_{m\times n}B_{n\times p} = (AB)_{m\times p}$

$(AB)_{m\times p}C_{p\times q} = (ABC)_{m\times q}$; and also

$B_{n\times p}C_{p\times q} = (BC)_{n\times q}$

$A_{m\times n}(BC)_{n\times q} = (ABC)_{m\times q}$

The associative property can be extended to any number of matrices to be multiplied together. As long as the dimensions of the matrices permit the operations to be performed, the matrices can be selected in pairs in any sequence. However, the **order** of the matrices in the product **cannot be changed** because matrix multiplication is not commutative.

For example,

$ABCDE = (AB)(CD)E = (AB)C(DE)$ and so on

4.2.4
Distributive Property of Matrices

Matrix multiplication is distributive over addition. By this is meant, as with real numbers, that if A, B, and C are three matrices with "suitable dimensions,"

$$A(B+C) = AB + AC$$

"Suitable dimensions" are that B and C have the same dimensions so that they can be added, and that the number of columns of A is the same as the number of rows of B and C so that the specified multiplication is possible.

This property means that the two matrices B and C can be added together before multiplication by A, and the result is the same as when A multiplies each of the matrices B and C and the products then are added together.

A verification of this property is left as an exercise for the student.

4.2.5
Some Additional Symbology

Recalling our designation of a generalized element of a matrix A as a_{ij} (the element in the ith row and the jth column), we shall use this symbology to give a generalized statement of the process of multiplication of two matrices.

Let A represent a matrix of m rows and n columns, and B represent a matrix with n rows and p columns. AB will be a matrix of m rows and p columns. $(ab)_{ij}$ is a generalized element of AB. From our previous discussion of matrix multiplication we recall that the element in the first row, first column of AB is obtained by summing the products when the elements of the first row of A multiply the elements of the first column of B in succession. That is,

$$(ab)_{11} = a_{11}b_{11} + a_{12}b_{21} + a_{13}b_{31} + \ldots + a_{1n}b_{n1}$$

Since there are n columns in A and n rows in B, each element in AB will be the sum of n individual element products. We can extend this notation to say that

$$(ab)_{ij} = a_{i1}b_{1j} + a_{i2}b_{2j} + a_{i3}b_{3j} + \ldots + a_{in}b_{nj} \qquad (4.2.1)$$

That is, the element in the ith row, jth column of AB is the sum of the n products when each of the n elements in the ith row of A multiplies the corresponding n elements in the jth column of B.

This product symbology is further condensed by the use of the summation symbol Σ. The symbol, Σ, is used to represent the sum of a number of terms in a sequence. The individual terms are identified by a subscript, which takes on successive counting number values as specified by the symbol.

For example, $\sum_{k=1}^{n}$ means to take the sum of the n terms with successive subscripts designated by the k.

$$\sum_{k=1}^{n} a_k \text{ means } a_1 + a_2 + a_3 + \ldots + a_n$$

The symbol, $\sum_{k=1}^{n} a_k$, is read "the sum of a-sub-k from $k=1$ to $k=n$."

Where the context makes the precise meaning clear, it is sometimes read "the summation over k of a_k."

Applying this symbology to the element of the product matrix, we have

$$(ab)_{ij} = \sum_{k=1}^{n} a_{ik} b_{kj}$$

The "k" in the summed terms appears where the successive counting numbers appear in Equation 4.2.1 above.

This type of symbol is used almost exclusively in written material involving matrices because it is so compact. Its very compactness, however, makes it difficult to read. To understand what is said, the reader must note carefully all of the information contained in the symbol.

EXERCISE 4.2

1. Find the following products:

a. $\begin{bmatrix} 3 & 2 \end{bmatrix} \begin{bmatrix} -1 \\ 1 \end{bmatrix}$

b. $\begin{bmatrix} 4 & 1 & 3 \end{bmatrix} \begin{bmatrix} -1 \\ 1 \\ 1 \end{bmatrix}$

c. $\begin{bmatrix} 2 & 3 \\ -1 & 4 \end{bmatrix} \begin{bmatrix} -1 & 0 \\ 2 & -3 \end{bmatrix}$

d. $\begin{bmatrix} 1 & -1 \\ 2 & 4 \\ -3 & 1 \end{bmatrix} \begin{bmatrix} 2 & 1 & 0 \\ -3 & 3 & 1 \end{bmatrix}$

e. $\begin{bmatrix} 2 \\ -1 \\ -2 \end{bmatrix} \begin{bmatrix} 1 & 3 & 3 \end{bmatrix}$

2. In each of the following cases, use the indicated dimensions of the matrices to determine whether AB and/or BA exist. Find the dimension of each product that exists.

a. $A_{2\times 3}$, $B_{2\times 3}$
b. $A_{2\times 3}$, $B_{3\times 2}$
c. $A_{1\times 4}$, $B_{4\times 4}$
d. $A_{3\times 3}$, $B_{3\times 1}$
e. $A_{7\times 8}$, $B_{3\times 7}$

3. $A = \begin{bmatrix} 1 & -2 \\ 3 & 1 \\ 4 & 2 \end{bmatrix}$ $\quad B = \begin{bmatrix} 2 & -3 \\ 1 & 4 \\ 7 & 0 \end{bmatrix}$ $\quad C = \begin{bmatrix} 1 & 2 \\ -1 & 3 \end{bmatrix}$

$$D = \begin{bmatrix} 4 & 3 & -1 \\ 2 & 0 & -1 \end{bmatrix} \quad E = \begin{bmatrix} 1 & 0 & 3 \\ -1 & 2 & 1 \\ -2 & 1 & 1 \end{bmatrix}$$

Find each product, AB, BA, BD, DB, and so on that is defined for the above matrices.

4. For the matrices of Exercise 3,

a. Find A^2 (i.e., AA), B^2, C^2, D^2, and E^2 if they exist.

b. What must be true about a matrix, A, in order that A^2 shall exist?

5. Let $A = \begin{bmatrix} 2 & -1 & -3 \\ 1 & 0 & 1 \\ 4 & -2 & 2 \end{bmatrix} \quad B = \begin{bmatrix} 5 & -1 \\ 2 & 2 \\ 3 & -2 \end{bmatrix} \quad C = \begin{bmatrix} 2 & 0 & 1 & 4 \\ 3 & -2 & -1 & 2 \\ 5 & 1 & -1 & 3 \end{bmatrix}$

$$D = \begin{bmatrix} -1 & -3 \\ 2 & 4 \\ 3 & 2 \\ -5 & 1 \end{bmatrix} \quad E = \begin{bmatrix} 3 & 1 & -4 & 5 \\ 2 & 2 & -1 & 3 \end{bmatrix}$$

a. Find $AB + CD$.

b. Find $AC - BE$.

6. Let $A = \begin{bmatrix} 1 & -1 \\ 2 & 0 \end{bmatrix} \quad B = \begin{bmatrix} 3 & -2 \\ 1 & 4 \end{bmatrix} \quad C = \begin{bmatrix} 2 & -1 & 3 \\ 1 & 0 & 2 \end{bmatrix}$

Show that $(AB)C = A(BC)$.

7. $A = \begin{bmatrix} 2 & 1 \\ -3 & 4 \end{bmatrix} \quad B = \begin{bmatrix} 4 & 3 & -1 \\ 2 & -2 & 0 \end{bmatrix} \quad C = \begin{bmatrix} -2 & 1 & 3 \\ -2 & 2 & 1 \end{bmatrix}$

Show that $A(B + C) = AB + AC$.

8. Let $A = \begin{bmatrix} 4 & 3 & -1 \\ 2 & 1 & 4 \end{bmatrix} \quad B = \begin{bmatrix} 3 & 1 \end{bmatrix} \quad C = \begin{bmatrix} 4 \\ 2 \\ -1 \end{bmatrix}$

a. Multiply these three matrices together in such a way that the final product is a single number.

b. What other dimensions can the product (in suitable order) of these same matrices have?

9. Given the matrices

$$A = \begin{bmatrix} 2 & 3 & -1 \\ 4 & 7 & 2 \\ -4 & 6 & 3 \end{bmatrix} \quad I = \begin{bmatrix} 1 & 0 & 0 \\ 0 & 1 & 0 \\ 0 & 0 & 1 \end{bmatrix}$$

$$T = \begin{bmatrix} 1 & 0 & 0 \\ 0 & 0 & 1 \\ 0 & 1 & 0 \end{bmatrix} \quad P = \begin{bmatrix} 0 & 0 & 1 \\ 1 & 0 & 0 \\ 0 & 1 & 0 \end{bmatrix}$$

Find the products

a. AI and IA. Compare each of these products with A.

b. AT and TA. Compare each of these products with A.

c. AP and PA. Compare each of these products with A.

d. What is the effect of multiplying a matrix times T? How is this different from multiplying T times the matrix?

10. Find two 2×2 matrices A and B such that $AB = [0]$, but $A \neq [0]$ and $B \neq [0]$. (There are many possibilities for A and B.)

11. Suppose in finding the matrices of Exercise 10 we let

$$A = \begin{bmatrix} 1 & 3 \\ 2 & 4 \end{bmatrix}$$

We now wish to find a matrix, B, such that $AB = [0]$.

Let $B = \begin{bmatrix} x_1 & y_1 \\ x_2 & y_2 \end{bmatrix}$

Write the conditions (in the form of algebraic equations) that the elements of B must satisfy in order that $AB = [0]$.

12. Let $A = \begin{bmatrix} a_1 & b_1 & c_1 \\ a_2 & b_2 & c_2 \\ a_3 & b_3 & c_3 \end{bmatrix} \quad X = \begin{bmatrix} x \\ y \\ z \end{bmatrix} \quad C = \begin{bmatrix} 2 \\ 1 \\ 3 \end{bmatrix}$

a. Find AX. What are its dimensions?
b. Use Definition 4.1.1 to write the conditions under which $AX = C$.

4.3 SYSTEMS OF LINEAR EQUATIONS

There are a great many situations, both in business and in the social studies, where quantities are combined and where the combination must satisfy several requirements at the same time. For example, consider a possible situation facing a marketer of coffee. He packages a blend of three different kinds of coffee: Brazilian standard, Brazilian select, and Colombian. His tasters tell him that the best flavor is obtained when there is twice as much Brazilian coffee as Colombian. They specify, however, that the Brazilian select is better than the Brazilian standard. The competitive situation is such that 60¢ per pound is the maximum price that he can pay for the blended coffee. Colombian coffee costs 65¢ per pound; Brazilian select costs 72¢ per pound; and Brazilian standard, 50¢ per pound. In what proportions should the three kinds of coffee be blended?

To set up this problem in mathematical terms let us consider a "typical" pound of blended coffee. Let x represent the amount of Colombian coffee in the blend; y, the amount of Brazilian select; and z, the amount of Brazilian standard. Thus,

$$x + y + z = 1$$

That is, the sum of the amounts of each kind of coffee totals one pound.

The cost of this pound of coffee will be the sum of $.65x$, the cost of x pounds of Colombian; plus $.72y$, the cost of the Brazilian select; plus $.50z$, the cost of the Brazilian standard. As we have specified, this cost should not exceed 60¢. Since this marketer wishes to market the best product he can, we have

$$.65x + .72y + .50z = .60$$

as the algebraic statement of the cost restriction.

The taste requirement says that $y + z$, the total amount of Brazilian coffee in the blend should equal $2x$, twice the amount of Colombian coffee. In algebraic terms

$$y + x = 2x$$

This latter statement can be rearranged, using the rules of algebra, to read

$$2x - y - z = 0$$

We now have an algebraic statement for the requirements to be satisfied in making up the coffee blend:

$$x + y + z = 1$$
$$.65x + .72y + .50z = .60$$
$$2x - y - z = 0$$

(The student will be given the opportunity to find the actual amounts of each kind of coffee to be used in Exercise 4.5, Problem 11).

A set of equations of this kind is called a **system** of **linear equations**. The equations are **linear** because all of the "unknowns," the variables, in the equations have no expressed exponent—the exponent is understood to be one. That is, the equations include no terms of the type, x^2, y^3, or $1/z$, and so on. Systems of linear equations, such as the one above, occur frequently in business and the social sciences. Our illustration has three variables, but in other situations there may be more than three. We shall see how matrices can be used, both to express such systems and to aid in their solution.

Before taking up the solutions to systems of linear equations and their interpretations, let us review carefully the way such systems are set up.

An equation is a statement written in the symbols, the **language**, of algebra. Each of the equations we obtained in the example above about the coffee blend was a statement, which expressed one of the requirements, or restrictions, imposed upon the blending of coffee. To visualize these restrictions in a form suitable for translating into the symbols of algebra, the student should place himself in the position of starting to mix the coffee. The first question is, "How much shall I mix?" In the example we chose one pound to serve as a model. We wished to formulate a recipe for a typical pound of blended coffee. **Any other quantity would serve equally well.** The result will be the recipe for blending whatever amount is chosen. If someone wishes to make a different amount, he can make up, for example, one-half of the recipe, or twice the recipe.

The next question is, "How much of each kind of coffee do I use?" Since the answer to this question is not specifically known, we assign symbols to represent these quantities and proceed, using these symbols, just as if we knew what specific quantities they represent. We then state the requirements we have to meet in terms of these representative symbols. The first statement tells how much coffee we are going to blend. The thing we need to recognize here is that if we take x pounds of one kind of coffee and add to it y pounds of a second kind and z pounds of a third kind, the total weight of coffee is going to be $x + y + z$ pounds of coffee. We then say that the total weight of coffee, $x + y + z$ pounds, is going to be, that is, is going to "equal," one pound.

The second statement relates to the cost of coffee. We know that if one pound of coffee costs 65¢, then x pounds will cost $65x$¢ or \$.65x. We form similar expressions for the cost of the other two kinds of coffee. The cost restriction says that the total cost is to be 60¢ per pound. Note that if we had chosen to make a recipe for P pounds of coffee, the total cost would have been $60P$¢, or \$.60P. That is, in the second equation the term to the right of the equal sign would have been multiplied by the amount of coffee that we had decided to blend.

The third statement is the taste requirement. As we have seen, it specifies that the total amount of Brazilian coffee equals twice the amount of Colombian. That is, $z + y = 2x$.

In this example we have the same number of specifications as there are unknown quantities to determine. In real situations this does not always happen. There may be more or fewer restrictions. In a later section we shall investigate the effect this will have on our solutions.

In this discussion we are purposely trying to avoid listing procedures to be followed in different "typical situations." In the actual use of mathematics the real-life situations rarely, if ever, match the "textbook types" in any recognizable way. The application of mathematics involves setting forth in an orderly manner in complete statements (sentences) all that we know and wish to know about a situation. Then these statements are translated into the symbols of mathematics. Once the symbolization has been accomplished, the mechanical operation of algebra can be used to help interpret and understand the statements and their implications. For example, in our discussion of the coffee blend above we used the statement,

"The amount of Brazilian coffee is twice the amount of Colombian,"

as one of the conditions we wished to meet. The amount of Brazilian coffee is known to be the combined amount of standard quality and select quality. Since y was the symbol used to represent the select quality, and z, the standard quality, this total is $y + z$. This quantity becomes the "subject" of our algebraic sentence. The verb "is" is translated as "=." The other quantity involved is the amount of Colombian coffee, x. However, we are specifying equality between the amount of Brazilian coffee and **twice** the amount of Colombian. A careful analysis of this type makes it clear that we must multiply x by 2. There is a temptation to think that since the amount of Brazilian coffee is larger than the amount of Colombian, that this is the quantity to be multiplied by 2. However, a careful analysis **before** we start using algebraic symbols (which somehow seem to lose their meaning while we are manipulating them) makes the proper algebraic statement clear.

The following exercises present different situations in which the student must interpret the meaning and translate it into algebraic symbols. They should be viewed not as problems in a math book for which "answers" must be found, but from the point of view of a person confronted with the situation described. What is he expected to know about the situation? What is he interested in learning? After completing the interpretation and translation, the student should read the algebraic statements carefully to verify that they correctly state the situation posed in the exercise. Correct translation of situations into mathematical terms (formation of mathematical models) is usually the most difficult part of the study of mathematics. However, it must be mastered before mathematics is of any use to us. Mastery of the operations of mathematics is essential, but it is a wasted talent if we do not know when and where to use it.

EXERCISE 4.3

1. Write equations to express the following:
   a. The quantity x is three larger than the quantity y.
   b. The quantity y is five more than twice x.
   c. The quantity y is four times the quantity x added to two.
   d. Twice y is three less than three times x.

2. Assign symbols to the unknown quantities and write equations to express the following:
   a. The sum of twice one quantity and three times another is 10.
   b. One quantity reduced by seven is four times a second quantity increased by two.
   c. A quantity of one type of counter, each worth five points, and another quantity of a second type of counter, each worth 12 points, produce a total of 120 points.
   d. The sum of three times a quantity reduced by two is the same as six added to five times another quantity.

Write the restrictions or descriptive conditions in the following situations in the form of systems of linear equations. **Write each restriction or condition as a separate equation.** In each case identify the meaning of each symbol; that is, identify the numerical quantity that each of the "unknowns" represents. The **statements** of the conditions are the "answers" requested, not the specific numbers that are the "solutions" to these statements.

3. In a factory two machines make the same part, but the control switch on the machines is adjusted so that one of the machines always makes four more parts per shift than the other. If the total output in a shift is 100 parts, how many does each machine produce?

4. If the machines of question 1 are adjusted so that, instead of producing four more parts than the other, one machine runs 10% faster than the other (i.e., makes 10% more parts per shift) and the total number of parts made in a shift is 210, how many does each machine produce?

5. A company uses two types of shipping cartons for its two products. One type of carton holds two of product A and three of product B. The other carton holds two of product A and one of product B. How many of each type carton should they use to ship ten of product A and twelve of product B?

6. An office uses two card sorting machines, the newer of which sorts cards at a 50% faster rate than the older one. When both machines are operating, they sort 500 cards per minute. How many cards can each machine sort by itself?

7. A gasoline distributor has a mixing pump that permits customers to choose their own blend of gasoline. Two basic grades of gasoline are available; one that sells for 32¢ per gallon; the other, for 40¢ per gallon. The customer sets a dial on the pump to choose the price, between 32¢ and 40¢, that he wishes to pay. The pump mixes appropriate amounts of each grade of gasoline so that the resulting mixture is worth the price selected. Presumably, the higher the price selected, the higher the quality of the gasoline blend obtained.

   A customer sets the dial at 37¢ per gallon and buys 10 gallons. How much of each grade of gasoline should the pump deliver so that the mixture is "worth" the price paid?

8. A dietitian is planning a special diet to be eaten by a volunteer in an experiment. The diet is to consist of carefully measured quantities of three different foods, which are labeled X, Y, and Z. The total weight of food to be eaten is 24 ounces per day. The diet also must include 74 units of protein and 86 units of vitamin C per day.

   Each ounce of the foods has the following amounts of protein and vitamin C:

	X	Y	Z
Protein	3	4	2
Vitamin C	2	5	3

   Formulate a set of equations expressing the conditions to be met for each day's food.

9. A man is attending a 4th of July party, and he wants to bring the hostess a patriotic bouquet. The florist has red and white carnations and blue asters. The carnations cost 20¢ each and the asters 25¢ each. The man decides to buy $5 worth. He asks the florist to give him twice as many white flowers as blue ones, and as many red ones as the number of white and blue together. How many of each does he buy?

10. A hardware chain makes up small packages of bolts in assorted sizes for its retail stores. One assortment (Assortment A) contains five 1-inch bolts, four $1\frac{1}{2}$-inch bolts, and three 2-inch bolts. Assortment B has four bolts of each of the three sizes; and Assortment C has two 1-inch bolts, four $1\frac{1}{2}$-inch, and six 2-inch. At the end of a large packaging run, the package machine operator finds he has 500 1-inch bolts, 480 $1\frac{1}{2}$-inch bolts, and 460 2-inch bolts left over. He wants to place them all in packages. How many packages of each assortment should he make up to use all the bolts?

11. A man makes sets of decorative shelves and sells them to gift shops. He uses sheets of plywood as the basic material for the shelves. He makes three different styles of shelves: A, B, and C. He buys the plywood in three different sizes of sheets, size 1, size 2, and size 3.

From a size 1 sheet he can make 1 style A, 3 style B, and 2 style C shelves.

A size 2 sheet makes 2 style A, 3 style B, and 3 style C.

2 style A, 1 style B, and 5 style C can be made from a size 3 sheet.

Whenever possible, he likes to fill an order for shelves from complete sheets of plywood and have no material left over. How many sheets of each size does he need to make exactly 16 sets of style A, 20 of style B, and 20 of style C?

12. A refinery blends gasoline from three main components, CP, CH, OT. The octane rating of CP is 100; of CH, 96; and of OT, 84. Assume that the octane rating of the final blend is proportional to the octane ratings of the components. How much of each component should be used to obtain a final blend with an octane rating of 92?

13. A manufacturer makes can lids by stamping them out of sheets of metal in a large press. He makes lids for three sizes of cans: pint, quart, and gallon. He has three dies used to form the lids; one makes eight pint-size, six quart-size, and two gallon-size lids from a sheet of metal. The second die forms 10 pint-size, eight quart-size, and two gallon-size lids per sheet; and the third, six pint-size, six quart-size and four gallon-size lids. He receives an order for 800 pint-size, 600 quart-size, and 300 gallon-size lids. He has 100 sheets of metal. How many sheets should he use with each die?

14. In the production department of a company two employees, A and B, make two assemblies, M and N. A can make four M's per hour and three N's per hour. B can make two M's per hour and five N's. The company needs exactly 28 M's and 29 N's per day for its production schedule. If A and B each work an eight-hour shift, how should they divide their time between M's and N's to meet exactly the production requirement?

4.4 SOLUTIONS TO SYSTEMS OF LINEAR EQUATIONS

In the preceding section we discussed the use of systems of linear equations to set forth the conditions to be met by variable quantities in different types of situations. Our task now is to develop a procedure for finding one or more sets of values for the variables that meet the conditions specified. The student has probably already learned one or two ways to do this for systems of two equations with two variables. The method we shall discuss here may appear more cumbersome to the student than the process he has already learned. This is the method, however, that is most readily programmed into a computer. Since the situations in business and behavioral science leading to the use of linear-equation systems usually involve a number of variables, the computations with any method are too lengthy and tedious to be done "manually." Thus, although we shall limit our examples and exercises to systems with only two or three variables, the method used to "solve" these systems can be readily extended to many variables with a computer.

To "solve" a system of equations is to find sets of values, one for each variable (unknown), that "satisfy" each of the equations in the system. The variables are assigned an order (1st, 2nd, 3rd, etc.). Thus, a "solution" to the system takes the form of a vector with as many components as there are variables in the system. The set of all the vectors that satisfy all the equations in the system is called the **solution set** of the system.

For example, in the system of equations,

$$\begin{aligned} x + 3y + 2z &= 1 \\ 2x + 4y + 6z &= 6 \\ 3x + 8y + 3z &= 1 \end{aligned}$$

the set of numbers $(2, -1, 1)$ used to replace x, y, and z, respectively, make each of the equations a true statement. To verify that this is true we replace the variable in each equation with these values. The first equation becomes

$$2 + 3(-1) + 2(1) = 2 - 3 + 2 = 1$$

The second and third equations become, respectively,

$$\begin{aligned} 2(2) + 4(-1) + 6(1) &= 4 - 4 + 6 = 6 \\ 3(2) + 8(-1) + 3(1) &= 6 - 8 + 3 = 1 \end{aligned}$$

Thus, all three statements are true for the specified values for the variables, and the vector $[2, -1, 1]$ is a **solution** for the system. In this case the solution set contains only this one vector.

It is necessary for the solution to make **all** of the equations in the system true. There are an unlimited number of vectors that will satisfy the first equation: $[3, 2, -4]$, $[1, -2, 3]$, and $[0, -1, 2]$, for example. However, none of these satisfies either of the other two equations. We need a process, therefore, that will enable us to find vectors to satisfy all of the equations in a system.

Let us study the equations in a system more carefully. Note that each equation is a **statement of equality between two numbers**. In the first equation, $x + 3y + 2z$ represents a **single number** expressed as a **sum**. The equation states the value of this number is one. The left side of the equation states that the number is obtained from three numbers (represented by x, y, and z) by taking the first (x) and adding it to three times the second (y) and then adding two times the third (z). The student should keep in mind this basic and simple structure of the equations. The process we shall develop is based on two fundamental properties of equal numbers:

i. If two equal numbers are multiplied by the same number, the resulting products are equal.

ii. If the same number or equal numbers are added to equal numbers, the resulting sums are equal.

To illustrate, let us consider the first equation above. We state that the number, $x + 3y + 2z$ is equal to one. Thus,

$$\begin{aligned} &2(x + 3y + 2z) = 2(1)\text{; or} \\ &2x + 6y + 4z = 2 \end{aligned}$$

If we replace the variables with the values, $(2, -1, 1)$, as before, we have

$$2(2) + 6(-1) + 4(1) = 4 - 6 + 4 = 2$$

We can also state that

$$-2(x + 3y + 2z) = -2(1); \text{ or}$$
$$-2x - 6y - 4z = -2$$

Let us consider now the second equation of the system. In this we state that $2x + 4y + 6z$ is a number equal to six. Using property **ii** above, we can add the two equal numbers of the first equation, $-2x - 6y - 4z$ and -2, to the equal numbers of the second equation, $2x + 4y + 6z$ and 6, to obtain two new equal numbers:

$$(-2x - 6y - 4z) + (2x + 4y + 6z) = -2 + 6$$

Simplifying, we have

$$-2y + 2z = 4$$

Note that neither of these two numbers involves x. This does not mean that we have removed x from the system, but only that we have obtained a relationship between the y and the z values of the vector that we are seeking. If we replace the y and z in this equation with the values of the solution vector, $[2, -1, 1]$, we have

$$-2(-1) + 2(1) = 2 + 2 = 4$$

We can combine the first equation with the third equation also in such a way as to obtain a relationship between the y and z in the systems. Let us multiply each of the equal numbers in the first equation by -3. We have

$$-3(x + 3y + 2z) = -3(1) \text{ or}$$
$$-3x - 9y - 6x = -3$$

If we add these two equal numbers, respectively, to the two equal numbers of the third equation, we have

$$-3x - 9y - 6z + (3x + 8y + 3z) = -3 + 1 \text{ or}$$
$$-y - 3z = -2$$

We note that, using values from the solution vector $[2, -1, 1]$, we have

$$-(-1) - 3(1) = 1 - 3 = -2$$

The student has probably surmised by now that a continuation of this process combining different pairs of equations, will eventually lead to a set of equations of the form

$$x = 2$$
$$y = -1$$
$$z = 1$$

That is, we can obtain three equations, each stating the equality between a single variable and a number. The solution vector for the system is now, quite clearly, $[2, -1, 1]$. Our only difficulty is that we have no organized pattern for selecting which equations to combine. If the job of finding the solutions is to be turned over to a computer, we need a well-organized and definite sequence of steps to follow. This is the basis for the following procedure, called the **Gauss elimination method** for obtaining solutions to a system of linear equations.

Let us return to our example system:

$$\begin{aligned} x + 3y + 2z &= 1 \qquad (1) \\ 2x + 4y + 6z &= 6 \qquad (2) \\ 3x + 8y + 3z &= 1 \qquad (3) \end{aligned}$$

Our objective is to replace this system with an equivalent system of the form,

$$\begin{aligned} x \qquad &= a \qquad (1') \\ y \quad &= b \qquad (2') \\ z &= c \qquad (3') \end{aligned}$$

We proceed as follows:

Step 1 Make certain that the coefficient of x in the first equation is one. If it is not, there are two alternatives:

a. If one of the other equations in the system has one as the coefficient for x, the order of the equations may be changed to put that equation first.

For example, if we had written the system originally as

$$\begin{aligned} 2x + 4y + 6z &= 6 \qquad (1) \\ x + 3y + 2z &= 1 \qquad (2) \\ 3x + 8y + 3z &= 1 \qquad (3) \end{aligned}$$

we could have interchanged the first two equations.

b. Each coefficient in the number on the left and also the number to the right of the equal sign can be divided by the coefficient of x in the first equation. (We are multiplying both equal numbers in the first equation by the reciprocal of the x-term.)

Let us choose alternative (b), and divide each number in the first equation by two. The system of equations becomes

$$\begin{aligned} x + 2y + 3z &= 3 \qquad (1a) \\ x + 3y + 2z &= 1 \qquad (2a) \\ 3x + 8y + 3z &= 1 \qquad (3a) \end{aligned}$$

Step 2 a. The two equal numbers of the first equation are used as "operators" to "remove" the x-term from each of the other equations. **All** of the numbers in the **first** equation are multiplied by the **algebraic opposite** of the x-coefficient in the **second** equation. The new first equation is then added to the second equation by adding the respective coefficients term by term. For example, the coefficient of the x-term in the second equation above is one. Its algebraic opposite is -1. Thus, we multiply each number in Equation 1a by -1 to obtain

$$-x - 2y - 3z = -3 \qquad (1a')$$

We add the numbers of this equation term by term to the numbers of Equation $2a$:

$$\begin{array}{rl} -x - 2y - 3z = -3 & \text{(1a')} \\ x + 3y + 2z = 1 & \text{(2a)} \\ \hline 0x + y - z = -2 & \text{(2b)} \end{array}$$

This equation is used as the new second equation.

b. The two equal numbers of the first equation are multiplied by the **algebraic opposite** of the x-coefficient in the **third**. The numbers of this new first equation are added, as above, term by term to the numbers of the third equation.

$$\begin{array}{rl} -3x - 6y - 9z = -9 & \text{(1a'')} \\ 3x + 8y + 3z = 1 & \text{(3a)} \\ \hline 0x + 2y - 6z = -8 & \text{(3b)} \end{array}$$

We retain the original first equation and the two new equations $2b$ and $3b$ to obtain a second system of equations:

$$\begin{array}{rl} x + 2y + 3z = 3 & \text{(1b)} \\ y - z = -2 & \text{(2b)} \\ 2y - 6z = -8 & \text{(3b)} \end{array}$$

The next step is to use the **second** equation as an operator to remove the y-terms from the other equations.

Step 3 Make certain that the coefficient of y in the second equation is one. If not, divide **each number** in the **second** equation by the coefficient of y. (Multiply by the reciprocal of the y-coefficient.) In our example the coefficient of y is already 1, so we proceed.

Step 4 a. Multiply each number in the **second** equation by the opposite of the y-coefficient in the **first** equation. Add the numbers of the resulting equation term by term to the respective numbers in the first equation. (Note that the x-coefficient in the first equation is not changed in this operation.)

In our example the coefficient of y in Equation $1b$ is 2. Thus, we multiply each number in Equation $2b$ by -2, and obtain

$$\begin{array}{rl} -2y + 2z = 4 & \text{(2b')} \\ x + 2y + 3z = 3 & \text{(1b)} \\ \hline x \qquad + 5z = 7 & \text{(1c)} \end{array}$$

We have a new first equation—one in which there is no y-term.

b. Multiply each number in the **second** equation by the algebraic opposite of the y-coefficient in the **third** equation. Add the numbers in the resulting equation to the respective numbers in the **third** equation. The resulting equation is a new third equation. In our example the y-coefficient in Equation $3b$ is 2. We multiply each number in Equation $2b$ by -2 and add:

$$\begin{array}{rl} -2y + 2z = 4 & \text{(2b'')} \\ 2y - 6z = -8 & \text{(3b)} \\ \hline -4z = -4 & \text{(3c)} \end{array}$$

We now have a new set of equations:

$$\begin{aligned} x \quad\quad + 5z &= 7 \quad\quad (1c)\\ y - \ \ z &= -2 \quad\quad (2c)\\ -4z &= -4 \quad\quad (3c) \end{aligned}$$

Step 5 Make certain the coefficient of z in the third equation is 1. If not, divide **each number** in the **third** equation by the coefficient of z.

In our example the coefficient of z in Equation $3c$ is -4. Thus, we divide each number in Equation $3c$ by -4 to obtain a new set of equations:

$$\begin{aligned} x \quad\quad + 5z &= 7 \quad\quad (1c)\\ y - \ \ z &= -2 \quad\quad (2c)\\ z &= 1 \quad\quad (3d) \end{aligned}$$

We now use the equal numbers of Equation $3d$ to remove the z-terms from the other two equations.

Step 6 a. Multiply each number in the new **third** equation by the algebraic opposite of the z-coefficient in the **first** equation. Add the resulting numbers term by term to the respective numbers in the first equation to obtain a new first equation. (When there are three variables, the result will give the value for x in the solution.)

In the example the coefficient of the z-term in Equation $1c$ is 5. Thus, we multiply each number in Equation $3d$ by -5 and add the result, term by term, to Equation $1c$:

$$\begin{array}{lrr} x \quad + 5z = & 7 & (1c)\\ \quad\quad - 5z = & -5 & (3d')\\ \hline x \quad\quad\quad = & 2 & (1d) \end{array}$$

b. Multiply each number in the third equation (Equation $3d$) by the algebraic opposite of the z-coefficient in Equation $2c$. Add the resulting numbers term by term to the respective numbers in Equation $2c$ to obtain a new second equation. (When there are three variables, the result will give the value for y.)

In our example the coefficient of z in Equation $2c$ is -1. Thus, we multiply each number in Equation $3d$ by $-(-1)$ or $+1$; and add:

$$\begin{array}{lrr} y - z = & -2 & (2c)\\ \quad\ \ z = & 1 & (3d'')\\ \hline y \quad\ \ = & -1 & (2d) \end{array}$$

The new set of equations gives the solution to the system of equations:

$$\begin{aligned} x &= 2 \quad\quad (1d)\\ y &= -1 \quad\quad (2d)\\ z &= 1 \quad\quad (3d) \end{aligned}$$

If the system has more equations and/or variables, the process described is extended to eliminate, if possible, all the variables but one in each equation, and have each equation retain a different variable. Those students familiar with computer programming will recognize the process as directly compatible with loop operations in a computer.

This process for finding solutions to systems of linear equations can be confusing to a student at first. We have "added and subtracted equations" and then discarded some and retained others. How do we know which to keep and which to discard? Why do we perform the operations in the manner described, and not some other way? Can we add the algebraic opposite of Equation 2 to Equation 1 instead of vice versa? There is, of course, no one sequence of operations that is the **only** one that will lead to the desired result. As long as no errors in arithmetic are made, **every equation** we obtain by combining multiples of the equal numbers of other equations is **equally valid for that system**. The decision on which operations to perform and in what order is entirely one of convenience. The particular sequence of operations outlined above was chosen because no other single sequence is shorter or simpler for general use, although there are others that are equally direct. Its principal virtue is that it can be readily programmed into a computer. As we shall see in the next two sections, it is readily adaptable to use with matrices and will lead to greater insight into the nature and meaning of the solutions to systems of linear equations.

The student should work a number of the following exercises to develop proficiency in performing the operations with the equations. Many of the systems encountered involve working with unusual fractions. Unfortunately, the nature of the process makes it impossible to avoid this.

EXERCISE 4.4

Find the solution sets for the following systems of linear equations, using the method described by this section.

1. $2x + 2y = 13$
$x + 2y = 8$

2. $.25x + .5y = 2$
$2x + 3y = 14$

3. $2x + 3y = 7$
$5x - 7y = 3$

4. $3x - 2y = 4$
$5x - 4y = 6$

5. $3x + y = 10$
$2x - 3y = 2$

6. $x + y - z = 4$
$2x + 5y + 4z = 5$
$2x + 3y + 2z = 13$

7. $\begin{aligned} x + y - 2z &= 1 \\ 3x + y - 4z &= 7 \\ x - 2y + 3z &= 5 \end{aligned}$

8. $\begin{aligned} x - 3y + 2z &= 5 \\ 2x - 4y + z &= 11 \\ 3x + 2y - z &= 11 \end{aligned}$

9. $\begin{aligned} x + y + 3z &= 8 \\ 3x - y + z &= -12 \\ x + 6y + z &= 29 \end{aligned}$

10. $\begin{aligned} x - 6y - z &= -3 \\ 3x - 2y + z &= 7 \\ 4x - 8y - 3z &= 16 \end{aligned}$

11. $\begin{aligned} x - 2y + 4z &= 13 \\ x + y + z &= -5 \\ 2x - y + z &= 0 \end{aligned}$

12. $\begin{aligned} x - 3y - 7z &= 7 \\ 2x - y + 5z &= 1 \\ x - 2y - 4z &= 2 \end{aligned}$

13. $\begin{aligned} 2x + 3y - z &= -1 \\ 3x + 2y - 4z &= 11 \\ 2x - 5y + 3z &= 3 \end{aligned}$

14. $\begin{aligned} 2x_1 + x_2 - x_3 &= 5 \\ x_1 - x_2 + 2x_3 &= -10 \\ x_1 + 2x_2 + 4x_3 &= 3 \end{aligned}$

15. $\begin{aligned} x_1 + 2x_2 + 3x_3 &= 8 \\ 3x_1 + 9x_2 &= 9 \\ 6x_1 + x_2 + 2x_3 &= 5 \end{aligned}$

16. $\begin{aligned} x_1 + 4x_2 + 3x_3 &= 1 \\ 2x_1 + 5x_2 + 4x_3 &= 4 \\ x_1 - 3x_2 - 2x_3 &= 5 \end{aligned}$

4.5 USE OF MATRICES IN SYSTEMS OF LINEAR EQUATIONS

We shall now develop a method for using matrices to express a system of linear equations. This use of matrices has implications that extend far beyond the limited applications that we shall make here. The branch of mathematics dealing with matrices, vectors, and related concepts is called linear algebra. It provides powerful tools for dealing with a surprising number of different mathematical concepts. As we shall see, the arithmetic computations involved often are tedious and cumbersome, so much so that until recently matrix algebra was limited to relatively few uses. The computations, however, can be quickly and accurately performed by a computer. Thus, in recent years the study of matrices and linear algebra has become an important part of our mathematical training.

Let us consider more carefully the form of a system of linear equations. We note that the numbers to the left of the equal sign in each equation have the same general form—they are sums of terms consisting of a coefficient and a variable. If we use the expression $ax + by + cz$ as a general illustration, a, b, and c are the respective coefficients for the variables x, y, and z. This form should be familiar to us—we should recognize it as the product of the two vectors $[a, b, c]$ and $[x, y, z]$. Note further that if we form a matrix of the coefficients in a system of equations and multiply it by the vector $[x, y, z]$ we have

$$\begin{bmatrix} 1 & 3 & 2 \\ 2 & 4 & 6 \\ 3 & 8 & 3 \end{bmatrix} \begin{bmatrix} x \\ y \\ z \end{bmatrix} = \begin{bmatrix} (1x + 3y + 2z) \\ (2x + 4y + 6z) \\ (3x + 8y + 3z) \end{bmatrix}$$

That is, the entire set of numbers on the left side of the equal signs can be represented as the product of a matrix of coefficients times the vector of "unknowns." Further, the result of this multiplication is a column **vector**, whose components are, respectively, $x+3y+2z$, $2x+4y+6z$, and $3x+8y+3z$. The system of equations from the previous section is

$$\begin{aligned} x + 3y + 2z &= 1 \\ 2x + 4y + 6z &= 6 \\ 3x + 8y + 3z &= 1 \end{aligned}$$

We have just seen that the numbers to the left of the equal signs can be regarded as a column vector. The set of numbers to the right side of the equal signs can also be considered as a column vector. Recall that two vectors are equal if, and only if, their components are respectively equal. Thus, the vector produced by the multiplication

$$\begin{bmatrix} 1 & 3 & 2 \\ 2 & 4 & 6 \\ 3 & 8 & 3 \end{bmatrix} \begin{bmatrix} x \\ y \\ z \end{bmatrix}$$

will be equal to the column vector

$$\begin{bmatrix} 1 \\ 6 \\ 1 \end{bmatrix}$$

if, and only if, $x + 3y + 2z = 1$ (the first components are equal); and $2x + 4y + 6z = 6$ (the second components are equal); and $3x + 8y + 3x = 1$ (the third components are equal). These three conditions are the three equations of the original system. Thus, the system of equations can be considered as a proposition for equality between two vectors.

It is customary in matrix algebra to represent an "unknown" or "variable" vector by a letter, such as u, v, x, and y, with a line over it. That is, "unknown" vectors are designated as $\overline{u}$, $\overline{v}$, $\overline{x}$, and $\overline{y}$. "Constant" vectors, or "known" vectors, are designated with a similar symbol using letters from the beginning of the alphabet, such as, $\overline{a}$, $\overline{b}$, and $\overline{c}$.

If we use the following symbols,

$$A = \begin{bmatrix} 1 & 3 & 2 \\ 2 & 4 & 6 \\ 3 & 8 & 3 \end{bmatrix} \qquad \bar{x} = \begin{bmatrix} x \\ y \\ z \end{bmatrix} \qquad \bar{c} = \begin{bmatrix} 1 \\ 6 \\ 1 \end{bmatrix}$$

the system of equations that we have been working with is represented in matrix notation by

$$A\bar{x} = \bar{c}$$

The system has been condensed into a very compact set of symbols. One of the problems students have in studying matrix algebra is the very compactness of the symbology. We must constantly keep in mind the meaning that is condensed into these apparently simple symbols.

4.5.1 Matrix Solutions to Systems of Linear Equations

Let us now use the concept of a matrix to shorten the process of obtaining solutions to sets of linear equations. Consider the system

$$\begin{aligned} 2x_1 + x_2 + x_3 &= 0 \\ -2x_1 + 3x_2 \quad\quad &= 1 \\ 3x_1 + 5x_2 + 2x_3 &= -1 \end{aligned} \tag{4.5.1}$$

The process developed in the previous section for finding the solution set for this system involves arithmetic operations with the numbers in the equations. Using the idea behind matrices that the position of a number in an array determines what it represents, let us form a matrix for this system using only the numbers:

$$\left[\begin{array}{rrr|r} 2 & 1 & 1 & 0 \\ -2 & 3 & 0 & 1 \\ 3 & 5 & 2 & -1 \end{array}\right] \tag{4.5.2}$$

Each row in the matrix contains the numbers in one of the equations of the system. The first **column** contains the coefficients for x_1; the second, the coefficients for x_2; the third, the coefficients for x_3; and the fourth, the quantities to which each of the unknown sums is equal. Note that a zero has been placed in the matrix as the coefficient for x_3 in the second equation. When we work with sums in the form used in the equations, we customarily omit terms that are always zero. In a matrix, however, we must have a number in each position.

The matrix that we have formed is called the **augmented matrix** for the system of equations. The word "augmented" distinguishes this matrix from the matrix of the coefficients, which we also use. The dotted line between the third and fourth columns shows the location of the equal signs in the system of equations. This line is included to designate the matrix as an augmented matrix representing a system of equations instead of a matrix of coefficients with three rows and four columns.

We can now perform the same type of arithmetic operations on the numbers in the rows of the matrix that we performed on the numbers in the equations in the previous section. These operations are usually called, appropriately, **row operations**. Our objective is to perform a sequence of row operations so that the matrix looks like this:

$$\left[\begin{array}{ccc|c} 1 & 0 & 0 & a \\ 0 & 1 & 0 & b \\ 0 & 0 & 1 & c \end{array}\right] \qquad (4.5.3)$$

A matrix of the form (4.5.3) is called the **reduced matrix** for the system of linear equations (4.5.1). The successive row operations may not lead directly to a reduced matrix, but we can expect to produce one of the following types of patterns:

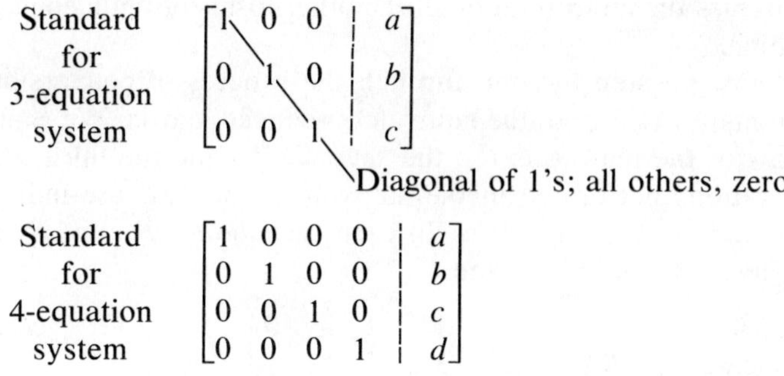

Other possibilities

Reduced matrices:

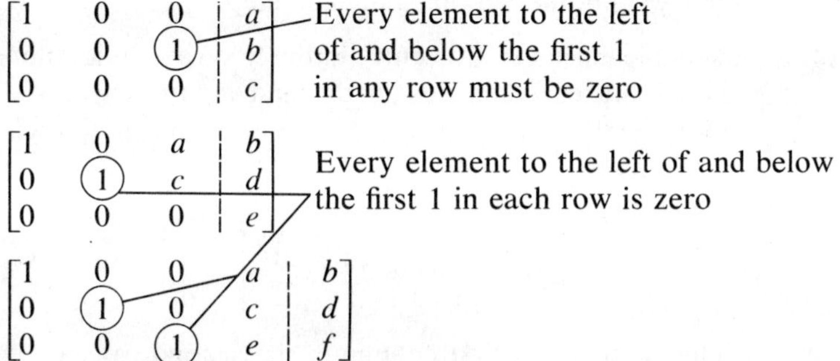

Every element to the left
of and below the first 1
in any row must be zero
Every element to the left of and below
the first 1 in each row is zero

Not reduced matrices

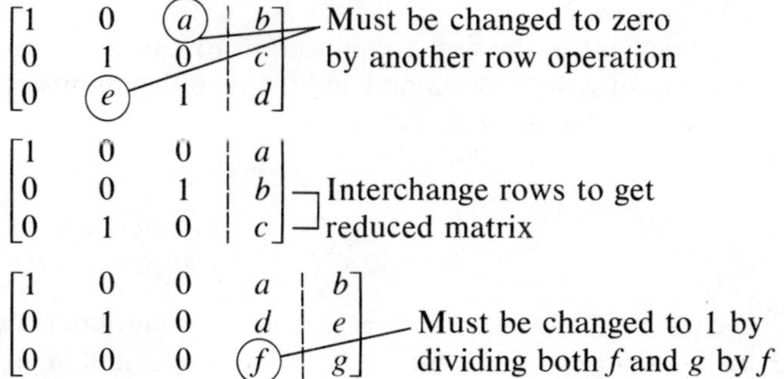

The numbers a, b, and c in the standard reduced matrix (4.5.3) above are the solution set, representing the values for x_1, x_2, and x_3, respectively. This matrix represents the system

$$\begin{aligned} x_1 + 0x_2 + 0x_3 &= a \\ 0x_1 + x_2 + 0x_3 &= b \\ 0x_1 + 0x_2 + x_3 &= c \end{aligned}$$

which we usually write in the form

$$\begin{aligned} x_1 &= a \\ x_2 &= b \\ x_3 &= c \end{aligned}$$

We shall discuss the interpretation of the other forms of reduced matrices later in this section.

Let us now go step by step through the process of conversion from the augmented matrix (4.5.2) to the equivalent reduced matrix. We shall designate the first row of the matrix as R_1; the second, R_2; and the third, R_3. The particular operations performed in the steps in the process are indicated in the column to the left of the matrices. In each case the R_1, R_2, and R_3 refer to the rows as listed in the **previous** step.

$$\left.\begin{matrix}\text{Original}\\ \text{Augmented}\\ \text{Matrix}\end{matrix}\right\} \quad \begin{matrix} R_1 \\ R_2 \\ R_3 \end{matrix} \left[\begin{array}{rrr|r} 2 & 1 & 1 & 0 \\ -2 & 3 & 0 & 1 \\ 3 & 5 & 2 & -1 \end{array}\right]$$

We shall follow the procedure outlined in the previous section for use with the complete system of equations. However, we shall use only the augmented matrix and omit writing the variables in each step.

1. The coefficient for x_1 (the entry in the first row, first column) is not 1. None of the other entries in the first column is 1. We, therefore, multiply each element in the first row by the reciprocal of 2; that is, $\frac{1}{2}$. We have

$$\tfrac{1}{2}R_1 \text{ (from above)} \quad \left[\begin{array}{rrr|r} 1 & \frac{1}{2} & \frac{1}{2} & 0 \\ -2 & 3 & 0 & 1 \\ 3 & 5 & 2 & -1 \end{array}\right] \quad \begin{matrix} \text{new } R_1 \\ \\ \\ \end{matrix}$$

2. a. The entry in the first column of the second row is -2 (coefficient of x_1 in the second equation). We multiply each element of the **first** row by $-(-2)$ or $+2$ and add these elements column by column to row 2.

b. The element in the first column, third row is 3. We multiply each element of row 1 by -3 and add the resulting numbers column by column to the elements of row 3.

$$\begin{matrix} \\ R_2 + 2R_1 \\ R_3 - 3R_1 \end{matrix} \quad \left[\begin{array}{rrr|r} 1 & \frac{1}{2} & \frac{1}{2} & 0 \\ 0 & 4 & 1 & 1 \\ 0 & \frac{7}{2} & \frac{1}{2} & -1 \end{array}\right] \quad \begin{matrix} \\ \text{new } R_2 \\ \text{new } R_3 \end{matrix}$$

3. The element in the second row, second column (coefficient of x_2 in second equation) is 4. We multiply each element in row 2 by $\frac{1}{4}$.

$$\tfrac{1}{4}R_2 \qquad \left[\begin{array}{ccc|c} 1 & \tfrac{1}{2} & \tfrac{1}{2} & 0 \\ 0 & 1 & \tfrac{1}{4} & \tfrac{1}{4} \\ 0 & \tfrac{7}{2} & \tfrac{1}{2} & -1 \end{array}\right] \qquad \begin{array}{l} \\ \text{new } R_2 \\ \\ \end{array}$$

4. a. The element in the second column of the first row is $\frac{1}{2}$. We multiply each element in the second row by $-\frac{1}{2}$ and add the resulting numbers column by column to the elements of the first row.
   b. The element in the second column of the third row is $\frac{7}{2}$. We multiply each element in the second row by $-\frac{7}{2}$ and add the resulting numbers column by column to the elements of the third row.

$$\begin{array}{l} R_1 - \tfrac{1}{2}R_2 \\ \\ R_3 - \tfrac{7}{2}R_2 \end{array} \qquad \left[\begin{array}{ccc|c} 1 & 0 & \tfrac{3}{8} & -\tfrac{1}{8} \\ 0 & 1 & \tfrac{1}{4} & \tfrac{1}{4} \\ 0 & 0 & -\tfrac{3}{8} & -\tfrac{15}{8} \end{array}\right] \qquad \begin{array}{l} \text{new } R_1 \\ \\ \text{new } R_3 \end{array}$$

5. The element in the third row, third column, is $-\frac{3}{8}$. We multiply each element in row 3 by $-\frac{8}{3}$ (reciprocal of $-\frac{3}{8}$).

$$\begin{array}{l} \\ \\ -\tfrac{8}{3}R_3 \end{array} \qquad \left[\begin{array}{ccc|c} 1 & 0 & \tfrac{3}{8} & -\tfrac{1}{8} \\ 0 & 1 & \tfrac{1}{4} & \tfrac{1}{4} \\ 0 & 0 & 1 & 5 \end{array}\right] \qquad \begin{array}{l} \\ \\ \text{new } R_3 \end{array}$$

6. a. The element in the third column of the first row is $\frac{3}{8}$. We multiply each element of the third row by $-\frac{3}{8}$ and add the resulting numbers column by column to the elements of the first row. (Note that by this time the zeros in the third row restrict the effect of this operation to the third and fourth columns of the augmented matrix.)
   b. The element in the third column of the second row is $\frac{1}{4}$. We multiply each element of the third row by $-\frac{1}{4}$ and add the resulting numbers column by column to the elements of the second row.

$$\begin{array}{l} R_1 - \tfrac{3}{8}R_3 \\ R_2 - \tfrac{1}{4}R_3 \\ \\ \end{array} \qquad \left[\begin{array}{ccc|c} 1 & 0 & 0 & -2 \\ 0 & 1 & 0 & -1 \\ 0 & 0 & 1 & 5 \end{array}\right] \qquad \begin{array}{l} \text{new } R_1 \\ \text{new } R_2 \\ \\ \end{array}$$

We have now converted the augmented matrix to its reduced form. We see that

$$x_1 = -2$$
$$x_2 = -1$$
$$x_3 = 5$$

is the solution set for this system of equations. The student should verify this by substituting these values into each of the original equations and checking whether they become true statements.

4.5.2 Additional Forms of Solutions

In the examples we have used in this and the preceding section, we have worked with systems of equations in which there have been the same number of equations as variables. In addition, in each case we have obtained a **unique** solution, a single set of replacement values for the variables. In practice, there are other possibilities. There may be more or fewer equations than variables. Even when the number of variables and equations is the same, the solution set may contain other than a single set of values, or there may be no solution set at all.

Example 4.5.1

Let us consider the system

$$\begin{aligned} x_1 + 2x_2 + 2x_3 &= 5 \\ 2x_1 + 3x_2 + x_3 &= 4 \\ x_1 + 4x_2 + 8x_3 &= 17 \end{aligned} \tag{4.5.4}$$

The augmented matrix for this system is

$$\left[\begin{array}{ccc|c} 1 & 2 & 2 & 5 \\ 2 & 3 & 1 & 4 \\ 1 & 4 & 8 & 17 \end{array}\right]$$

We can apply operations to the rows as indicated:

$$\begin{array}{l} \\ R_2 - 2R_1 \\ R_3 - R_1 \end{array} \left[\begin{array}{ccc|c} 1 & 2 & 2 & 5 \\ 0 & -1 & -3 & -6 \\ 0 & 2 & 6 & 12 \end{array}\right]$$

$$\begin{array}{l} \\ (-1)R_2 \\ \end{array} \left[\begin{array}{ccc|c} 1 & 2 & 2 & 5 \\ 0 & 1 & 3 & 6 \\ 0 & 2 & 6 & 12 \end{array}\right]$$

$$\begin{array}{l} R_1 - 2R_2 \\ \\ R_3 - 2R_2 \end{array} \left[\begin{array}{ccc|c} 1 & 0 & -4 & -7 \\ 0 & 1 & 3 & 6 \\ 0 & 0 & 0 & 0 \end{array}\right]$$

The reduced matrix has all zeros in the third row. Quite clearly we cannot eliminate the coefficients for x_3 from the first two rows in the usual way. If we convert this matrix into a system of linear equations, which are equivalent to the original system, we have

$$\begin{aligned} x_1 - 4x_3 &= -7 \\ x_2 + 3x_3 &= 6 \\ 0x_1 + 0x_2 + 0x_3 &= 0 \end{aligned} \tag{4.5.5}$$

The third equation states that $0 = 0$, which is a true statement. In essence, our system of three equations has been reduced to two, **which are equivalent** to the original system. When this occurs the restriction on the system represented by the third equation has already been included in the statement of the other two restrictions. We say that the three equations are **not independent** of each other; or, more elegantly, the three equations are **linearly dependent**. This means that any of the three equations can be obtained by combining suitable multiples of the other two. In our example above each coefficient of the third equation can be obtained by subtracting two times the corresponding coefficient in the second equation from five times the coefficient of the first equation. That is, $5(1) - 2(2) = 1$; $5(2) - 2(3) = 4$; and $5(2) - 2(1) = 8$.

What effect does this situation have on the solution set of the system? Let us rewrite the two equations in the system (4.5.5) above as

$$\begin{aligned} x_1 &= 4x_3 - 7 \\ x_2 &= -3x_3 + 6 \end{aligned} \tag{4.5.6}$$

The significance of the system (4.5.6) is that we **can choose any number we wish** as the **replacement for** x_3. The values for x_1 and x_2 are then computed from the equations in the system (4.5.6). Let us suppose, for example, that we find three to be a convenient value for x_3. Then

$$\begin{aligned} x_1 &= 4 \cdot 3 - 7 = 12 - 7 = 5 \\ x_2 &= -3 \cdot 3 + 6 = -9 + 6 = -3 \end{aligned}$$

Thus, the set of values, $(5, -3, 3)$ for x_1, x_2, x_3, respectively, will be a solution for the system. We check this by replacing x_1, x_2, and x_3 with these values in the original system:

$$\begin{aligned} &5 + 2(-3) + 2(3) = 5 \\ &2(5) + 3(-3) + 3 = 4 \\ &5 + 4(-3) + 8(3) = 17 \end{aligned}$$

Each of these statements is true. Thus, the set of numbers $(5, -3, 3)$ is a solution. Other solutions can be obtained by choosing other value for x_3.

The complete solution set for this system includes all the possible choices for x_3, and the resulting values for x_1 and x_2. Thus, there are an unlimited number of solutions. We write this as

$$\begin{aligned} x_1 &= 4x_3 - 7 \\ x_2 &= -3x_3 + 6 \\ x_3 &= \text{any number} \end{aligned}$$

In vector terms this is written

$$\bar{x} = \begin{bmatrix} -7 \\ 6 \\ 0 \end{bmatrix} + k \begin{bmatrix} 4 \\ -3 \\ 1 \end{bmatrix}$$

where k is any constant.

Example 4.5.2

Let us consider the system

$$\begin{aligned} x_1 + x_2 + x_3 &= 4 \\ x_1 + 3x_2 + 2x_3 &= 6 \\ 3x_1 + 5x_2 + 4x_3 &= 10 \end{aligned}$$

The augmented matrix for this system is

$$\left[\begin{array}{ccc|c} 1 & 1 & 1 & 4 \\ 1 & 3 & 2 & 6 \\ 3 & 5 & 4 & 10 \end{array}\right]$$

Using row operations in the manner previously employed, we have

$$\begin{array}{l} \\ R_2 - R_1 \\ R_3 - 3R_1 \end{array} \left[\begin{array}{ccc|c} 1 & 1 & 1 & 4 \\ 0 & 2 & 1 & 2 \\ 0 & 2 & 1 & -2 \end{array}\right]$$

$$\begin{array}{l} \\ \frac{1}{2}R_2 \\ \\ \end{array} \left[\begin{array}{ccc|c} 1 & 1 & 1 & 4 \\ 0 & 1 & \frac{1}{2} & 1 \\ 0 & 2 & 1 & -2 \end{array}\right]$$

$$\begin{array}{l} R_1 - R_2 \\ \\ R_3 - 2R_2 \end{array} \left[\begin{array}{ccc|c} 1 & 0 & \frac{1}{2} & 3 \\ 0 & 1 & \frac{1}{2} & 1 \\ 0 & 0 & 0 & -4 \end{array}\right]$$

The third row of the reduced matrix this time has zeros in each of the first three columns, but the element in the fourth column is −4. Since the element in the third column is zero, we cannot remove the third column elements from either row 1 or row 2. When we convert the matrix into a system of linear equations, we have

$$\begin{aligned} x_1 \qquad + \tfrac{1}{2}x_3 &= 3 \\ x_2 + \tfrac{1}{2}x_3 &= 1 \\ 0 &= -4 \end{aligned}$$

The third equation makes the statement that zero is equal to −4, a fallacy. We have converted our original system of equations into an equivalent system in which one part is always false. Thus, the three parts of the original system are mutually inconsistent, and there is no set of numbers that will satisfy all three of the original equations. This result means either that it is not possible to meet all of the conditions represented by the original equations, or that one or more of the conditions was improperly translated into the symbols of algebra.

Example 4.5.3

Let us consider the system

$$\begin{aligned} x_1 + 2x_2 - x_3 &= 5 \\ x_1 + 2x_2 + 3x_3 &= 3 \\ x_1 + x_2 - 2x_3 &= 8 \\ 2x_1 - x_2 + x_3 &= k \end{aligned}$$

The augmented matrix for this system is

$$\left[\begin{array}{ccc|c} 1 & 2 & -1 & 5 \\ 1 & 2 & 3 & 3 \\ 1 & 1 & -2 & 8 \\ 2 & -1 & 1 & k \end{array}\right]$$

Using row operations as before to obtain the reduced matrix:

$$\begin{array}{l} \\ R_2 - R_1 \\ R_3 - R_1 \\ R_4 - 2R_1 \end{array} \left[\begin{array}{rrr|l} 1 & 2 & -1 & 5 \\ 0 & 0 & 4 & -2 \\ 0 & -1 & -1 & 3 \\ 0 & -5 & 3 & k - 10 \end{array}\right]$$

We have hit our first snag. We have zeros in the first two columns of row two, but a nonzero element in the second column of row 3. As we have seen above, this situation cannot exist in a reduced matrix. To correct the situation we merely interchange rows 2 and 3. This is an allowable operation, since there is no set order in which the equations of our original system must be listed. If we had listed equation (3) in the second position originally, this problem would not arise.

Our "partially reduced" augmented matrix now looks like this:

$$\left[\begin{array}{rrr|l} 1 & 2 & -1 & 5 \\ 0 & -1 & -1 & 3 \\ 0 & 0 & 4 & -2 \\ 0 & -5 & 3 & k - 10 \end{array}\right]$$

We now follow the standard procedure, noting the operations performed at the left:

$$\begin{array}{l} \\ (-1)R_2 \\ \\ \\ \end{array} \left[\begin{array}{rrr|l} 1 & 2 & -1 & 5 \\ 0 & 1 & 1 & -3 \\ 0 & 0 & 4 & -2 \\ 0 & -5 & 3 & k - 10 \end{array}\right]$$

$$\begin{array}{l} R_1 - 2R_2 \\ \\ \\ R_4 + 5R_2 \end{array} \left[\begin{array}{rrr|l} 1 & 0 & -3 & 11 \\ 0 & 1 & 1 & -3 \\ 0 & 0 & 4 & -2 \\ 0 & 0 & 8 & k - 10 - 15 \end{array}\right]$$

$$\begin{array}{l} \\ \\ \frac{1}{4}R_3 \\ \\ \end{array} \left[\begin{array}{rrr|l} 1 & 0 & -3 & 11 \\ 0 & 1 & 1 & -3 \\ 0 & 0 & 1 & -\frac{1}{2} \\ 0 & 0 & 8 & k - 25 \end{array}\right]$$

$$\begin{array}{l} R_1 + 3R_3 \\ R_2 - R_3 \\ \\ R_4 - 8R_3 \end{array} \left[\begin{array}{rrr|l} 1 & 0 & 0 & \frac{19}{2} \\ 0 & 1 & 0 & -\frac{5}{2} \\ 0 & 0 & 1 & -\frac{1}{2} \\ 0 & 0 & 0 & k - 21 \end{array}\right]$$

Note that row 4 has zeros in each of the first three columns. In this system there are three variables, but **four** conditions. Whenever the number of conditions exceeds the number of variables, there will **always** be rows in which the elements in the columns to the left of the "equal line" are all zeros. The entry in the last column is the critical one. In this case we represented the original number by k. Note that if we had replaced k by 21 in the original system, the fourth row of the reduced matrix would represent the statement $0 = 0$, which is true. In this case, and this case **only**, the four original conditions are consistent with each other, and we have a solution. For any other replacement value for k, say 30, the fourth row represents the statement $0 = 9$, which is a fallacy. The original conditions are not consistent, and there is no solution.

Let us summarize the possible solutions for a system of linear equations.

1. Single solutions

When the matrix of elements to the left of the "equal-line" for the **reduced** matrix has a diagonal of 1's with all the other elements zero, there is a single solution to the system; namely, the set of numbers to the right of the "equal-line."

$$\left[\begin{array}{ccc|c} 1 & 0 & 0 & a \\ 0 & 1 & 0 & b \\ 0 & 0 & 1 & c \\ 0 & 0 & 0 & 0 \end{array}\right] \leftarrow$$

If the number of rows exceeds the number of variables, the "excess" rows are **all** zeros

2. Multiple solutions

When the matrix of elements to the left of the "equal-line" in the reduced matrix of the system has more columns than there are rows with nonzero elements, there are multiple solutions. If this matrix has n columns and m rows, and n is greater than m, values for at least $n - m$ of the variables may be chosen at will. The values of the remaining variables must then be **computed** using these chosen values.

$$\left[\begin{array}{ccccc|c} 1 & 0 & 0 & a & b & c \\ 0 & 1 & 0 & d & e & f \\ 0 & 0 & 1 & g & h & i \end{array}\right]$$

In this example values may be chosen at will for two of the variables, and the other three are then computed, using the chosen values.

If we have a reduced matrix of the type shown above, but in which there is an entire column of zeros, the variable represented by that column has no effect on the system. That variable can be assigned any value at all, and the other variables are not affected.

$$\left[\begin{array}{cccc|c} 1 & 0 & 0 & 0 & a \\ 0 & 1 & 0 & 0 & b \\ 0 & 0 & 0 & 1 & c \end{array}\right]$$

All zeros (third column)

In the system represented by this matrix, $x_1 = a$; $x_2 = b$; $x_4 = c$; x_3 can have **any** value. In effect this means that the conditions represented by the system made no restrictions on the value of x_3, even though the quantity represented by x_3 may have been included in the statement of the conditions.

3. No solution

When the matrix to the left of the "equal-line" in a reduced matrix has a row of zeros, but the element to the right of the "equal-line" in that row is not zero, the conditions for the system are inconsistent and there is no solution.

EXERCISE 4.5

Set up the augmented matrices for the following systems of equations, and find their solution sets, using the method of this section.

1. $x_1 - 2x_2 = 7$
$2x_1 - 3x_2 = 10$

2. $2x_1 - 5x_2 = 10$
$3x_1 + 2x_2 = -4$

3. $3x_1 + x_2 = 8$
$5x_1 - 2x_2 = -5$

4. $x_1 + x_2 + 5x_3 = 8$
$2x_1 + x_2 - x_3 = 3$
$3x_1 + 2x_2 + 5x_3 = 12$

5. $x_1 + x_2 + 2x_3 = 4$
$2x_1 + 4x_2 + 5x_3 = 12$
$x_1 - 3x_2 + 3x_3 = 5$

6. $x_1 + 3x_2 + 2x_3 = -6$
$3x_1 + 2x_2 - x_3 = 3$
$4x_1 + x_2 - 3x_3 = 9$

7. $x_1 + 3x_2 - x_3 = 5$
$2x_1 + x_2 + 3x_3 = 3$
$3x_1 + 2x_2 + 4x_3 = 8$

8. $x_1 + 2x_2 + 3x_3 = 2$
$3x_1 + x_2 - x_3 = -4$
$3x_1 + 2x_2 + x_3 = 2$

9. $x_1 + 2x_2 + 4x_3 = 6$
$2x_1 + x_2 - x_3 = 3$
$2x_1 + 3x_2 + 5x_3 = 9$

10. $x_1 + 3x_2 - 7x_3 = 2$
$4x_1 + x_2 + 5x_3 = -3$
$2x_1 + x_2 + x_3 = 1$

11. Find desired coffee blend for situation described in Section 4.3.

12. Use the methods of this section to find solutions, where they exist, for problems 3 to 14 of Exercise 4.3.

Interpret the solutions according to the situations described in those cases where a unique solution does not exist.

4.6
INVERSE MATRIX

4.6.1
Identity Matrix

In the previous sections we encountered matrices of the form

$$\begin{bmatrix} 1 & 0 & 0 \\ 0 & 1 & 0 \\ 0 & 0 & 1 \end{bmatrix}$$

Such matrices are called **identity** matrices. An **identity matrix** is a **square** matrix in which **each element** in the **diagonal** from the upper left corner to the lower right corner is the number **one**. Each of the other elements in the matrix is **zero**. There is an identity matrix for each dimension of square matrix. In ordinary usage the symbol I is used to designate an identity matrix. It implies the identity matrix whose dimension is appropriate for the situation. For example, if A and B are both 4×4 matrices and $AB = I$, the I in the equation represents the 4×4 identity matrix. Where the context does not make clear the dimension I is to have, a subscript is used. For example, I_2 means the 2×2 identity matrix and I_3, the 3×3 identity matrix.

The primary property of an identity matrix, which gives it its name, is that for any matrix A, for which the product is defined,

$$IA = AI = A$$

If A is a square matrix, the I in the product IA has the same dimension as the I in the product AI. If A is not square, the I in the product IA will have a different dimension from the I in the product AI. For example, if A is a 2×3 matrix, such as

$$A = \begin{bmatrix} 1 & 2 & 3 \\ 4 & 5 & 6 \end{bmatrix}$$

the product, IA, is defined for I_2:

$$\begin{bmatrix} 1 & 0 \\ 0 & 1 \end{bmatrix} \begin{bmatrix} 1 & 2 & 3 \\ 4 & 5 & 6 \end{bmatrix} = \begin{bmatrix} 1 & 2 & 3 \\ 4 & 5 & 6 \end{bmatrix}$$

For the product, AI, however, the I is I_3:

$$\begin{bmatrix} 1 & 2 & 3 \\ 4 & 5 & 6 \end{bmatrix} \begin{bmatrix} 1 & 0 & 0 \\ 0 & 1 & 0 \\ 0 & 0 & 1 \end{bmatrix} = \begin{bmatrix} 1 & 2 & 3 \\ 4 & 5 & 6 \end{bmatrix}$$

4.6.2
The Inverse of a Matrix

The inverse of a matrix, A, is defined as follows:

Definition 4.6.1

> **Inverse Matrix:** A matrix, B, such that $BA = AB = I$ is the **inverse** of **matrix** A. The usual symbol for the inverse matrix of A, is A^{-1}.

It is not our purpose here to concern ourselves with the details of matrix algebra, so we shall state without proof some of the important properties of matrices and their inverses.

A matrix must be **square** to have an inverse. However, **not all square matrices have inverses**. In the language of Chapter 2, being a square matrix is a **necessary, but not** a **sufficient** condition for a matrix to have an inverse.

If A^{-1} is the inverse of the matrix A, then A is the inverse of A^{-1}. We have $A^{-1}A = AA^{-1} = I$.

If a matrix has an inverse, that inverse is unique. That is, if $BA = I$ and $CA = I$, then B **must equal** C.

Thus far we have spoken in general terms about matrices and their inverses, "if they exist." The question immediately arises, of course, how can we tell if a matrix does or does not have an inverse? Unfortunately, the best way to answer the question is to go through the process for finding the inverse. If it can be found, the matrix has an inverse; otherwise, not. A square matrix that has no inverse is called a **singular** matrix. There are ways to test a matrix for singularity, but they involve about the same amount of work as the process for finding the inverse. If a person uses a test of this kind, and the matrix turns out not to be singular, he still has the job of finding the inverse. We shall, therefore, determine whether a matrix has an inverse by the process of computing the inverse. We let

$$A = \begin{bmatrix} a_{11} & a_{12} & a_{13} \\ a_{21} & a_{22} & a_{23} \\ a_{31} & a_{32} & a_{33} \end{bmatrix}$$

and

$$X = \begin{bmatrix} x_1 & x_2 & x_3 \\ y_1 & y_2 & y_3 \\ z_1 & z_2 & z_3 \end{bmatrix}$$

From the definition of an inverse matrix, X is the inverse of A, if, and only if,

$$XA = AX = I$$

If we work out the product of the matrices, using the relationship $AX = I$, we have

$$\begin{bmatrix} a_{11} & a_{12} & a_{13} \\ a_{21} & a_{22} & a_{23} \\ a_{31} & a_{32} & a_{33} \end{bmatrix} \begin{bmatrix} x_1 & x_2 & x_3 \\ y_1 & y_2 & y_3 \\ z_1 & z_2 & z_3 \end{bmatrix} = \begin{bmatrix} 1 & 0 & 0 \\ 0 & 1 & 0 \\ 0 & 0 & 1 \end{bmatrix}$$

The set of computations to obtain the first column of the product I is

$$\begin{aligned} a_{11}x_1 + a_{12}y_1 + a_{13}z_1 &= 1 \\ a_{21}x_1 + a_{22}y_1 + a_{23}z_1 &= 0 \\ a_{31}x_1 + a_{32}y_1 + a_{33}z_1 &= 0 \end{aligned}$$

We can find thc values, x_1, y_1, z_1, by solving these three equations simultaneously. If we write out the computations for each of the other two columns of I, we have

$$\begin{aligned} a_{11}x_2 + a_{12}y_2 + a_{13}z_2 &= 0 \\ a_{21}x_2 + a_{22}y_2 + a_{23}z_2 &= 1 \\ a_{31}x_2 + a_{32}y_2 + a_{33}z_2 &= 0 \end{aligned}$$

and

$$\begin{aligned}a_{11}x_3 + a_{12}y_3 + a_{13}z_3 &= 0\\ a_{21}x_3 + a_{22}y_3 + a_{23}z_3 &= 0\\ a_{31}x_3 + a_{32}y_3 + a_{33}z_3 &= 1\end{aligned}$$

The augmented matrix for the first set of equations is

$$\left[\begin{array}{ccc|c} a_{11} & a_{12} & a_{13} & 1\\ a_{21} & a_{22} & a_{23} & 0\\ a_{31} & a_{32} & a_{33} & 0\end{array}\right]$$

For the second and third sets of equations

$$\left[\begin{array}{ccc|c} a_{11} & a_{12} & a_{13} & 0\\ a_{21} & a_{22} & a_{23} & 1\\ a_{31} & a_{32} & a_{33} & 0\end{array}\right]$$

$$\left[\begin{array}{ccc|c} a_{11} & a_{12} & a_{13} & 0\\ a_{21} & a_{22} & a_{23} & 0\\ a_{31} & a_{32} & a_{33} & 1\end{array}\right]$$

The process of solution for these **three** systems of equations involves the **same** row operations, since they all are the same set of coefficients. Therefore, let us perform all three solutions at the same time, as follows:

$$\left[\begin{array}{ccc|ccc} a_{11} & a_{12} & a_{13} & 1 & 0 & 0\\ a_{21} & a_{22} & a_{23} & 0 & 1 & 0\\ a_{31} & a_{32} & a_{33} & 0 & 0 & 1\end{array}\right]$$

After we perform the row operations required to obtain the reduced matrix, the numbers in the first column to the right of the"equal-line" will be the values for x_1, y_1, and z_1. In the second column we will have the values for x_2, y_2, z_2; and in the third column, the values for x_3, y_3, z_3. Thus, the matrix to the right of the "equal-line" will be the inverse matrix for A.

If in the process of performing the row operations for obtaining the reduced matrix we find that there is no solution, then A has no inverse.

Example 4.6.1

Let us find the inverse of the matrix

$$A = \begin{bmatrix} 1 & 2 & 3\\ 2 & 0 & -2\\ 1 & -2 & 4\end{bmatrix}$$

We set up the special augmented matrix

$$\left[\begin{array}{ccc|ccc} 1 & 2 & 3 & 1 & 0 & 0\\ 2 & 0 & -2 & 0 & 1 & 0\\ 1 & -2 & 4 & 0 & 0 & 1\end{array}\right]$$

and use row operations to find the reduced matrix. Each row operation is applied to all of the columns to the right of the "equal-line."

$$\begin{array}{l} \\ R_2 - 2R_1\\ R_3 - R_1\end{array}\left[\begin{array}{ccc|ccc} 1 & 2 & 3 & 1 & 0 & 0\\ 0 & -4 & -8 & -2 & 1 & 0\\ 0 & -4 & 1 & -1 & 0 & 1\end{array}\right]$$

$$-\tfrac{1}{4}R_2 \quad \left[\begin{array}{ccc|ccc} 1 & 2 & 3 & 1 & 0 & 0 \\ 0 & 1 & 2 & \frac{1}{2} & -\frac{1}{4} & 0 \\ 0 & -4 & 1 & -1 & 0 & 1 \end{array}\right]$$

$$\begin{matrix} R_1-2R_2 \\ \\ R_3+4R_2 \end{matrix} \left[\begin{array}{ccc|ccc} 1 & 0 & -1 & 0 & \frac{1}{2} & 0 \\ 0 & 1 & 2 & \frac{1}{2} & -\frac{1}{4} & 0 \\ 0 & 0 & 9 & 1 & -1 & 1 \end{array}\right]$$

$$\tfrac{1}{9}R_3 \quad \left[\begin{array}{ccc|ccc} 1 & 0 & -1 & 0 & \frac{1}{2} & 0 \\ 0 & 1 & 2 & \frac{1}{2} & -\frac{1}{4} & 0 \\ 0 & 0 & 1 & \frac{1}{9} & -\frac{1}{9} & \frac{1}{9} \end{array}\right]$$

$$\begin{matrix} R_1+R_3 \\ R_2-2R_3 \\ \\ \end{matrix} \left[\begin{array}{ccc|ccc} 1 & 0 & 0 & \frac{1}{9} & \frac{7}{18} & \frac{1}{9} \\ 0 & 1 & 0 & \frac{5}{18} & -\frac{1}{36} & -\frac{2}{9} \\ 0 & 0 & 1 & \frac{1}{9} & -\frac{1}{9} & \frac{1}{9} \end{array}\right]$$

Therefore, the inverse of the matrix A is the matrix

$$A^{-1} = \begin{bmatrix} \frac{1}{9} & \frac{7}{18} & \frac{1}{9} \\ \frac{5}{18} & -\frac{1}{36} & -\frac{2}{9} \\ \frac{1}{9} & -\frac{1}{9} & \frac{1}{9} \end{bmatrix}$$

Verification that $AA^{-1} = I$ is left as an exercise for the student.

Example 4.6.2

We consider now the matrix

$$B = \begin{bmatrix} 2 & 3 & 1 \\ -3 & 1 & -7 \\ 1 & 5 & -3 \end{bmatrix}$$

To find the inverse of B, if it exists, we set up the special augmented matrix,

$$\left[\begin{array}{ccc|ccc} 2 & 3 & 1 & 1 & 0 & 0 \\ -3 & 1 & -7 & 0 & 1 & 0 \\ 1 & 5 & -3 & 0 & 0 & 1 \end{array}\right]$$

and perform the following row operations:

$$\tfrac{1}{2}R_1 \quad \left[\begin{array}{ccc|ccc} 1 & \frac{3}{2} & \frac{1}{2} & \frac{1}{2} & 0 & 0 \\ -3 & 1 & -7 & 0 & 1 & 0 \\ 1 & 5 & -3 & 0 & 0 & 1 \end{array}\right]$$

$$\begin{matrix} \\ R_2+3R_1 \\ R_3-R_1 \end{matrix} \left[\begin{array}{ccc|ccc} 1 & \frac{3}{2} & \frac{1}{2} & \frac{1}{2} & 0 & 0 \\ 0 & \frac{11}{2} & -\frac{11}{2} & \frac{3}{2} & 1 & 0 \\ 0 & \frac{7}{2} & -\frac{7}{2} & -\frac{1}{2} & 0 & 1 \end{array}\right]$$

$$\tfrac{2}{11}R_2 \quad \left[\begin{array}{ccc|ccc} 1 & \frac{3}{2} & \frac{1}{2} & \frac{1}{2} & 0 & 0 \\ 0 & 1 & -1 & \frac{3}{11} & \frac{2}{11} & 0 \\ 0 & \frac{7}{2} & -\frac{7}{2} & -\frac{1}{2} & 0 & 1 \end{array}\right]$$

$$\begin{matrix} R_1-\frac{3}{2}R_2 \\ \\ R_3-\frac{7}{2}R_2 \end{matrix} \left[\begin{array}{ccc|ccc} 1 & 0 & 2 & \frac{1}{11} & -\frac{3}{11} & 0 \\ 0 & 1 & -1 & \frac{3}{11} & \frac{2}{11} & 0 \\ 0 & 0 & 0 & -\frac{16}{11} & -\frac{7}{11} & 1 \end{array}\right]$$

We note that Row 3 consists of all zeros to the left of the "equal-line." Considering the matrix to the left of the "equal-line" and the third column on the right side, we have

$$\left[\begin{array}{ccc|c} 1 & 0 & 0 & 0 \\ 0 & 1 & -1 & 0 \\ 0 & 0 & 0 & 1 \end{array}\right]$$

This is the augmented matrix that gives us the elements of the third column of the inverse. As we have seen in the previous section, however, when the augmented matrix has a row of zeros to the left of the "equal-line" and a number other than zero in that row to the right of the "equal-line," there is no solution to the system of equations. Thus, the matrix B has no inverse.

In summary, a 3×3 matrix A has an inverse if, and only if we can find solutions to the set of equations,

$$A\bar{x} = \begin{bmatrix} 1 \\ 0 \\ 0 \end{bmatrix} \qquad A\bar{y} = \begin{bmatrix} 0 \\ 1 \\ 0 \end{bmatrix} \qquad A\bar{z} = \begin{bmatrix} 0 \\ 0 \\ 1 \end{bmatrix}$$

For matrices of other dimensions, we have a similar set of equations to solve. There are as many equations as there are rows and columns in the matrix whose inverse we seek.

We can solve all of the systems of equations in the set at the same time if we use the special augmented matrix in which the identity matrix of the appropriate dimension is used in place of the usual column vector to the right of the "equal-line." In matrix symbology, the augmented matrix will have the form

$$[A \mid I]$$

After the row operations are performed on A to obtain the reduced matrix, we have the following:

$$[I \mid A^{-1}]$$

If it is not possible to obtain the identity matrix as the reduced matrix; that is, if we obtain one or more rows of zeros to the left of the "equal-line," the matrix has no inverse.

4.6.3 Use of Inverse Matrix in Solutions of Linear Systems

In a previous section we expressed a system of linear equations as a single matrix equation:

$$A\bar{x} = \bar{c}$$

A is a matrix whose elements are the coefficients of the variables in the system, $\bar{x}$ is a vector whose elements are the variables, and $\bar{c}$ is a vector whose elements are the quantities to which each of the sums of the unknowns in the system is equal. If A, $\bar{x}$, and $\bar{c}$ are assigned meaning as shown,

$$A = \begin{bmatrix} 1 & -1 & 2 \\ 2 & -1 & 2 \\ 1 & 2 & -1 \end{bmatrix} \qquad \bar{x} = \begin{bmatrix} x_1 \\ x_2 \\ x_3 \end{bmatrix} \qquad \bar{c} = \begin{bmatrix} 12 \\ 15 \\ 0 \end{bmatrix}$$

The matrix equation, $A\bar{x} = \bar{c}$ represents the system of linear equations

$$\begin{aligned} x_1 - x_2 + 2x_3 &= 12 \\ 2x_1 - x_2 + 2x_3 &= 15 \\ x_1 + 2x_2 - x_3 &= 0 \end{aligned}$$

If the matrix A has an inverse, we can multiply each of the vectors, $A\bar{x}$ and $\bar{c}$, by this matrix to form the equation,

$$A^{-1}A\bar{x} = A^{-1}\bar{c}$$

Since

$$A^{-1}A = I$$

we have

$$I\bar{x} = A^{-1}\bar{c}, \qquad \text{or} \qquad \bar{x} = A^{-1}\bar{c}$$

We can find A^{-1} by the process of Section 4.6.2.

$$\left[\begin{array}{rrr|rrr} 1 & -1 & 2 & 1 & 0 & 0 \\ 2 & -1 & 2 & 0 & 1 & 0 \\ 1 & 2 & -1 & 0 & 0 & 1 \end{array}\right]$$

$$\begin{array}{l} R_1 \\ R_2 - 2R_1 \\ R_3 - R_1 \end{array} \left[\begin{array}{rrr|rrr} 1 & -1 & 2 & 1 & 0 & 0 \\ 0 & 1 & -2 & -2 & 1 & 0 \\ 0 & 3 & -3 & -1 & 0 & 1 \end{array}\right]$$

$$\begin{array}{l} R_1 + R_2 \\ R_2 \\ R_3 - 3R_2 \end{array} \left[\begin{array}{rrr|rrr} 1 & 0 & 0 & -1 & 1 & 0 \\ 0 & 1 & -2 & -2 & 1 & 0 \\ 0 & 0 & 3 & 5 & -3 & 1 \end{array}\right]$$

$$\begin{array}{l} \\ \\ \frac{1}{3}R_3 \end{array} \left[\begin{array}{rrr|rrr} 1 & 0 & 0 & -1 & 1 & 0 \\ 0 & 1 & -2 & -2 & 1 & 0 \\ 0 & 0 & 1 & \frac{5}{3} & -1 & \frac{1}{3} \end{array}\right]$$

$$\begin{array}{l} R_1 \\ R_2 + 2R_3 \\ \end{array} \left[\begin{array}{rrr|rrr} 1 & 0 & 0 & -1 & 1 & 0 \\ 0 & 1 & 0 & \frac{4}{3} & -1 & \frac{2}{3} \\ 0 & 0 & 1 & \frac{5}{3} & -1 & \frac{1}{3} \end{array}\right]$$

Thus,

$$A^{-1} = \begin{bmatrix} -1 & 1 & 0 \\ \frac{4}{3} & -1 & \frac{2}{3} \\ \frac{5}{3} & -1 & \frac{1}{3} \end{bmatrix}$$

We now find the product $A^{-1}\bar{c}$.

$$\begin{bmatrix} -1 & 1 & 0 \\ \frac{4}{3} & -1 & \frac{2}{3} \\ \frac{5}{3} & -1 & \frac{1}{3} \end{bmatrix} \begin{bmatrix} 12 \\ 15 \\ 0 \end{bmatrix} = \begin{bmatrix} -12 + 15 \\ 16 - 15 \\ 20 - 15 \end{bmatrix} = \begin{bmatrix} 3 \\ 1 \\ 5 \end{bmatrix}$$

Thus, the solution set for the system is

$$x_1 = 3$$
$$x_2 = 1$$
$$x_3 = 5$$

Finding the inverse A^{-1} involves approximately the same amount of work as finding the solution set for the system using the method of Section 4.5. There is little to be gained by using the inverse matrix in the solution of a single system of equations. There are situations, however, in which the same coefficient matrix occurs in a number of linear-equation systems. In such a situation, finding the inverse of the coefficient matrix can simplify the task of finding the solutions to the systems of equations.

4.6.4 Matrix Equations

The situation arises in using matrices in which it is desired to solve an equation of the type

$$AX = B$$

where A, B, and X are all square matrices of the same dimension. A and B are known matrices and X is the matrix, which when multiplied by A, will give B as a product. The problem is to find the matrix X if it exists.

As we have seen earlier, if we multiply the matrix A by its inverse, we obtain the identity matrix. The solution of our problem can be obtained as follows:

If $\quad AX = B$

$A^{-1}AX = A^{-1}B$ ←(multiplying both sides of the equation by the same matrix, A^{-1})

or $\quad IX = A^{-1}B$

or $\quad X = A^{-1}B.$

The matrix X is found by multiplying the product matrix B by the inverse of A. If A has no inverse, we cannot find the solution by this method. (If A is singular, i.e., has no inverse, there may be either no solution or an unlimited number of solutions for X, analogous to the situations for systems of equations discussed in Section 4.5.1.)

Example 4.6.3

In the matrix equation

$$AX = B$$

let

$$A = \begin{bmatrix} 1 & -2 & 1 \\ -2 & 1 & 1 \\ 1 & 3 & -2 \end{bmatrix} \qquad B = \begin{bmatrix} -1 & 3 & -7 \\ -7 & 0 & 11 \\ 10 & -5 & 4 \end{bmatrix}$$

We wish to find the matrix X.

First, we find the inverse of A:

$$\left[\begin{array}{rrr|rrr} 1 & -2 & 1 & 1 & 0 & 0 \\ -2 & 1 & 1 & 0 & 1 & 0 \\ 1 & 3 & -2 & 0 & 0 & 1 \end{array}\right]$$

$$\begin{array}{l} \\ R_2 + 2R_1 \\ R_3 - R_1 \end{array} \left[\begin{array}{rrr|rrr} 1 & -2 & 1 & 1 & 0 & 0 \\ 0 & -3 & 3 & 2 & 1 & 0 \\ 0 & 5 & -3 & -1 & 0 & 1 \end{array}\right]$$

$$\begin{array}{l} \\ -\frac{1}{3}R_2 \\ \\ \end{array} \left[\begin{array}{rrr|rrr} 1 & -2 & 1 & 1 & 0 & 0 \\ 0 & 1 & -1 & -\frac{2}{3} & -\frac{1}{3} & 0 \\ 0 & 5 & -3 & -1 & 0 & 1 \end{array}\right]$$

$$\begin{array}{l} R_1 + 2R_2 \\ \\ R_3 - 5R_2 \end{array} \left[\begin{array}{rrr|rrr} 1 & 0 & -1 & -\frac{1}{3} & -\frac{2}{3} & 0 \\ 0 & 1 & -1 & -\frac{2}{3} & -\frac{1}{3} & 0 \\ 0 & 0 & 2 & \frac{7}{3} & \frac{5}{3} & 1 \end{array}\right]$$

$$\begin{array}{l} \\ \\ \frac{1}{2}R_3 \end{array} \left[\begin{array}{rrr|rrr} 1 & 0 & -1 & -\frac{1}{3} & -\frac{2}{3} & 0 \\ 0 & 1 & -1 & -\frac{2}{3} & -\frac{1}{3} & 0 \\ 0 & 0 & 1 & \frac{7}{6} & \frac{5}{6} & \frac{1}{2} \end{array}\right]$$

$$\begin{array}{l} R_1 + R_3 \\ R_2 + R_3 \\ \\ \end{array} \left[\begin{array}{rrr|rrr} 1 & 0 & 0 & \frac{5}{6} & \frac{1}{6} & \frac{1}{2} \\ 0 & 1 & 0 & \frac{1}{2} & \frac{1}{2} & \frac{1}{2} \\ 0 & 0 & 1 & \frac{7}{6} & \frac{5}{6} & \frac{1}{2} \end{array}\right]$$

Thus,

$$A^{-1} = \begin{bmatrix} \frac{5}{6} & \frac{1}{6} & \frac{1}{2} \\ \frac{1}{2} & \frac{1}{2} & \frac{1}{2} \\ \frac{7}{6} & \frac{5}{6} & \frac{1}{2} \end{bmatrix}$$

We find X by multiplying A^{-1} times B; that is, $X = A^{-1}B$.

$$X = \begin{bmatrix} \frac{5}{6} & \frac{1}{6} & \frac{1}{2} \\ \frac{1}{2} & \frac{1}{2} & \frac{1}{2} \\ \frac{7}{6} & \frac{5}{6} & \frac{1}{2} \end{bmatrix} \begin{bmatrix} -1 & 3 & -7 \\ -7 & 0 & 11 \\ 10 & -5 & 4 \end{bmatrix} = \begin{bmatrix} 3 & 0 & -2 \\ 1 & -1 & 4 \\ -2 & 1 & 3 \end{bmatrix}$$

The student should verify the multiplication of $A^{-1}B$, and then multiply A times X to verify that this product is equal to B.

Example 4.6.4

We have seen earlier that matrix multiplication is not commutative; that is, in general $AB \neq BA$. Thus, if the equation in Example 4.6.3 had been $YA = B$. We expect the matrix Y to be different from the X found in that example.

The equation, $YA = B$, can be solved using A^{-1}, but by performing the multiplication in reverse order:

$$\begin{aligned} YA &= B \\ YAA^{-1} &= BA^{-1} \\ YI = Y &= BA^{-1} \end{aligned}$$

Using the matrices from Example 4.6.3 we have,

$$Y = BA^{-1} = \begin{bmatrix} -1 & 3 & -7 \\ -7 & 0 & 11 \\ 10 & -5 & 4 \end{bmatrix} \begin{bmatrix} \frac{5}{6} & \frac{1}{6} & \frac{1}{2} \\ \frac{1}{2} & \frac{1}{2} & \frac{1}{2} \\ \frac{7}{6} & \frac{5}{6} & \frac{1}{2} \end{bmatrix}$$

$$= \begin{bmatrix} -\frac{15}{2} & -\frac{9}{2} & -\frac{5}{2} \\ 7 & 8 & 2 \\ \frac{21}{2} & \frac{5}{2} & \frac{9}{2} \end{bmatrix}$$

We see that Y is a matrix whose elements are completely different from those of X, the matrix we found in Example 4.6.3.

There are many uses for inverse matrices other than the two shown in this section. We shall employ one of them in a later section in our discussion of Markov chains.

EXERCISE 4.6

1. Perform the multiplication to verify that the matrix obtained in Example 4.6.3 is the inverse of the matrix A in that example.

2. Find the inverses, if they exist, for the following matrices:

a. $\begin{bmatrix} 1 & 3 \\ 2 & 4 \end{bmatrix}$

b. $\begin{bmatrix} 2 & 3 \\ -4 & -6 \end{bmatrix}$

c. $\begin{bmatrix} 1 & 2 \\ 2 & 1 \end{bmatrix}$

d. $\begin{bmatrix} -2 & 1 & 0 \\ 1 & -2 & 1 \\ 0 & 1 & -2 \end{bmatrix}$

e. $\begin{bmatrix} 0 & 1 & -2 \\ -1 & 1 & 3 \\ 2 & -3 & 0 \end{bmatrix}$ (See Hint)

f. $\begin{bmatrix} 1 & 2 & 3 \\ 3 & 4 & 5 \\ 1 & 1 & 1 \end{bmatrix}$

g. $\begin{bmatrix} 1 & 2 & -1 \\ 1 & 1 & -2 \\ 2 & 4 & 6 \end{bmatrix}$

h. $\begin{bmatrix} 1 & 2 & -1 & 3 \\ 2 & 3 & -4 & 6 \\ 2 & 5 & 0 & 7 \\ 1 & 2 & 0 & 2 \end{bmatrix}$

(**Hint:** Be sure to set up the augmented matrix before interchanging or otherwise combining the rows.)

3. Use the inverse matrices found in Problem 2 to solve the following systems of equations:

a. $x + 3y = 7$
$2x + 4y = 2$

b. $x + 2y = 6$
$2x + y = 6$

c. $$\begin{aligned} -2x + y \quad &= 4 \\ x - 2y + z &= 3 \\ y - 2z &= 7 \end{aligned}$$

d. $$\begin{aligned} y - 2z &= -7 \\ -x + y + 3z &= 2 \\ 2x - 3y \quad &= -2 \end{aligned}$$

e. $$\begin{aligned} x + 2y - z &= 4 \\ x + y - 2z &= 3 \\ 2x + 4y + 6z &= -10 \end{aligned}$$

f. $$\begin{aligned} x_1 + 2x_2 - x_3 + 3x_4 &= 6 \\ 2x_1 + 3x_2 - 4x_3 + 6x_4 &= -1 \\ 2x_1 + 5x_2 \quad + 7x_4 &= 4 \\ x_1 + 2x_2 \quad + 2x_4 &= 7 \end{aligned}$$

4. Solve the following matrix equations:

a. $$\begin{bmatrix} 1 & 3 \\ 2 & 1 \end{bmatrix} X = \begin{bmatrix} 2 & 1 \\ -1 & 4 \end{bmatrix}$$

b. $$X \begin{bmatrix} 1 & 3 \\ 2 & 1 \end{bmatrix} = \begin{bmatrix} 2 & 1 \\ -1 & 4 \end{bmatrix}$$

c. $$\begin{bmatrix} 7 & -11 & 16 \\ -3 & 5 & -7 \\ 1 & -2 & 3 \end{bmatrix} X = \begin{bmatrix} 4 & 1 & -3 \\ 2 & -2 & 5 \\ -1 & 3 & 7 \end{bmatrix}$$

d. $$\begin{bmatrix} 1 & 2 & 3 \\ 2 & 3 & 4 \\ 2 & -1 & 0 \end{bmatrix} X = \begin{bmatrix} 3 & -1 & -3 \\ 2 & 1 & -7 \\ -5 & 3 & 2 \end{bmatrix}$$

e. $$X \begin{bmatrix} 1 & 2 & 3 \\ 2 & 3 & 4 \\ 2 & -1 & 0 \end{bmatrix} = \begin{bmatrix} 3 & -1 & -3 \\ 2 & 1 & -7 \\ -5 & 3 & 2 \end{bmatrix}$$

5. Given that $AB = AC$, where A, B, and C are square matrices.
   a. Under what circumstances can we be sure that $B = C$? Compare this to the situation with real numbers in which $ab = ac$, but b $\neq c$.
   b. Consider the case in which

$$A = \begin{bmatrix} 1 & 2 & 1 \\ -2 & 2 & 1 \\ -1 & 4 & 2 \end{bmatrix} \qquad B = \begin{bmatrix} 3 & -2 & -1 \\ 1 & 2 & 2 \\ 2 & -1 & 1 \end{bmatrix} \qquad C = \begin{bmatrix} 3 & -2 & -1 \\ -1 & 1 & 1 \\ 6 & 1 & 3 \end{bmatrix}$$

   How does your answer to part *a* explain the products AB and AC?

5

linear inequalities and linear programming

5.1 LINEAR INEQUALITIES

In the previous two chapters we dealt with methods for determining specific values for quantities under specified conditions. The algebraic propositions used were **equations**. There are many situations, however, in which our interest is not so much with the particular value of a quantity but, instead, with restricting it within certain limits. "The cost of the materials cannot exceed 10 dollars, or we will be forced to raise the price." "The applicant must score above 70% on the aptitude test to qualify for an interview." In these statements the expressions used to denote quantity are "cannot exceed" and "score above." These are two of a great many such expressions in our language. They all translate into one of four algebraic relations: "greater than," "less than," "greater than or equal to," and "less than or equal to." They are called, for obvious reasons, **algebraic inequalities**.

5.1.1 Terminology

The relationships, "greater than" and "less than," are called **strong inequalities**, **strict inequalities**, or **absolute inequalities**. They specify that the two quantities in the relationship must be different. This contrasts with the **weak inequalities**, "greater than or equal to" and "less than or equal to," in which equality between the two quantities is possible. The weak inequalities are sometimes translated as "not less than" and "not greater than," respectively.

The strong inequalities are denoted by the symbols, "$>$" and "$<$." "$a > b$" states that "a is greater than b." "$a < b$" states, "a is less than b." The two relationships are not completely separate concepts. "$a > b$" can be reversed to become "$b < a$." (**Remark**: When the student reads a sentence such as the preceding one, he should read it as, "a is greater than b can be reversed to become b is less than a." The student should regard the algebraic symbols merely as **words** in his vocabulary.

Note that $a < b$ is not the negation for $a > b$. The possibility that $a = b$ must be included. The expression, "a is not greater than b," has the same meaning as, "a is less than **or equal** to b." The symbolic form of this expression is $a \leq b$. The addition of the bar below the inequality symbol denotes the possibility of equality between the two numbers. Similarly, "a is not less than b" is symbolized as $a \geq b$ —"a is greater than or equal to b." Each of these weak inequalities is the disjunction of two statements. $a \leq b$ is the disjunction of $a < b$ with $a = b$. $a \geq b$ is the disjunction of $a > b$ with $a = b$.

As with strong inequalities, the weak inequalities can be reversed. $a \leq b$ is the same as $b \geq a$.

The language used to express the strong inequalities is more or less standardized. $a > b$ is ordinarily expressed either as "a is greater than b" or "a is larger than b." $a < b$ is expressed as "a is smaller than b" or "a is less than b."

Expressions for the weak inequalities are more varied. Some of these are listed below:

$a \leq b$	a is less than or equal to b
	a is not greater than b
	a is at most b
	a is not more than b
$a \geq b$	a is greater than or equal to b
	a is not less than b
	a is at least as great as b
	a is at least equal to b

The expressions "at least" and "at most" are used frequently. The statement, "The machine weighs at least two tons," is translated into algebraic symbols as $W \geq 2$, where W represents the weight of the machine in tons.

5.1.2 The Algebra of Inequalities

A complete treatment of the algebra of inequalities would take far more space than we have available here. We shall discuss only those few properties of inequalities that we shall need in subsequent sections.

Our discussion thus far has assumed that the student has an intuitive understanding of the basic concepts, "is less than," and "is greater than." Inequalities are usually defined in algebra in terms of equality, which is one of the fundamental undefined terms of mathematics.

Definition 5.1.1

"$<$": For any two numbers, a and b, $a < b$ if, and only if, $a + k = b$, where k is a **positive number.**

">": For any two numbers, a and b, $a > b$ if, and only if, $a = b + k$, where k is a **positive number.**

An alternate definition for inequalities is based on the location of the numbers on a number line.

Definition (Alternate) 5.1.1A

"<" **and** ">": For any two numbers, a and b, $a < b$ if, and only if, the point representing a on a standard number line is to the **left** of the point representing b.

$a > b$ if, and only if, the point representing a on a standard number line is to the **right** of the point representing b.

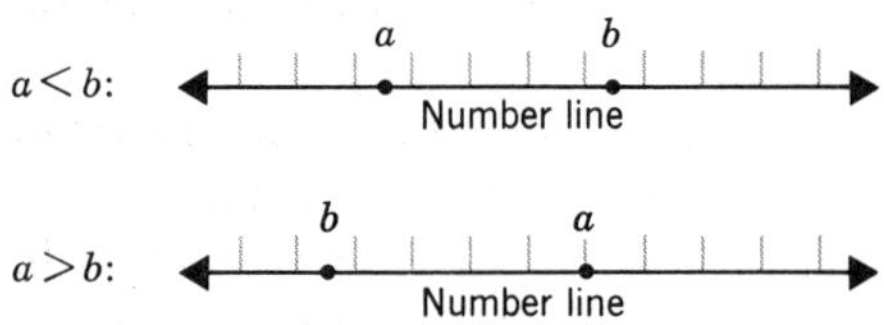

(The location of the numbers with respect to zero on the line is of no consequence.)

For the applications of inequalities we are considering in this text, this alternate definition and its implications is probably more useful.

In earlier sections we have "solved" algebraic propositions by performing certain operations on them to obtain equivalent propositions whose solution sets are more apparent. We can use the same general procedure and the same operations on inequalities as on equations, **with one important exception**. The algebraic tautologies defining these operations are listed here for convenient reference. The exceptional one is marked with a #.

Algebraic Tautologies for Inequalities

These tautologies are written for strong inequalities. They apply also to weak inequalities.

1. The same, or equal, quantities can be added (or subtracted) to both "sides" of an inequality, and the resulting inequality is also valid.

In symbols, if $a < b$, and $c = d$, then $(a + c) < (b + d)$.

If $a > b$ and $c = d$, then $(a + c) > (b + d)$.

(Subtracting a number is the same as adding that number with the algebraic sign changed.)

2. Both numbers in an inequality can be multiplied (or divided) by the same **positive** number and the resulting inequality is also valid.

In symbols, if $a > b$ and $k > 0$, then $ka > kb$.

If $a < b$ and $k > 0$, then $ka < kb$.

(Dividing by any number is the same as multiplying by its reciprocal.)

#3. If both numbers in an inequality are multiplied (or divided) by the same **negative** number, the resulting inequality is valid provided the **sense of the inequality is reversed**.

In symbols, if $a > b$ and $k < 0$, then $ka < kb$.

If $a < b$ and $k < 0$, then $ka > kb$.

The third of these tautologies should be noted carefully. It applies only to **multiplication** (or division) by negative numbers. Two or three examples should serve to assure the student of its validity. Consider the numbers 2 and 5. $2 < 5$. If we multiply each by -1, the resulting numbers are -2 and -5. Note that $-2 > -5$. The inequality symbol has been reversed. For the numbers, -1 and 2, $-1 < 2$. If we multiply each by -2, the results are 2 and -4; but $2 > -4$. The inequality symbol has been reversed. Consider $-3 > -6$. Multiplying both "sides" of the inequality by -3 yields $9 < 18$. The inequality symbol has been reversed.

5.1.3 Linear Inequalities

The applications in this text involve only linear inequalities. These are similar in form to the linear equations discussed in Chapters 3 and 4, but the equal sign is replaced with one of the inequality symbols. The usual form of the inequalities that we shall use is illustrated by the expression $3x + 2y \leq 10$. This is analogous to the form for linear functions discussed in Section 3.3.1. Notice that the expression $3x + 2y \leq 10$ describes a relation that is **not** a function. For any value assigned to x, there are countless values for y that satisfy the expression. Thus, the set of number pairs belonging to the relation will have many in which the same value for x is paired with different values for y.

In our applications we shall study situations in which linear inequalities can serve as mathematical models. Usually a set, or system, of linear inequalities is needed. We shall develop methods for finding the solution sets for these systems. Let us consider an example.

Example 5.1.1

A manufacturer of pet food uses two principal ingredients as the source of nutrients in his product: meat by-products and a protein-enriched cereal. The remainder of the food is water, a small amount of flavoring, and an inexpensive filler.

To meet nutritional standards, each pound of pet food must contain **at least** 18 units of protein, 20 units of carbohydrates, and 3 units of fats.

Each ounce of the meat by-product contains 3 units of protein, 4 units of carbohydrates, and 3 units of fats.

Each ounce of the cereal contains 6 units of protein, 5 units of carbohydrates, and 1/3 of a unit of fats.

These requirements constitute a system of inequalities involving the amounts of the two principal ingredients of the pet food. First, we shall see how to translate the verbal statements into their algebraic equivalents. In succeeding sections we shall see how to determine the amounts of these ingredients that the manufacturer can use to meet the requirements.

The amounts of the principal ingredients are the variables. Let x represent the number of ounces of meat by-product, and y, the number of ounces of cereal, in each pound of pet food.

The assignment of these meanings to the symbols provides three algebraic statements:

$$\begin{aligned} &x \geq 0 \\ &y \geq 0 \\ &x + y \leq 16 \end{aligned}$$

The first two expressions come from the physical fact that the manufacturer cannot use a negative amount of cither food. Although this is an obvious fact, it must be included in the algebraic statement of the problem, because it is an important limitation on the values of the numbers that we shall use. The third expression states that a pound of pet food cannot contain more than 16 ounces of ingredients.

We now must write the requirements for the different nutrients in algebraic terms. Each requirement becomes an algebraic inequality. Let us consider the protein requirement. The nutritional specification states that each pound of pet food must contain at least 18 units of protein. Each ounce of meat by-product provides 3 units of protein, and each ounce of cereal provides 6 units. Thus, the x ounces of meat by-product will provide $3x$ units, and the y ounces of cereal will provide $6y$ units. Together they supply $3x + 6y$ units. In accordance with the specification, this total must be at least 18. That is,

$$3x + 6y \geq 18$$

Similarly, the requirements for carbohydrates and fats become, respectively,

$$\begin{aligned} 4x + 5y &\geq 20 \\ 3x + \tfrac{1}{3}y &\geq \ \ 3 \end{aligned}$$

All of the requirements are expressed in the following system of inequalities:

$$\begin{aligned} &\left.\begin{aligned} x &\geq 0 \\ y &\geq 0 \\ x + \ y &\leq 16 \end{aligned}\right\} \text{Physical Requirements} \\ &3x + 6y \geq 18 \quad \text{Protein} \\ &4x + 5y \geq 20 \quad \text{Carbohydrates} \\ &3x + \tfrac{1}{3}y \geq 3 \quad \ \text{Fats} \end{aligned}$$

As we shall see, there are a great many combinations of amounts of the two foods that meet all of the requirements. This is a characteristic of this type of problem. Usually there is an additional consideration, such as minimizing cost or maximizing profit, which serves as a criterion for choosing which of the "possible solutions" to use. In a succeeding section we shall see how to apply these criteria.

In summary, a situation in which there are limiting specifications, or conditions, is translated into algebraic terms as follows:

1. The quantities whose values are to be determined are identified, and variables assigned to represent them.
2. Any physical limitations on these variables are listed. (Amounts of materials cannot be negative, etc.)
3. An algebraic inequality is formulated for **each** of the conditions to be met by the quantities represented by the variables.

It is important to keep clearly in mind precisely what each variable represents. When the expressions involving the variables are formed, the "units" of these expressions should be noted so that all the expressions added together, or compared, have the same units. For instance, in the foregoing example we found the **amount** of **protein** supplied by one ingredient, added it to the **amount** of **protein** supplied by the other ingredient, and formed an expression comparing this total with the **specified amount** of **protein**. Although students are undoubtedly familiar with the old cliche, "You can't add apples to oranges," one of the most common errors made in formulating algebraic equations and inequalities is that the quantities involved are not all quantities of the same thing. Each algebraic expression formed should be translated back into English to determine that it makes sense. Keep in mind that each inequality or equation is a **sentence** stating **one condition**, and only one, in the language of algebra.

EXERCISE 5.1

1. Given the sets of numbers, $A = \{1,2,3,4,5\}$, $B = \{5,6,7,8,9,10\}$, and $C = \{10,11,12,13,14\}$. Let $a \in A$, $b \in B$, and $c \in C$.

   Supply the **most informative inequality symbol** in each of the following:

   $a \quad b \qquad a \quad c \qquad c \quad b \qquad b \quad c \qquad b \quad a \qquad c \quad a$

2. For each of the following, draw a number line and position the numbers to illustrate the relationship. Assume that a, b, c, and d represent specific numbers.

   a. $a < b$
   b. $c > d$
   c. $a < b$ and $b < c$ (usually condensed to $a < b < c$)

d. $a < b$ and $a < c$ (Is there more than one relative positioning possible?)

e. $a > b$, $a > c$, $b < d$, and $c > d$. (Is more than one relative positioning possible?)

3. Indicate on a number line all of the possibilities (show a line segment, if needed) for the location of x in the following:

a. $x \geq 3$
b. $x < -2$
c. $2 \leq x \leq 4$
d. $-1 < x < 5$

4. Use a number line, and make a graph showing the possible values for x in each of the following:

a. $x \leq 5$
b. $x > -3$
c. $-3 \leq x \leq 0$
d. $2 \leq x \leq 6$

5. Use the algebraic tautologies of Section 5.1.2 to find the solution sets for the following:

a. $x + 3 \geq 2$
b. $2x - 4 \leq x + 1$
c. $3x + 7 \geq x - 1$
d. $x + 5 \leq 3x - 7$
e. $15 - 5x \geq 2x + 1$

6. Find solution sets for the following:

a. $x - 3 \geq -2$
b. $3x - 5 \leq x + 3$
c. $4 - 3x \geq 10 - 5x$
d. $3 - 2x \leq 3x + 8$
e. $2x + 7 \geq 5x - 2$

For the following problems, set up the system of inequalities that describes the conditions to be met.

7. A sand and gravel company has a machine to separate sand from rock and then crush the rock for concrete. The quarry has two areas to supply the sand-rock mixture. In one area (Area A) each cubic yard of the quarried mixture contains $1\frac{1}{2}$ tons of sand and $\frac{1}{2}$ ton of rock. The mixture from the second area (Area B) contains 1 ton of sand and 1 ton of rock per cubic yard. The machine requires an input of at least 60 tons of sand and 40 tons of rock per hour to keep operating. What are the possible amounts of quarried mixture from the two areas that will keep the machine supplied?

8. A woman has been making handcrafted belts as a hobby. A specialty dress shop wants her to make them for sale. She decides to make two styles of belts, one with silver ornaments and one with glass beads. The materials for the silver belt cost \$3 each, and the materials for the beaded belts cost \$1.50 each. The woman has \$60 per week that she can spend on materials. The silver belts take 1 hour to make, and the beaded belts take $1\frac{1}{2}$ hours. The woman can devote up to 36 hours each week to making the belts. What are the quantities of each kind of belt that she can promise to deliver?

9. A manufacturer produces two styles of a patented clothes rack. There are three operations involved in the manufacture: cutting the materials, assembling the racks, and applying the finish. For the standard model it takes $\frac{3}{4}$ hour for cutting, 1 hour for assembly, and 1 hour for finishing. The deluxe model requires 1 hour for cutting, 1 hour for assembly, and 2 hours for finishing. In his shop each day he has 30 man-hours available for cutting, 36 man-hours for assembly, and 50 man-hours for finishing. How many of each type of rack can he produce each day?

10. A chemical company produces a fertilizer that it guarantees to contain at least 20% nitrogen, 12% phosphorus, and 8% potassium. It makes the fertilizers by blending mixtures of chemicals with the following compositions:

	Mixture *A*	Mixture *B*
Nitrogen	20%	40%
Phosphorus	20%	12%
Potassium	10%	10%
Inerts	50%	38%

What are the possible amounts of each mixture that the company can use to provide fertilizer with at least the guaranteed amounts of each component? (**Suggestions:** Compute the amount of each component in one pound of mixture. Set up expressions for the number of pounds of each mixture that will produce the amount of each component in 100 pounds of fertilizer.)

5.2 GRAPHING INEQUALITIES

In systems of inequalities involving only two variables, a great deal of information can be obtained from a graph of the system. In Sections 3.2 and 3.3 we studied the Cartesian coordinate system and the graphs of linear functions. The graph of an inequality uses the same coordinate system and graphing procedure used with linear equations, but there are a few important differences.

To start, the inequality symbol in the algebraic expression to be graphed is replaced by an equal sign, converting the inequality into an equation. The graph for this equation is drawn in the standard manner. For a linear inequality, this graph is a straight line, which divides the plane of the graph into two "half-planes."

Recall that the purpose of a graph is to exhibit the number pairs which belong to a relation—that is, which "satisfy" the algebraic expression defining the relation. For an equation, all of the number pairs belonging to the relation lie along the curve in the graph. The coordinates of each point on the curve, when substituted for the variables in the equation, produce a true statement. For an inequality, the points satisfying the inequality lie in a region, or area, of the graph instead of on a line or curve.

Example 5.2.1

Consider the inequality $y \leq x + 3$. To graph this inequality we first draw the graph of the equation, $y = x + 3$. This is a straight line,

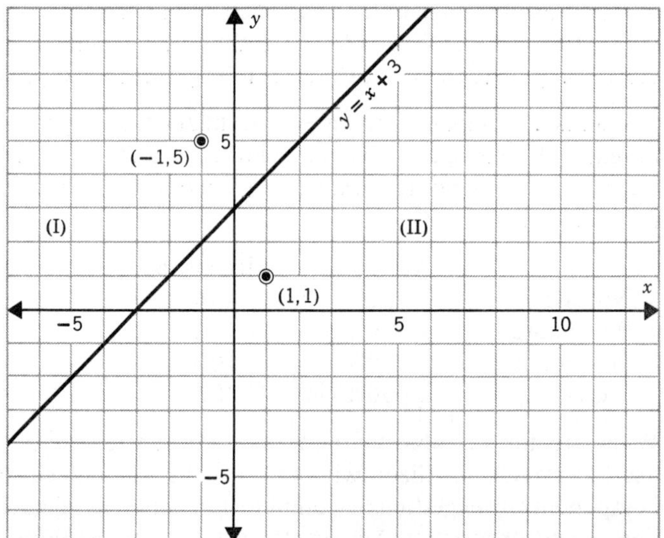

Figure 5.2.1

as shown in Figure 5.2.1. It divides the plane of the graph into two regions, labeled in the figures as (I) and (II). We choose a point anywhere in either of the regions, **but not on the line**. Suppose we have chosen (1, 1). The coordinates of this point are substituted into the inequality, $y \leq x + 3$. We have

$$1 \leq 1 + 3 \quad or \quad 1 \leq 4$$

which is a true statement. Thus, the point (1, 1) belongs to the relation, $y \leq x + 3$. **All of the other points on the same side of the line—in Region (II)—also belong to this relation.**

Suppose we had chosen the point (−1,5). When we substitute the coordinates into the inequality, we have

$$5 \leq -1 + 3 \quad or \quad 5 \leq 2$$

This is not a true statement. The point (−1,5) does not belong to the relation $y \leq x + 3$. None of the other points on the same side of the line—in Region (I)—belong to the relation either.

The graph of the relation includes all of the points in the half-plane shown shaded in Figure 5.2.2. Since the relation $y \leq x + 3$ includes the equation, $y = x + 3$, the points on the dividing line are also included.

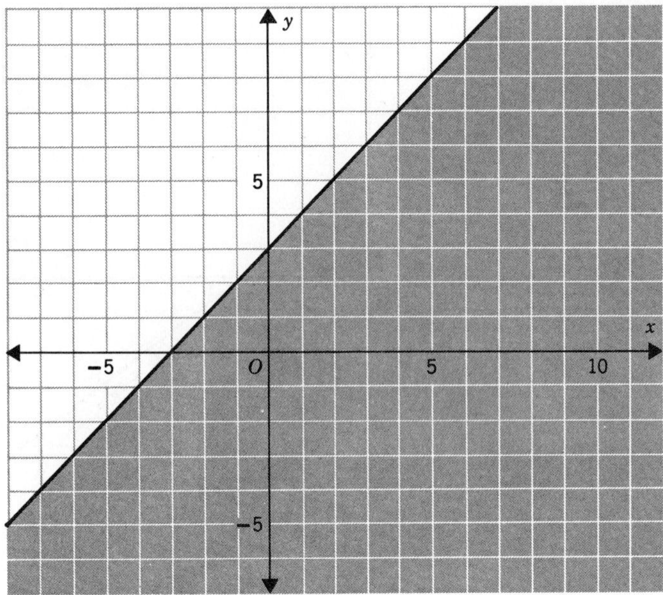

Figure 5.2.2

The procedure for graphing an inequality can be summarized as follows:

1. Replace the inequality symbol with an equal sign.
2. Graph the resulting equation.
3. Choose any convenient point **not on the graph of the equation**. (The point (0,0) is the easiest to work with if the graph of the equation does not pass through it.) Substitute the coordinates of the point chosen into the inequality.
4. If the resulting statement is true, all of the points in the region on the same side of the curve as the chosen point satisfy the inequality. If the resulting statement is false, the points in the region on the **other side** of the curve from the chosen point satisfy the inequality.

This procedure takes care of all of the points in the plane **except those on the curve drawn.** These belong to the relation defined by the equation formed when the inequality symbol is replaced by an equal sign. If the inequality is a **weak inequality**—if the inequality symbol is $\leq$ or $\geq$ – the points on the curve also belong to the relation. In this case the curve is drawn with a **solid line**. If the relation is defined by a **strict inequality**—the symbol is $<$ or $>$ – the points on the curve do not belong to the relation. In this case the curve is drawn with a **dotted**, or **broken line**.

Example 5.2.2

Consider the relation defined by the inequality, $3x + 4y > 12$. A graph for the equation, $3x + 4y = 12$, is drawn, as shown in Figure 5.2.3. Since we are considering a strict inequality, the line is drawn broken. We choose (0,0) as a test point.

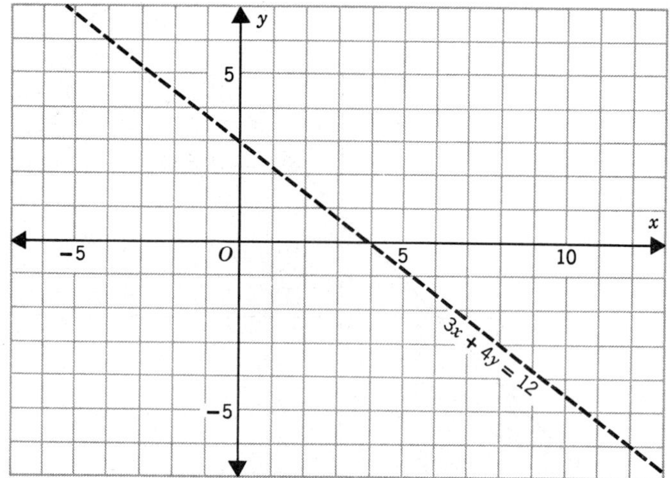

Figure 5.2.3

We have

$$3(0) + 4(0) > 12, \qquad \text{or} \qquad 0 > 12$$

This is not a true statement, and the points satisfying the inequality lie on the **other side** of the line from (0,0). The region including these points is shown shaded in Figure 5.2.4.

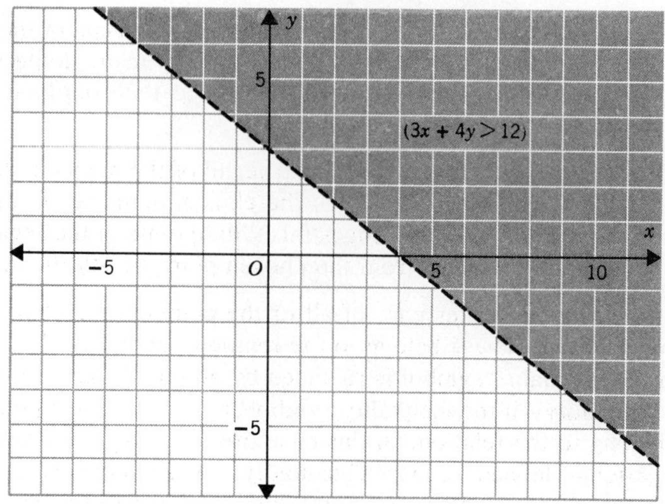

Figure 5.2.4

Let us now examine more carefully the reason that all of the points on one side of a line, and only those points, satisfy an inequality. Let us consider the linear inequality, $y \leq \frac{1}{2}x - 2$. (Recall that any linear inequality can be expressed in a similar form.) For any given value of x, say 6, the inequality becomes $y \leq \frac{1}{2}(6) - 2$, or $y \leq 1$. Thus, for $x = 6$ the points satisfying this inequality are those whose y-coordinate is less than or equal to one.

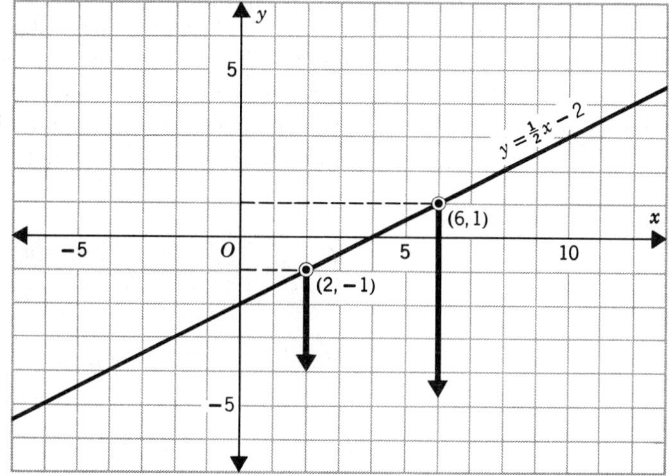

Figure 5.2.5

These points lie on the vertical line, $x = 6$, below $y = 1$, as shown in Figure 5.2.5. Note that the point (6, 1) is on the line, $y = \frac{1}{2}x - 2$. Any point with an x-coordinate of 6, whose y-coordinate is greater than one, is above the line; and none of these points satisfy the inequality. When $x = 2$, the inequality becomes $y \leq \frac{1}{2}(2) - 2$, or $y \leq -1$. The points with x-coordinates of 2 that satisfy the inequality are those with y-coordinate less than or equal to -1. The point $(2, -1)$ is on the line $y = \frac{1}{2}x - 2$, and the others are on the line $x = 2$ starting at $(2, -1)$ and extending downward. The points above the line have y-coordinates greater than -1 when $x = 2$, and they do not satisfy the inequality. Similar reasoning for the other values for x lead to the conclusion that the points satisfying the inequality all lie on one side of the line representing the equation formed from the inequality.

5.2.1 Graphs for Systems of Inequalities

In a system of inequalities the interest is in the points that satisfy **all** of the inequalities. If we consider each inequality as a relation composed of a **set** of points, the points of interest lie in the **intersection** of all the sets in the system. A graph can be regarded as a sort of quantified Venn diagram. The intersection of all the sets is the area in which all of the sets overlap. The points in this area comprise the **solution set** for the system.

Example 5.2.3

Consider the system

$$y \leq 2x + 7$$
$$2x + 3y \geq 6$$

Let us first plot the inequality, $y \leq 2x + 7$. We obtain the graph shown in Figure 5.2.6. The shaded portion represents the points in the relation. Let us now add the relation $2x + 3y \geq 6$ to the graph. The result is shown in Figure 5.2.7. The horizontal shading represents the points in the relation $2x + 3y \geq 6$; and the vertical shading, those in $y \leq 2x + 7$. The crosshatched area includes the points of the solution set. Note that the point $(-2, 6)$ when substituted into $y \leq 2x + 7$ produces the statement, $6 \leq -4 + 7$, which is not true.

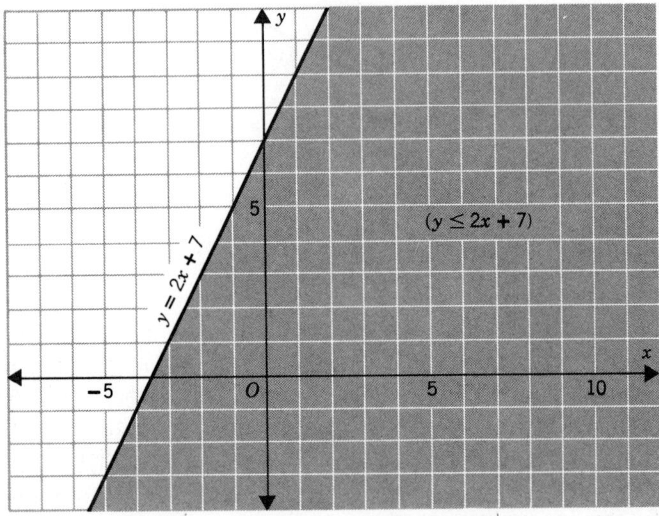

Figure 5.2.6

This point does not satisfy this condition, although when substituted into the inequality, $2x + 3y \geq 6$, it produces the true statement, $-4 + 18 \geq 6$. Similarly, the point $(0, 1)$ satisfies the condition, $y \leq 2x + 7$; but it does not satisfy $2x + 3y \geq 6$. The point $(1, 5)$ satisfies both conditions.

We observe for the system of inequalities in Example 5.2.3 that when the x-coordinate of the point $(1, 5)$, a point that satisfies both conditions, is placed in the inequality $y \leq 2x + 7$, we have $y \leq 9$. When 1 is substituted for x in $2x + 3y \geq 6$, we have, after rearranging, $y \geq \frac{4}{3}$. Thus, that portion of the line lying between the lines $y = 2x + 7$ and $2x + 3y = 6$ (the thick line in Figure 5.2.7) contains all of the points satisfying both conditions for the situation in which $x = 1$. The complete set of points satisfying both conditions is made up of similar intervals for other x-values. The two boundary lines intersect (have

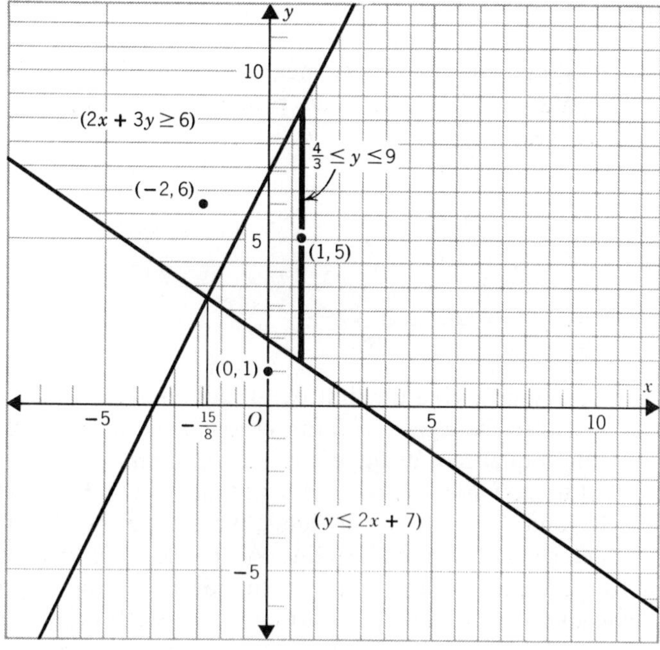

Figure 5.2.7

the same y-value) when $x = -\frac{15}{8}$. For x-values less than $-\frac{15}{8}$, there are no points that satisfy both inequalities.

Let us consider now the effect of adding a third condition to those specified in Example 5.2.3. Let us add the condition, $2x + y \leq 6$. Figure 5.2.8 shows this condition added to the graph of Figure 5.2.7. Note that the solution set consists of the points within and on a closed figure.

If the added restriction is the same inequality with the symbol reversed, $2x + y \geq 6$, the effect is to remove the triangle of points from the solution set. The solution set is now the open region to the right of the dividing lines, as shown in Figure 5.2.9.

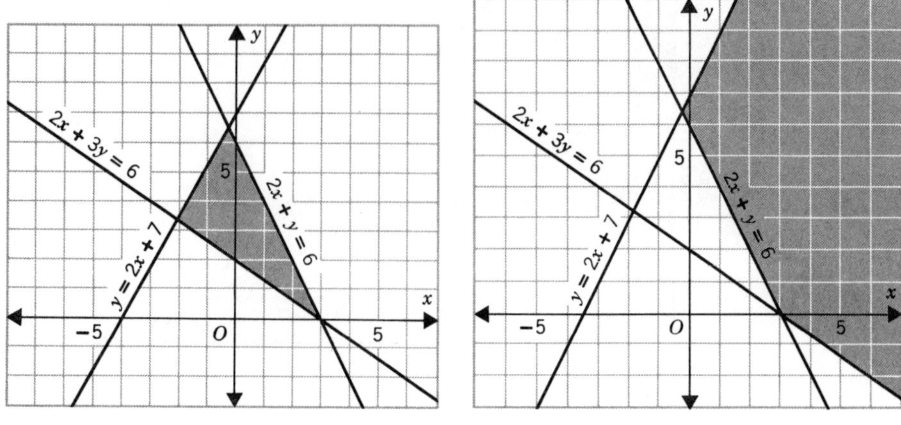

Figure 5.2.8

Figure 5.2.9

Example 5.2.4

Let us graph the system of inequalities obtained as the algebraic representation of the pet-food problem of Example 5.1.1. Recall that we had the system

$$\begin{aligned} x &\geq 0 \\ y &\geq 0 \\ x + y &\leq 16 \\ 3x + 6y &\geq 18 \\ 4x + 5y &\geq 20 \\ 3x + \tfrac{1}{3}y &\geq 3 \end{aligned}$$

The points satisfying the first two requirements lie above or on the x-axis and to the right or on the y-axis. That is, they are the points in the first quadrant of the coordinate system.

Let us now add the requirement that $x + y \leq 16$. If we let $x = 0$, $y = 16$; and if $x = 16$, $y = 0$. Thus, the two points $(0, 16)$ and $(16, 0)$ will determine the boundary line for this set. Let us choose a point not on the line to determine which side of the line represents this inequality. The easiest point to consider is the origin $(0, 0)$, which satisfies the inequality. Thus, the first three specifications determine the triangular area shown in Figure 5.2.10. Any point in or on that triangle satisfies all three conditions.

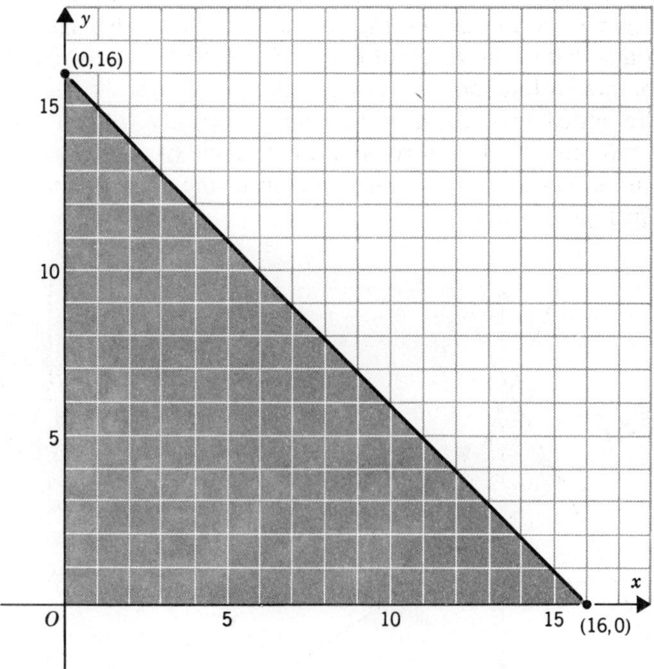

Figure 5.2.10

The set of points specified for the next condition, $3x + 6y \geq 18$ is shown in Figure 5.2.11. The line $3x + 6y = 18$ is the border for the set. When $x = 0$, $y = 3$; and when $y = 0$, $x = 6$. Thus, the points $(0, 3)$ and $(6, 0)$ determine the boundary line.

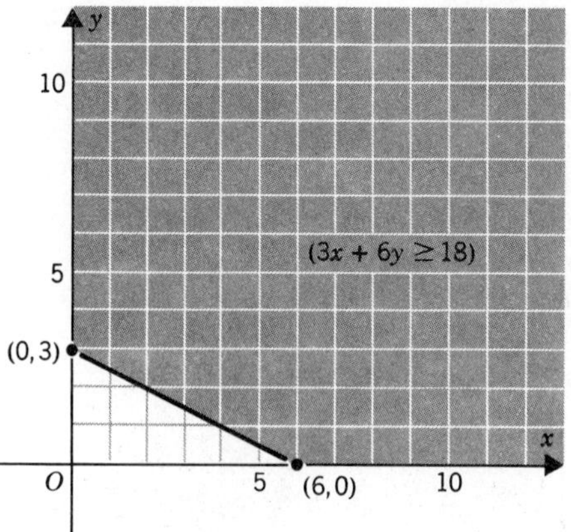

Figure 5.2.11

In this case, if we use the point $(0,0)$ as a test point, we find that it does not satisfy the inequality $(0 + 0 \not\geq 18)$. Thus, the set specified is on the side of the line away from $(0,0)$.

In similar fashion we can draw the graphs for the conditions, $4x + 5y \geq 20$ and $3x + \frac{1}{3}y \geq 3$. These are shown together in Figure 5.2.12.

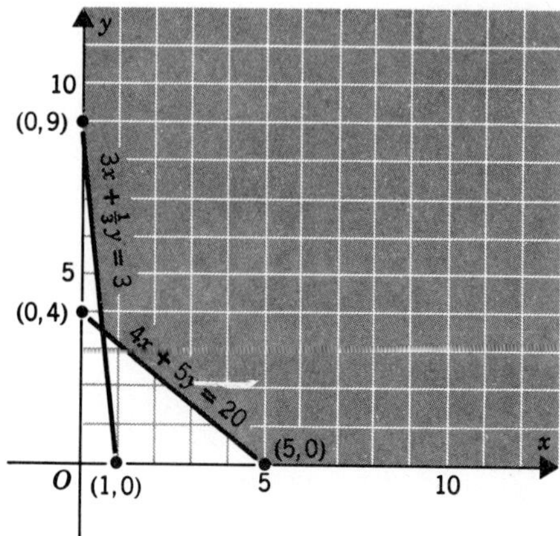

Figure 5.2.12

Figure 5.2.13 shows all six conditions together on the same graph. The shaded area represents the set of points satisfying all of the conditions.

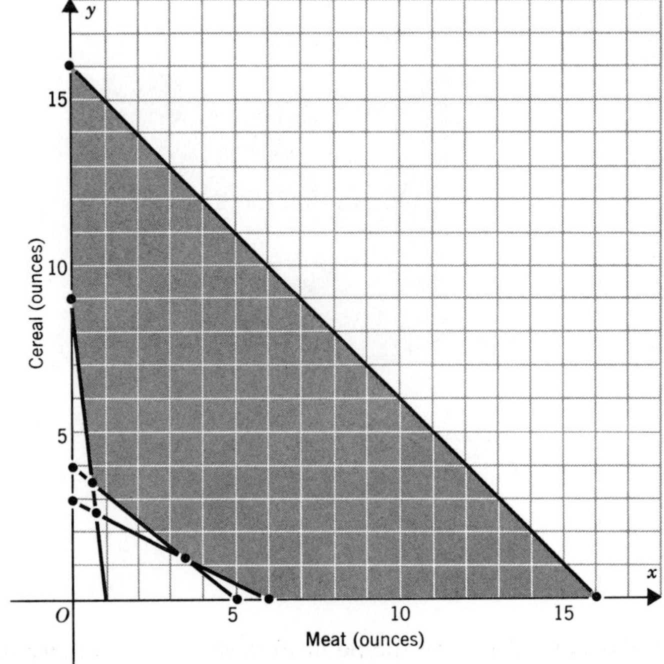

Figure 5.2.13

Any point in or on the polygon shown represents values for the quantities x and y that satisfy all of the conditions specified. In the next section we shall see how to select the particular point or points that represent the optimum selection with respect to some particular criterion.

EXERCISE 5.2

Graph the following inequalities:

1. $y \geq \frac{3}{2}x - 4$

2. $4x + 5y \leq 10$

3. $2x - 5y < 20$

4. $y \geq x^2 - 3x + 2$

Graph the solution sets for the following systems:

5. $$\begin{aligned} x + 2y &\le 4 \\ x - y &\le 1 \end{aligned}$$

6. $$\begin{aligned} x + 2y &\le 4 \\ x + y &\ge 1 \end{aligned}$$

7. $$\begin{aligned} x + 2y &\ge 4 \\ x - y &\ge 4 \\ x &\le 5 \end{aligned}$$

8. $$\begin{aligned} x + 3y &\ge -3 \\ 2x - y &\ge -13 \\ 3x + 2y &\le 12 \end{aligned}$$

9. $$\begin{aligned} x + y &\ge 2 \\ 4x + 3y &\le 12 \\ x \ge 1, &\quad y \ge 1 \end{aligned}$$

10. $$\begin{aligned} x \ge 0, &\quad y \ge 0 \\ x + y &\ge 1 \\ 3x + 2y &\ge 6 \end{aligned}$$
Interpret the results.

11. $$\begin{aligned} x \ge 0, &\quad y \ge 0 \\ y &\ge 2x + 1 \\ 2y &\le x - 1 \end{aligned}$$
Interpret the results.

12. Graph the system of inequalities that represents the situation described in Problem 7 of Exercise 5.1.

13. Graph the system of inequalities that represents the situation described in Problem 8 of Exercise 5.1.

14. Graph the system of inequalities that represents the situation described in Problem 9 of Exercise 5.1.

15. Graph the system of inequalities that represents the situation described in Problem 10 of Exercise 5.1.

5.3
LINEAR PROGRAMMING—PART I

The situations whose mathematical models are a set of linear inequalities usually involve a consideration for which the "best," or optimal solution is sought. In business situations it is desired to obtain solutions that produce, for example, the highest profit, the lowest cost, or the fastest completion of a task. As we have seen in the previous sections, when a system of linear inequalities has any solutions at all, it usually has a great many. We need now some means by which we can decide which of these solutions is the best for our purposes.

For example, the pet-food manufacturer of Example 5.2.4 can use any mixture from all cereal to all meat by-product and still meet the nutritional requirements. There is no indication in the solution set of the inequalities (the nutritional requirements) which mixture is "best" for him. A new consideration must be introduced to help make this decision. In this example it is likely that the manufacturer prefers to use the mixture that costs the least. That is, he wishes to **minimize his cost**. The cost of the ingredients becomes the key quantity in choosing a solution.

The quantity to be optimized is one whose value depends on the same variables used in the set of inequalities. It is related to these variables by a function, called the **objective function**. Finding the optimum solution consists of determining which one of the solutions for the set of inequalities produces the optimum value when substituted for the variables in the objective function.

In the pet-food problem, for example, if the meat by-product costs 2¢ per ounce and the cereal, 1¢ per ounce, the objective function—the cost function—becomes

$$C(\text{¢ per pound}) = 2x + 1y$$

(Recall that x represents the number of ounces of meat by-product to be used in each pound of pet food and y, the number of ounces of cereal.) Referring

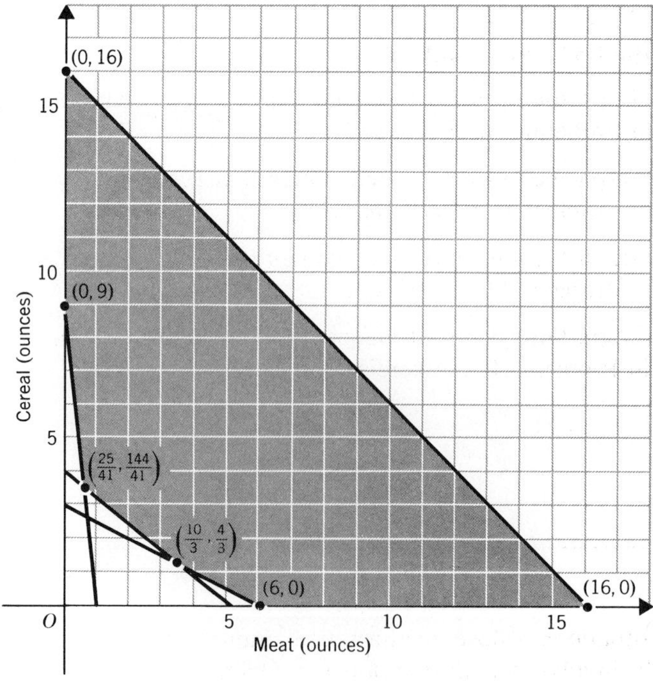

Figure 5.3.1

to Figure 5.3.1, (the graph of the solution set from Example 5.2.4 is shown again in Figure 5.3.1) the task of minimizing the cost is that of finding the point in the shaded region (the solution set) for which the product of two times the x-coordinate added to the y-coordinate gives the smallest number.

At first glance it may appear that this task is still beyond our capabilities. There are a great many points in the solution set of the inequalities. However, systems of linear inequalities such as we are using here have a useful property, which simplifies our task a great deal. We note that since we are dealing with **linear** inequalities, the region containing the solutions is bounded by **straight lines**. It is, therefore, a **polygon** with specific points—"corners," or vertices—where the sides meet. It turns out (we shall not attempt to prove it here) that if there is an optimum value for the objective function, it will occur at one of the vertices, or along one edge, of the "solution polygon." Thus, the objective function need be evaluated only for the values of the variables at the corner points of the solution set. There will be relatively few of these corner points, and the best value for the key quantity can be chosen by direct comparison among them.

Example 5.3.1

Let us consider again the pet-food problem. Suppose that the meat by-product costs 2¢ per ounce, and the cereal 1¢ per ounce. What amount of each ingredient should the manufacturer use to meet the nutritional requirements at the least cost?

We form an equation from which the cost of the ingredients in each pound of pet food can be computed:

$$2x + y = C$$

This equation is the **objective function** for the problem. It relates the key quantity in the situation (in this case the cost) to the variables. As before, x is the amount of meat by-product and y is the amount of cereal in ounces used in each pound of pet food; and C is the cost of these ingredients. The solution set for the nutritional requirements is shown in Figure 5.3.1. The value for each of the variables is shown at each of the corners, or **vertices**, of the solution polygon. These values are obtained by solving for the intersections of the boundary lines of the polygon. The vertex, (25/41, 144/41), is the intersection of the two lines, $3x + \frac{1}{3}y = 3$ and $4x + 5y = 20$. This intersection is the point that "satisfies" both these equations simultaneously. It is, therefore, the solution of the system

$$3x + \tfrac{1}{3}y = 3$$
$$4x + 5y = 20$$

The solution is found using the method of Chapter 4. The vertex, (10/3, 4/3), is the solution of the system

$$4x + 5y = 20$$
$$3x + 6y = 18$$

The other four vertices occur where either x or y is zero. The value for the other variable at each of these vertices is computed by substituting the zero value into the equation of the boundary line.

The problem of choosing a solution—deciding how much of each ingredient to use—is now reduced to determining for which of the vertex points the cost is the least. We use the objective function to compute the cost for each of these points.

At (16,0):	$C = 2(16) + 0 = 32¢$
At (6,0):	$C = 2(6) + 0 = 12¢$
At (10/3, 4/3):	$C = 2(10/3) + (4/3) = 8¢$
At (25/41, 144/41):	$C = 2(25/41) + (144/41) = 4.7¢$
At (0,9):	$C = 2(0) + 9 = 9¢$
At (0,16):	$C = 2(0) + 16 = 16¢$

It is clear that the minimum cost occurs at the vertex, (25/41, 144/41). Thus, the pet food will contain 25/41 ounces of meat by-product and 144/41 ounces of cereal in each pound. The remainder is water and filler.

Let us summarize the methods used thus far:

The purpose of linear programming is to obtain the optimum value for the "key quantity" in a situation, consistent with a set of "operating conditions," by making an appropriate choice for the variables in the situation.

The "operating conditions" are a system of linear inequalities defining the restrictions the given situation places on the values for the variables.

The solution set for the system of inequalities is the set of all the feasible, or possible, values for the variables: the values that meet all the restrictions. The particular values for which the key quantity has its optimum value are the values for one of the "corner points" of the solution set.

The relationship by which the key quantity can be computed from the values of the variables is called the **objective function**.

The optimum "solution" is found by computing the value of the key quantity at each of the "corner points" of the solution set, using the objective function. The optimum value among these is determined by direct comparison. The largest value is chosen if a maximum is sought; the smallest, if a minimum is sought. This optimum value is the optimum for the entire solution set.

5.3.1 Linear Programming—Graphical Solutions

The optimum solution for the objective function can be found graphically if the system can be plotted on a Cartesian coordinate system. In the situations we are considering here the objective function is a linear function and, thus, can be written in the form

$$ax + by = K$$

K is the "key quantity"—the quantity to be maximized or minimized. a and b are the unit values for the variable quantities. In the pet-food problem, for example, a was 2¢, the cost per ounce of meat by-product; b was 1¢, the cost per ounce of cereal. K was the total cost of the ingredients per pound of pet food.

Since the objective function is linear, its graph will be a straight line. In each situation the values for a and b remain the same. These values determine the slope of the line. For each different value for K, a different line will result—all of which have the **same slope**. At every point on any of these lines the coordinates of the point give the values for the **variables** for which the key quantity, K, has the same value. Finding the optimum solution consists of finding the line for which K has the best value, and for which there is at least one point on the line in the solution polygon of the inequalities. The coordinates of that common point are the optimum values for the variables.

Example 5.3.2

Let us consider the following system of inequalities:

$$x \geq 0, \qquad x \leq 6$$
$$y \geq 0, \qquad y \leq 7$$
$$3x + 2y \geq 12$$
$$x + 2y \geq 8$$

The solution polygon for this system is shown in Figure 5.3.2.

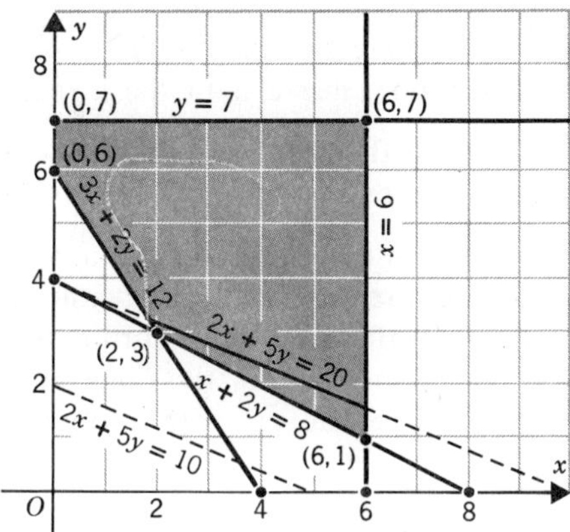

Figure 5.3.2

The coordinates for the vertices were found by solving the boundary-line equations simultaneously in pairs. Let us find a minimum value for K in the objective function, $K = 2x + 5y$. We start by choosing two convenient "test values" for K, say 10 and 20. These values enable us to establish the slope of the objective function and to see the effect that different values for K have on the location of the objective function line. The lines are shown dotted in Figure 5.3.2. We note that these two lines are parallel, and that no points on the line for which $K = 10$ are within the solution set for the inequalities. There are, however, points in the solution set on the line for $K = 20$. These are the points along the solid portion of the line $2x + 5y = 20$.

By comparing the locations of the two objective function lines, we see that the line for the smaller value for K is closer to the origin.

This suggests that moving the objective function line closer to the origin is associated with smaller values for K. Thus, if we draw lines parallel to the line $2x + 5y = 20$ closer to the origin, but still intersecting the solution polygon for the inequalities, we have smaller values for K. The line for the smallest possible value for K passes through the point $(6, 1)$. The coordinates of this point are the optimal values for x and y. The student can verify that this is so by computing K for each of the vertices of the solution polygon.

Let us summarize the process for finding the optimum value for an objective function graphically.

1. Plot the solution set for the system of inequalities.
2. Choose two convenient values for the quantity to be optimized; use them in the objective function to form the equations for two "objective lines." Plot these lines on the graph.
3. Use the relative location of these two lines to determine the direction to move the lines to produce the desired effect on the quantity being optimized. For maximizing, move in the direction from the smaller value to the larger and vice versa.
4. Lay a straight edge on one of the objective lines and, keeping it parallel to the line, move it in the direction determined in Step 3.
5. The last point in the solution polygon touched by the straight edge as it is moved gives the optimum values for the variables.
6. If the objective lines are parallel to any of the sides of the solution polygon, one entire side of the polygon may be last touched by the straight edge. In this case the coordinate for any point on the edge will yield the same value for the objective quantity. Thus, any point on the edge can be chosen as the "operating point" for the system.

Example 5.3.3

Let us consider a situation in which the conditions to be met are given by the system of inequalities:

$$\begin{aligned} x \geq 0, \quad & y \geq 0 \\ x + y &\leq 11 \\ x + 2y &\leq 18 \\ 3x + 2y &\leq 30 \end{aligned}$$

Suppose we wish to maximize the quantity, P, given by the objective function

$$P = 5x + 8y$$

The solution set for the given condition is shown in Figure 5.3.3.

Let us choose 40 and 60 as the test values for P. The objective lines for these values are shown dotted in Figure 5.3.3. If we lay a straight edge along the line $5x + 8y = 40$ and move it toward the

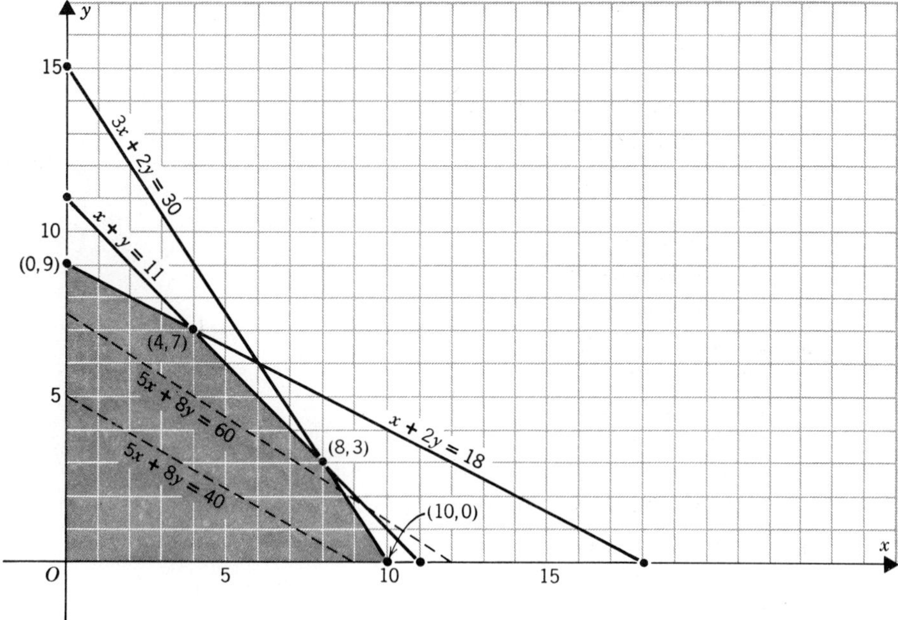

Figure 5.3.3

other line (we are seeking a maximum value for P), we see that the last point in the solution set touched by the straight edge is (4, 7). Thus, the optimum values for x and y are 4 and 7. The maximum value for P is

$$5(4) + 8(7) = 20 + 56 = 76$$

The student can verify that the value for P at each of the other vertices of the solution polygon is less than 76.

EXERCISE 5.3

For situations described by the following systems of inequalities, find the optimum solutions for the indicated objective functions:

1. $y \geq 0$
$x + 2y \leq 4$
$x - y \geq 1$
 a. Maximize $P = x + 3y$
 b. Minimize $C = 2x + y$

2. $x \geq 0, \qquad y \geq 0$
$x + 2y \leq 8$
$x - 2y \geq -4$
 a. Maximize $P = x + 3y$
 b. Maximize $C = x + y$

3. $y \le 4, \quad x \le 5$
$x + 2y \ge 4$
$x - y \le 4$
   a. Maximize $P = 3x - y$
   b. Minimize $P = 2x + 3y$

4. $x \ge 1, \quad y \ge 1$
$x + y \ge 3$
$4x + 3y \le 24$
   a. Maximize $P = 2x + 7y$
   b. Minimize $P = 5x - y$

5. $x \ge 0, \quad y \ge 0$
$2x + y \le 20$
$1.5x + 2y \le 20$
$x + 2y \le 18$
   a. Maximize $P = 2x + 3y$
   b. Minimize $R = 30 + 3x - 2y$

6. $x \ge 0, \quad y \ge 0$
$2x + 3y \ge 18$
$x + y \ge 7$
$x + 3y \ge 12$
   a. Minimize $S = 5x + 6y$
   b. Maximize $T = 80 - 2x - 5y$

7. $x \ge 1, \quad x \le 12$
$y \ge 3, \quad y \le 10$
$x + y \ge 7$
$7x + 5y \le 85$
   a. Maximize $F = 3x + 4y$
   b. Minimize $G = 5 + 5x - y$

8. $x \ge 0, \quad y \ge 0$
$x \le 5, \quad y \le 4$
$3x + 2y \le 18$
$3x + 6y \le 30$
   a. Maximize $M = 6x + 4y$
   b. Minimize $N = 40 - 2x - 4y$

9. Consider the situation of Problem 7, Exercise 5.1.
   a. It costs \$1 per cubic yard to dig and haul the sand-rock mixture from Area *A* to the machine, and \$4 per cubic yard to dig and haul the mixture from Area *B*. How much should be supplied to the machine from each area to minimize costs?
   b. As material continues to be removed from Area *A*, the cost rises. At what cost per cubic yard for quarrying Area *A* should the operation be changed, and what will the new operating condition be?

10. Consider the situation in Problem 8, Exercise 5.1. The woman can sell the silver-ornamented belts for \$10, and the beaded belts for \$6.50.
   a. Keeping in mind the cost of materials, how many belts of each kind should she make to maximize her profit?
   b. Keeping in mind the time required to make the belts, how many

of each kind should she make to obtain the maximum profit per hour worked?

11. Consider the situation in Problem 9, Exercise 5.1.
If the profit on a deluxe rack is $10 and the profit on the standard model is $8, how many should the manufacturer produce of each to maximize his profit?

12. Consider the situation in Problem 10, Exercise 5.1.
  a. If mixture *A* costs 5¢ per pound to produce, and mixture *B* costs 4¢ per pound, how much of each should be included in 100 pounds of fertilizer to meet the specifications at least cost?
  b. Suppose the price of mixture *B* starts to increase. At what price for *B* will the minimum cost occur for the use of equal amounts of *A* and *B*? For mixture *A* exclusively (i.e., none of mixture *B* at all)?

5.4 LINEAR PROGRAMMING—PART II

In the previous section we developed a procedure for finding optimum solutions in situations involving only two unknown quantities. Since many quantities to be optimized depend on more than two other quantities, we shall study briefly the basic procedures for finding solutions when there are more than two variables.

5.4.1 Graphical Solutions for More Than Two Variables

Although graphs, by their very nature are limited to two independent variables, there are many situations involving three, and sometimes four, variables where graphs can be used. The key concept here is **independent** variables. When there are fixed relationships among the variables, they are not independent. For example, suppose three ingredients are combined to form a fixed amount of mixture. Although there are three different variable quantities—the amounts of each of the three ingredients to be used—only two are independent. As soon as the amounts for two of the ingredients are chosen, the third amount becomes the amount needed to form the specified total quantity. To illustrate the procedure, let us consider a simple example.

Example 5.4.1

A small shop manufactures three styles of table lamps, *A*, *B*, and *C*. The total capacity of the shop is 25 lamps per day. The owner of the shop has decided that he must produce at least three of each style per day to keep his "product line" complete. Each style lamp requires special work, so that no more than 15 of any one style can be produced per day. The profit on Style *A* is $7; on Style *B*, $8; and Style *C*, $6. If he can sell all the lamps he produces, what number of each style should he produce to obtain the maximum profit?

Let x represent the number of Style A produced,
y, the number of Style B, and
z, the number of Style C.

The capacity of his plant requires that

$$x + y + z = 25$$

or

$$z = 25 - x - y$$

The minimum production requirements are

$$x \geq 3, \qquad y \geq 3, \qquad z \geq 3$$

The latter requirement can be expressed in terms of x and y:

$$25 - x - y \geq 3, \qquad \text{or} \qquad x + y \leq 22$$

The maximum product limits are

$$x \leq 15, \qquad y \leq 15, \qquad z \leq 15$$

The latter condition can be expressed in terms of x and y:

$$25 - x - y \leq 15, \qquad \text{or} \qquad x + y \geq 10$$

We can now form a graph showing the set of feasible values for x, y, and z. This is shown in Figure 5.4.1.

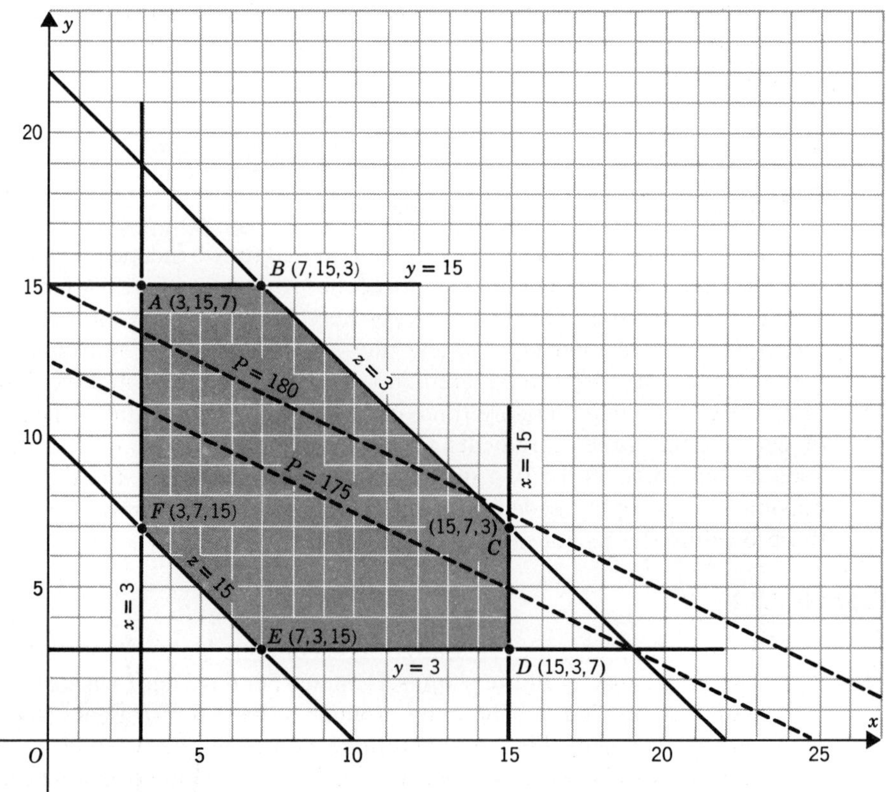

Figure 5.4.1

The vertices of the solution polygon, Points A, B, C, D, E, and F, are each shown with the values for x, y, and z. The x- and y-values were obtained from the graph. The z-value was obtained in each case from the relation, $z = 25 - x - y$.

The profit function is given by

$$P = 7x + 8y + 6z$$

Replacing it by its equivalent in terms of x and y, we have

$$P = 7x + 8y + 6(25 - x - y)$$

or

$$P = x + 2y + 150$$

The dashed lines on the graph show the profit function for profits of \$175 and \$180. The farther away from the origin for the objective line, the higher the profit becomes. Thus, Point$B$ is seen as the optimum operating point. That is, the shop should produce seven Style A, fifteen Style B, and three Style C lamps each day.

The student should compute the profit for each of the corner points to verify that Point B does yield the highest profit.

Let us consider now a more complex problem.

Example 5.4.2

A chemical processing plant receives 800 pounds per hour of a raw material, which is converted into three different chemicals, A, B, and C. The plant has three different processes for the conversion, each producing different amounts of the three chemicals. The market price for the chemicals produced changes frequently, and the amount of each chemical produced is changed to keep the profit as high as possible. The operation of the plant has the following requirements:

The machinery for each process must be kept in operation, and a minimum of 100 pounds per hour of raw material must be fed to each.

The maximum amount of raw material that can be handled by each process is 400 pounds per hour.

The following chart shows the amount of each of the products produced per pound of raw material by each process.

	Product		
	A	B	C
Process I	.1	.8	.1
Process II	.6	.2	.2
Process III	.2	.5	.3

The plant can sell all of each product that it produces, but it has contracts that require at least 180 pounds of A and 340 pounds of B to be produced each hour.

Let us first translate these requirements into algebraic expressions, and then find the set of solutions which meet all the conditions.

We let

$$x = \text{pounds per hour of raw material to Process I}$$
$$y = \text{pounds per hour of raw material to Process II}$$
$$z = \text{pounds per hour of raw material to Process III}$$

The total amount of raw material to be used gives the relationship,

$$x + y + z = 800$$

Thus, we can treat the problem as a two-variable problem by letting

$$z = 800 - x - y.$$

The minimum processing requirement gives

$$x \geq 100 \qquad y \geq 100 \qquad z \geq 100$$

The third requirement becomes

$$800 - x - y \geq 100 \qquad \text{or} \qquad x + y \leq 700$$

The maximum capacity of each process produces the relationship,

$$x \leq 400 \qquad y \leq 400 \qquad x + y \geq 400$$

(The student should verify the third condition by setting up the condition for z and making the substitution of z's value in terms of x and y.)

From the production table we see that the amount of chemical A produced is

$$.1x + .6y + .2z$$

The contract for chemical A calling for at least 180 pounds per hour provides the condition

$$.1x + .6y + .2z \geq 180$$

The student should verify that this can be restated as

$$-.1x + .4y \geq 20$$

Similarly, the contract for chemical B produces the requirement

$$.3x - .3y \geq -60$$

We have a set of eight inequalities that must be satisfied by the amounts of raw material fed to each process. A graph of this system of inequalities is shown in Figure 5.4.2.

The graph shows that the solution polygon has six vertices. The coordinates for these operating points are shown. The third number is the value for z, obtained from the relationship, $x + y + z = 800$.

Let us suppose the price for A is 9¢ per pound; for B, 11¢ per pound; and for C, 10¢. What amount of raw material should be fed to each process so that the gross sales value of the three products is a maximum?

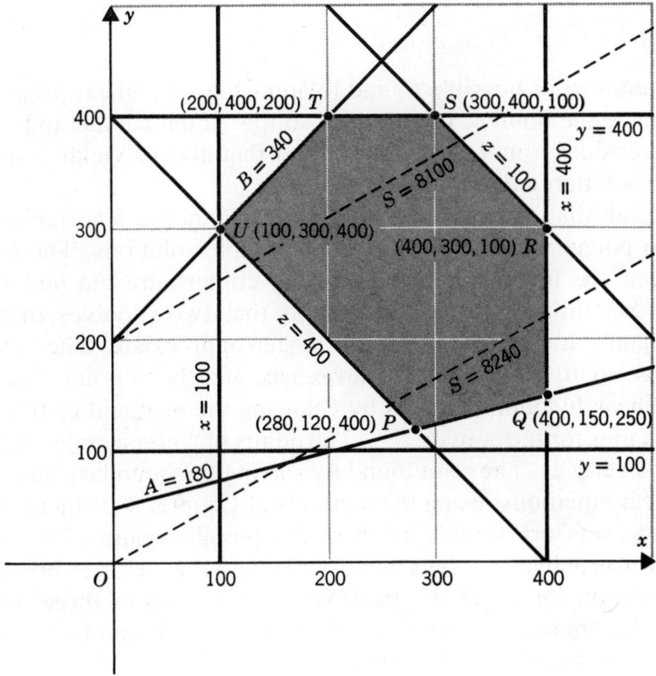

Figure 5.4.2

We note that the amount of Chemical A produced is

$$A = .1x + 6y + .2z$$

The amount of B produced is

$$B = .8x + .2y + .5z$$

The amount of C produced is

$$C = .1x + .2y + .3z$$

Thus, the objective function giving the total sales value is:

$$S = 9(.1x + .6y + .2z) + 11(.8x + .2y + .5z) \\ + 10(.1x + .2y + .3z)$$

Combining terms, we have

$$S(\text{¢ per hour}) = 10.7x + 9.6y + 10.3z$$

Replacing z by its value in terms of x and y, we have

$$S = 10.7x + 9.6y + 10.3(800 - x - y)$$

or

$$S = .4x - .7y + 8240$$

The student should verify that the maximum value for S occurs at Point Q – 400 pounds per hour of raw material being fed to Process I, 150 pounds per hour to Process II, and 250 pounds per hour to Process III.

Linear Programming—Algebraic Solutions

It is not always possible to find relationships in a linear programming problem to reduce the number of variables so that graphical methods can be used. We now consider a more general method that always yields a solution, assuming that a solution exists.

Recall that the optimum point occurs either at a corner or along one edge of the polygon bounding the set of feasible solutions. The boundaries of the polygon are linear functions, and the corners are the intersections of these linear functions. When there are more than two variables, of course, the terms "polygon" and "corner" are no longer appropriate. The sets containing the feasible solutions are called **convex sets**, and the "corners" are called **vertices.** The linear functions formed by changing the inequalities to equations are the boundaries for the convex sets. The points of "intersection" of these boundaries are the vertices. They are found by solving the boundary equations as systems of linear equations, using the methods of Chapter 4. If there are **two** variables, solution sets are found for all of the possible **pairs** of boundary equations. When a problem involves **three** variables, the vertices are found by finding the solution for all of the possible combinations of **three** equations each. In general, if there are n variables, the vertices are found by solving the combinations of boundaries n at a time.

As an example of how to proceed let us consider again the problem of Example 5.3.3, whose conditions are given by the system of inequalities

$$\begin{aligned} &x \geq 0 \qquad y \geq 0 \\ &x + y \leq 11 \\ &x + 2y \leq 18 \\ &3x + 2y \leq 30 \end{aligned}$$

We have already seen the graphical solution for this problem. In the algebraic solution we must form all of the possible pairs of boundary equations, and then find their interesections. Specifically, we have the following systems of equations:

1. $x = 0$, $y = 0$

2. $x = 0$, $x + y = 11$

3. $x = 0$, $x + 2y = 18$

4. $x = 0$, $3x + 2y = 30$

5. $y = 0$, $x + y = 11$

6. $y = 0$, $x + 2y = 18$

7. $y = 0$, $3x + 2y = 30$

8. $x + y = 11$, $x + 2y = 18$

9. $x + y = 11$, $3x + 2y = 30$

10. $x + 2y = 18$, $3x + 2y = 30$

The 10 points obtained as solutions are the 10 **possible vertices. They will not all meet all of the conditions**, however. **Each of these 10 points must be tested** to determine whether it meets all of the conditions. That is, the x- and y-values for each point must be substituted in **each** of the conditions to determine whether the condition is met. In this problem the 10 intersections of the 10 pairs of lines listed above are, respectively, (0,0), (0,11), (0,9), (0,15), (11,0), (18,0), (10,0), (4,7), (8,3), and (6,6). Of these, the ones that satisfy all of the conditions and, thus, qualify as possible solutions for the optimum are (0,0), (0,9), (10,0), (4,7), and (8,3).

In the graphical solution to this problem shown in Figure 5.3.3, note that the five possible solutions above are the vertices of the polygon containing the solutions to the system of inequalities. The other five intersections, (0,11), (0,15), (11,0), (18,0), and (6,6), all lie outside this polygon.

The next step in the algebraic solution is to evaluate the objective function for each of the vertex points. The optimum value is determined by comparison. In Example 5.3.3 the objective function was given as $P = 5x + 8y$.

At (0,0):	$P = 0 + 0 = 0$
At (0,9):	$P = 0 + 8(9) = 72$
At (10,0):	$P = 5(10) + 0 = 50$
At (4,7):	$P = 5(4) + 8(7) = 76$
At (8,3):	$P = 5(8) + 8(3) = 64$

The largest value for P is 76, which is its value at the point, (4,7).

If the same value is obtained for the objective function at two of the points, and this is the optimum value, any of the points on the line joining these two points can be used as the optimum solution. For example, in the preceding problem, if the objective function is $M = 3x + 3y$, we have for the values at each point.

At (0,0):	$M = 0 + 0 = 0$
At (0,9):	$M = 3(0) + 3(9) = 27$
At (10,0):	$M = 3(10) + 3(0) = 30$
At (4,7):	$M = 3(4) + 3(7) = 33$
At (8,3):	$M = 3(8) + 3(3) = 33$

The value for M is 33 at each of the two points, (4,7) and (8,3). This is also the largest value for M. Thus, the maximum value for M occurs at any point along the line segment joining (4,7), and (8,3). Any point on this line segment can be used as the optimum "operating point."

The solution in this example required finding the solution set for 10 pairs of equations. If there are several variables and conditions, it could be necessary to solve a large number of systems of linear equations. For example, in a situation with five variables and eight conditions, there are 56 systems of five equations each. For a situation involving six variables and 10 conditions, there would be 210 systems of six equations each! Clearly, optimum operating points cannot be found for such systems without the aid of a high-speed computer.

A special procedure, called the **simplex method**, reduces the amount of computations needed; but it, too, requires a computer when there are more than three or four variables. The method consists of two algorithms (computational procedures), one for finding maxima and one for minima, based on the fundamental principles developed in this and the preceding chapter. If the student is interested in learning how to apply the method, he is referred to any of the several texts devoted more particularly to linear programming.

As the foregoing discussion indicates, the computations involved in finding an optimum condition can be quite tedious. When we consider the alternatives, however, the task becomes less onerous. The rewards for finding an optimum operating condition in a given situation can be quite great. When we consider the many possible operating conditions, and the huge task of finding the best one by trial and error, the linear programming process becomes quite attractive. When a computer is available to perform the tedious part of the computation, linear programming is a particularly powerful process for determining optimum operating conditions in a complex situation.

EXERCISE 5.4

1. a. Graph the following system:

$$x + y + z = 16$$
$$x \geq 2 \qquad y \geq 4 \qquad z \geq 0$$
$$x \leq 10 \qquad y \leq 12 \qquad z \leq 8$$

b. Find the maximum value for

$$A = 3x + 5y + z$$

subject to the conditions of (*a*).

2. a. Find the solution set for the following system graphically:

$$x + y + z = 24$$
$$x \geq 2 \qquad y \geq 0 \qquad z \geq 6$$
$$x \leq 15 \qquad y \leq 10 \qquad z \leq 20$$

b. At what point in this system will the quantity,

$$C = 6x - 2y + 3z$$

have a minimum value?

3. a. Graph the solution set for the following system:

$$z = 4x + 3y$$
$$x \geq 0 \qquad y \geq 0 \qquad z \geq 0$$
$$x \leq 6 \qquad y \leq 8 \qquad z \leq 36$$

b. Find the maximum value for

$$A = 8x + 7y + 2z$$

4. a. Graph the solution set for the following system:

$$3x + 2y + z = 36$$
$$x \geq 0 \qquad y \geq 0 \qquad z \geq 0$$
$$x \leq 10 \qquad y \leq 12 \qquad z \leq 18$$

b. Find the operating point such that the quantity

$$C = 4x + 2y + 7z$$

has its maximum value.

5. Find the solution set for the following system graphically:

$$x + u = 10 \qquad y + v = 12$$
$$x \geq 0 \qquad y \geq 0 \qquad u \geq 0 \qquad v \geq 0$$
$$x + y \leq 16 \qquad u + v \leq 18$$

(**Hint**: Let x and y be the independent variables. Express u and v in terms of x and y, respectively. Convert the conditions for u and v to conditions involving x and y.)

6. a. Graph the following system:

$$x + u = 10 \qquad y + v = 12$$
$$x \geq 2 \qquad y \geq 2 \qquad u \geq 2 \qquad v \geq 3$$
$$3x + 2y \leq 30 \qquad u + 2v \leq 24$$

b. Find the maximum value for

$$A = 3x + 2y + 4u + 5v$$

7. a. Find graphically the solution set for the system in which

$$x + y + z = 20$$
$$x \geq 0 \qquad y \geq 0 \qquad z \geq 0$$
$$x + 6y - 2z \leq 80$$
$$4x + 2y + z \leq 68$$

b. Find the operating point for the above system such that the quantity $R = 4x + 3y + 6z$ has maximum value. What is this value?

c. For which operating point will the quantity $C = 3x + 2y + 5z$ be a minimum? What is the minimum value for C?

8. Consider the situation in Example 5.4.2. Suppose the price of C increases to 11¢ per pound. Find the operating point that will maximize the total sales value.

9. a. Find the vertices of the solution set for the following system of inequalities, using algebraic methods:

$$x \geq 0 \qquad y \geq 0 \qquad z \geq 0$$
$$3x + 2y + 5z \leq 21$$
$$x + 3y + 4z \leq 14$$

b. For which of these vertices does the objective function $C = 3x + 8y + 5z$ have a maximum value? For which vertex is it a minimum?

10. a. An appliance distributor maintains two warehouses, A and B, in an urban area. There are 20 refrigerators in Warehouse A and 24 in Warehouse B. Store M places an order for 16 refrigerators, and Store N orders 18. Set up a system of inequalities for the number to be shipped from each warehouse to each store, and construct a graph of the system. (**Hint**: Let x represent the number shipped from A to M, and y, the number from A to N. Express the number shipped from B to M and B to N in terms of x and y.)

b. It costs \$5 each to ship the refrigerators from A to M; \$6, from A to N; \$3, from B to M; and \$5, from B to N. What shipping arrangement will cost the least?

11. A smelter receives three grades of ore from which it extracts lead, zinc, and manganese. Grade A ore is 10% lead, 4% zinc, and 2% manganese. Grade B has 4% lead, 6% zinc, and 1% manganese. Grade C is 8% lead, 5% zinc, and 3% manganese. The smelting process handles 100 tons of ore per hour. The smelter can sell all the metal produced, but it has orders that require at least 6.8 tons of lead, 4.7 tons of zinc, and 1.8 tons of manganese to be produced each hour.

a. Set up the system of inequalities and equations that show the amounts of each grade of ore that can be fed to the smelter to produce the required amounts of metal. Construct a graph of this system.

b. Grade A ore costs \$5 per ton; Grade B, \$5; and Grade C, \$6. What amounts of each grade ore should be used to meet the operating requirements at minimum cost?

12. A shop makes three styles, A, B, and C, of one of its products. All three styles are finished by the same machine, which can handle 60 units per day. The company works this machine to capacity. The manufacture of each item requires three other operations: grinding, drilling, and machining. The number of hours for each operation required by a unit of each style is shown in the table:

	Grinding	Drilling	Machining
Style A	5 hours	2 hours	2 hours
Style B	2 hours	3 hours	3 hours
Style C	4 hours	4 hours	2 hours

There are 240 hours of grinding time, 200 hours of drilling time, and 160 hours of machining time available each day. There are orders that require at least 10 units each of Styles A and B be produced each day.

a. Set up the system of conditions for the operation of this shop in terms of the units of each style produced per day. Construct the graph of this system, and find the coordinates of the vertex points.

b. If the profit on each Style A is \$11; on each Style B, \$7; and on each style C, \$10; how many of each style should be made to maximize the profit? How much is the profit per day?

6

counting patterns

6.0 INTRODUCTION

In our discussions thus far we have developed several different mathematical concepts and then used them to formulate mathematical models for many different types of situations. Setting up mathematical models is the basic process in applying mathematics. These models are used to predict the effects that changes in the parameters and variables will have on the situation being described. In a preceding chapter, for example, we developed a mathematical model by which a pet-food manufacturer can determine the amounts of ingredients to use in his product to meet specified nutritional standards. We used the model further to compute the amount of each ingredient to use to keep the cost of the product as low as possible for a particular price schedule for the ingredients. If the prices for the ingredients change, the model can be used to compute readily any changes that might be desirable in the amounts of ingredients used to keep the cost as low as possible while still meeting the nutrition specifications.

Mathematical models in which the effect of changes can be predicted with with certainty are called **deterministic**. Most of the mathematical models in the physical sciences and many of those in business and the behavioral sciences are of this type. It is likely that the student has worked only with deterministic models in his uses of mathematics. In fact, many people regard mathematics as completely deterministic in nature.

It is ordinarily not possible to devise a deterministic mathematical model for a situation involving the actions or behavior of people. Since the activities of people are not governed by inexorable laws analogous to the laws of physics and chemistry, the kind of mathematics invented to formulate such laws is understandably unsuitable for making models for these activities. However, a kind of mathematics has been invented that takes uncertainties into account. It makes possible the formulation of mathematical models that are both useful and reasonably accurate in describing the phenomena of human behavior. Mathematical models of this type are called **stochastic**. They are based on the **probability**, or **relative likelihood**, of events. They do not describe **exactly** what will happen in a given situation but what is **likely** to happen.

The uncertainties inherent in the structure of a stochastic model impose limitations on their use. One of the most important, and one that is frequently ignored, is that a stochastic model cannot be used to make any valid predictions about the action of an **individual**. It can be applied reliably only to large groups, and the predictions that it provides take the form of telling how many members of the group can be expected to take each of the possible courses of action in the situation described. For example, a mathematical model can be formulated describing the frequency of job changing in a given occupation. This model could predict with reasonable accuracy how many people in that occupation will change jobs in the coming year. It cannot predict with any accuracy at all whether or not a specific individual in that occupation will change jobs.

It is not our purpose to go beyond a few examples of applying stochastic mathematical models to business or the behavioral sciences. The use of these models in these fields is relatively new, and new methods and procedures are being developed regularly. Many applications have already been made, however, and we wish to study the **mathematical structure** of the models being used.

The information with which stochastic models are formulated comes essentially from statistical studies. The formulation of the models themselves uses the **mathematics of probability**. In this text our interest is in **finite probability**; that is, the concepts of probability applicable to situations in which there is a **finite** or **countable set** of **discrete events** possible. Our first task is to develop ways for counting these possible events. The counting patterns that we shall study can be applied to many different situations in which the simple enumeration method, "one, two, three, . . . ," is inadequate.

6.1 FUNDAMENTAL COUNTING PRINCIPLES

Perhaps the first reaction of a student to a chapter on counting is that he learned to count when he was very young and has not forgotten. However, the usual enumeration method of making an association of individual items with the natural numbers is useful only in counting small groups. Consider the following relatively simple situations:

A person is studying the management structure and personnel inter-relationships in a company. He has 10 junior executive positions included in his study and 10 people to fill these positions. He would like to consider all of the different possibilities for assigning the 10 people to the 10 jobs. How many different possibilities are there?

A psychologist is giving an association test with 10 words and 10 colors. How many different sets of associations are possible?

A sociologist is studying personal interactions in a group. There are 10 couples in the group, and the sociologist would like to try all of the different sets of pairings of the 10 men and 10 women. How many are possible?

The student probably recognizes that each of these situations follows the same pattern. If we try to list the possibilities in order to count them, we are likely to become discouraged. There are 3,628,800 in each case. Clearly, we need a formula by which we can compute the number of possibilities using the key number, 10, which occurs in all three examples. The three foregoing examples fit one of several counting patterns that we shall study in this chapter. We shall develop enumeration formulas that will meet the **basic counting criterion**:

> In each situation with more than one possible result, or outcome, the enumeration formula for the number of possible results shall count each of them **exactly once**, no more, no less.

The student will recall that application of this criterion led to the formulas developed in Chapter 1 for determining the number of elements in the union of two or more sets.

6.1.1 Fundamental Principle of Counting

We shall consider first the counting of events or arrangements in sequence. This need not be a sequence in time or position, but merely a set of separate considerations relating to the same situation. For example, suppose a person is buying a refrigerator. He visits a store that carries three different brands of refrigerators. Each brand comes in four different sizes, each of which is available in five different colors. How many different refrigerators are there to choose from?

The **fundamental principle of counting** is that, in a sequence, if the first event, or choice, has m possibilities and the second has n, the two taken together have $m \cdot n$ possibilities. This principle comes from the fact that there will be n possibilities for the second event for each possibility for the first event. For the first possibility of the first event there will be n for the second event. For the second possibility of the first event, there will be n more for the second, or a total of $2n$, and so on up to the $m \cdot n$ possibilities for the two events. This principle can be extended to sequences with as many parts as desired. The total number of possibilities is the product of the number of possibilities for each part. If there are three parts with the first having m possibilities; the second, n; and the third, p; the three together have $m \cdot n \cdot p$ possibilities. Thus, in the refrigerator purchase mentioned above there are $3 \cdot 4 \cdot 5 = 60$ different refrigerators to choose from.

In an election involving two different offices, if there are four candidates for the first office and three for the second, there are $4 \cdot 3$ different possible outcomes for the election.

In a multiple-choice examination with 10 questions, each of which has four choices for the answer, there will be $4 \cdot 4 \cdot 4 \cdot 4 \cdot 4 \cdot 4 \cdot 4 \cdot 4 \cdot 4 \cdot 4$ or 4^{10} different ways of supplying answers to the questions.

The validity of the fundamental principle of counting can be illustrated by a **tree diagram**. A tree diagram depicts each of the possibilities graphically by using the branchlike nature of a sequence of events.

Example 6.1.1

Let us consider in more detail the example of the election with two offices to fill: Suppose the four candidates for the first office are Jones, Smith, Brown, and Green. The three candidates for the second office are Johnson, Ronson, and Bronson. The tree diagram for the outcomes of the election is shown in Figure 6.1.1. The 12

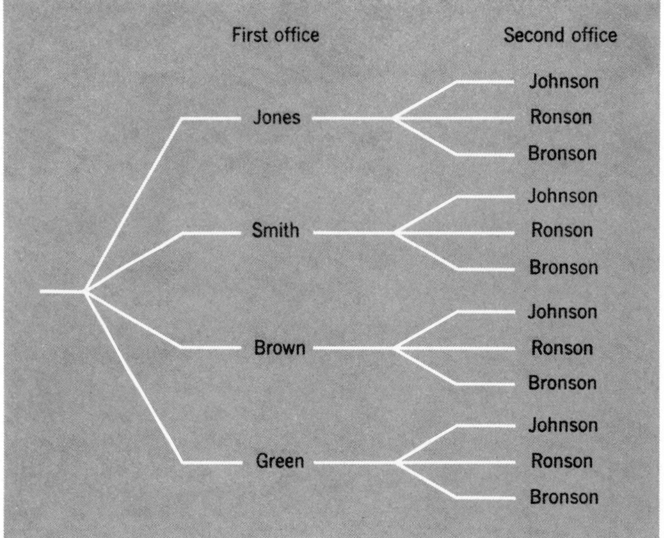

Figure 6.1.1

"branches" under the heading of "second office" show the 12 possible outcomes. The first three show the possibilities assuming Jones wins the first office; Jones–Johnson, Jones–Ronson, and Jones–Bronson. The second three show the possibilities assuming Smith wins the first office; and so on.

The tree diagram is the basic means for displaying a counting pattern. In a situation such as that in Example 6.1.1 the symmetry of the counting pattern makes it simpler to use the formula directly instead of a diagram. However, in other situations a tree diagram is quite useful in making certain that all of the possibilities have been counted once and only once. The diagram is of particular usefulness when all of the possibilities in one or more of the parts of a sequence are not compatible with all of the possibilities in the other parts.

Example 6.1.2

Let us consider the possible outcomes of a tennis match won by the first player to win three sets, (i.e., a best three out of five match). Each set is won by one player or the other; and it may require the playing of three, four, or five sets before one player wins the necessary three sets. Figure 6.1.2 shows the possible outcomes, with line segments representing the sets. The two players are A and B,

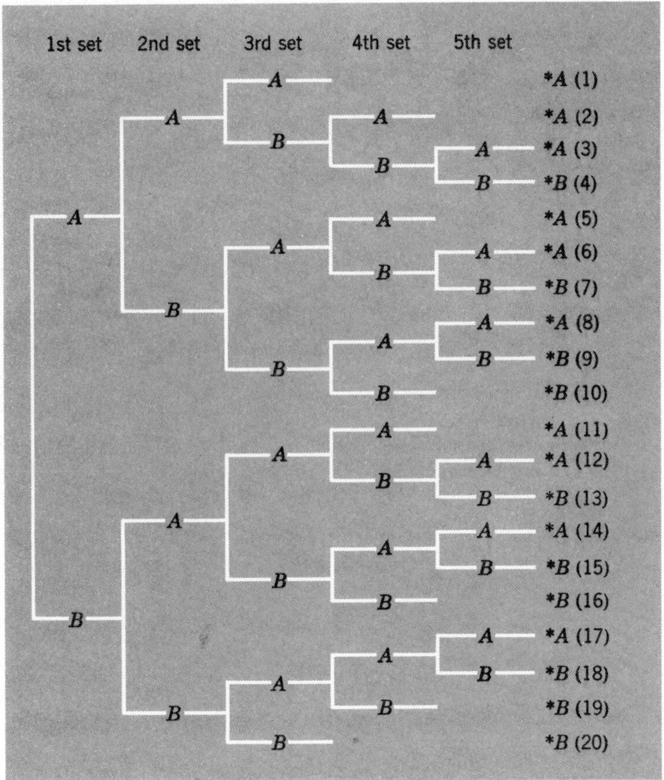

Figure 6.1.2

and the letter next to the line segment indicates the winner of the set. The asterisk and letter at the end of a sequence of line segments indicate the winner of the match. The diagram provides an orderly way of keeping track of all of the possibilities. In this situation we know that at least three sets must be played. The first three segments of the tree show the possibility that either player can win each set. Each line segment is followed by two segments: one represents the possibility that A wins; the other, that B wins. As an aid in drawing the diagram, note that the lower half follows the same pattern as the upper half with the winners of each set interchanged. At the end of three sets, it is necessary to consider the possibility that the match is ended; that is, that one or the other player has won three sets. These two possibilities are shown at the top and the bottom of the diagram—paths 1 and 20. Each of the other branches is followed by the two possibilities—that one player or the other wins the next set. After the fourth set it is again necessary to check whether the match is over. For paths 2, 5, 10, 11, 16, and

19, one of the players has won three sets; and the match is ended. The remaining branches are each followed by the possibilities that one player or the other wins. At this point the match is certain to be ended. We can now count the total number of possibilities. We see there are two possibilities for a three-set match; six, for a four-set match; and 12, for a five-set match. There are $12 + 6 + 2 = 20$ possibilities in all.

The two examples discussed here illustrate two general types of situations. In the first, Example 6.1.1, the outcomes in different steps of the sequence were unrelated, or independent of each other. In this case the total number of outcomes can be computed from the first principle of counting as the product of the number of possibilities in the first part and the number in the second part.

In the second example, Example 6.1.2, after the third set, the results in succeeding sets depend on the results of the preceding sets. If one of the players has won all three of the first sets, the fourth set will not be played. Similarly, if one of the players has won three out of the first four sets, the fifth set will not be played. Note also that the winner of the match wins the last set.

In the foregoing example the tree diagram provides a convenient method for keeping track of the different possibilities; but, of course, it does not represent the only method for solving the problem. We shall consider others in a later section. Later we shall also see how to count possibilities in special patterns that occur frequently. In all cases the purpose is to provide convenient and accurate procedures for counting in complex situations.

One of the confusing aspects of using counting patterns at first is the choice of the arithmetic operation used to combine the possibilities. Note that in Example 6.1.1 the two numbers, 4 and 3, representing the numbers of possibilities for the winners in the two offices, were **multiplied** to obtain the total number of possibilities. In Example 6.1.2 the three numbers, 2, 6, and 12, representing the number of possibilities for 3-set, 4-set, and 5-set matches, were **added** to obtain the total. The key concept in Example 6.1.1 is that there are four possibilities for the first office **and** three for the second. **Both** selections are to be made. Whenever one choice **and** another are to be made, the total number of possibilities is the **product** of the numbers in each. In Example 6.1.2 the match is over in 3 sets **or** 4 sets **or** 5 sets. The total of the possibilities involves **alternative** possibilities. Whenever the first choice involves one choice **or** another, the total number of possibilities is the **sum** of the numbers in each.

EXERCISE 6.1

1. An election is held to fill three offices. If there are three candidates for each of the first two offices and four candidates for the third, how many different results are possible?

2. A homeowner wants to install floor tiles in a new room that he has added to his house. The tiles are one foot square. The room is 12 feet by 15 feet. How many tiles will he need?

3. a. A citizens' group maintains a file of registered voters. In one precinct the forms made out for each voter are separated first into two groups by sex, M or F; then, by party preference into three groups, R, D, or no-party; and finally into six age groups, 18–20, 21–29, 30–39, 40–49, 50–59, and 60 or over. A separate file is set up for each of the possible groupings. How many files are needed?
   b. In a second precinct the workers use the same groupings, but they separate the forms first into the three groups by party, then into the six age classifications, and finally by sex. How many files are needed in this precinct?

4. Referring to Example 6.1.2, in how many of the possible outcomes does the winner of the first set also win the match? The winner of the second set? The winner of the first two sets?

5. An appliance dealer wishes to stock a complete line of refrigerators. The brand that he sells comes in four colors: white, avocado, gold, and coppertone. There are five sizes: 12 cubic feet, 14 cubic feet, 17 cubic feet, 19 cubic feet, and 21 cubic feet. An automatic ice-cube maker is a factory-installed option available for any model. The two smaller sizes, 12 cubic feet and 14 cubic feet, have single doors, which can be factory-installed to open from the left or the right. What is the minimum number of refrigerators that he must stock to have at least one of each model available?

6. A psychologist has a method for analyzing personality that involves rating a person in each of three different characteristics. There are seven different scores possible for each of these ratings. How many different personality types does the method provide?

7. A small restaurant advertises, "For Variety, Eat at Ben's Beanery, Over **200** Different Dinners Each Day." The menu lists

   Choice of Soup or Salad
   Choice of 3 different entrees
   Choice of 2 kinds of potatoes
   Choice of 3 different vegetables
   Choice of coffee, tea, or milk
   Dessert

   How many different desserts must be offered to make good on the advertising claim? What is the exact number of "different" dinners that this number of desserts will yield?

8. In an election there are four offices to fill. Three of the offices have three candidates, and one office has four candidates. The ballot also includes three issues on which the voter is to vote "yes" or "no." After the ballots were counted, one of the tally clerks stated that each voter had voted on every item and that no two ballots were marked exactly alike. What is the maximum number of people who voted in this election?

9. The "World's Largest Automobile Dealer" advertises that he has in stock for immediate delivery a complete stock of Fowlare motor cars with or without any of the factory-installed options. The Fowlares are

produced in three model lines: the Lunger, the Astrocat, and the Dynamonster. There are five body styles available: a two-door sedan, a four-door sedan, a hard-top coupe, a convertible, and a station wagon; and each can be obtained in a choice of eight decorator colors. There are three engine sizes, an econosix, a standard eight, and a supereight. With the econosix and the standard eight there is a choice of two transmissions, a three-speed manual shift or a turbo-slide automatic. With the supereight engine a third transmission, the dynaslush automatic, is also available. For the hardtop coupés and the convertible, a four-speed floor-mounted transmission is another option. Any model can be fitted with optional air conditioning, power brakes, and power steering. How many different cars must the dealer keep in stock to make good on his advertising claim? (**Suggestion:** Use a tree diagram for the nonsymmetric part: transmissions, body style, and engine size. Then apply the fundamental principle of counting.)

10. An organization has an executive committee of three members. If one of these three is designated as president, one as vice-president, and one as secretary, how many different sets of officer designations are possible? (**Hint**: Use a tree diagram and make the designations in sequence.)

11. If the organization of Exercise 10 has an executive committee of five members instead of three, how many different sets of officer designates are possible?

12. Extend the tree diagram of Example 6.1.2 to find the number of possibilities for a best four out of seven match. (**Hint**: Do only the top half of the diagram and multiply by two.)

13. A merchant has an item packaged in sets of five, two, or one. In how many different ways can he make up an order for 12 of the items? (**Suggestion**: Make a tree diagram. To avoid repetitions make each branch follow the same package-size sequence; that is, start with large packages and progress to smaller ones, or vice versa.)

14.

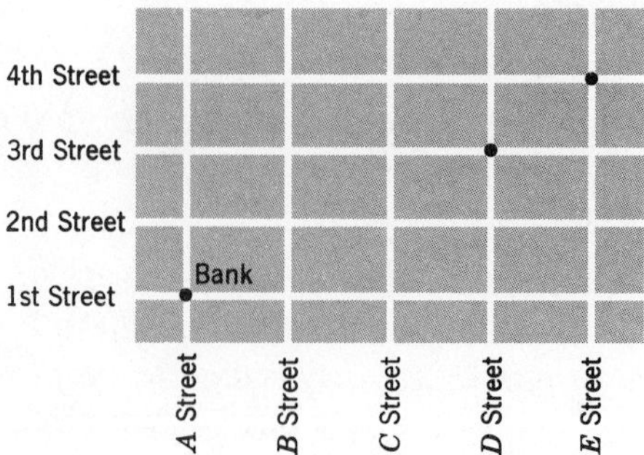

a. A bank messenger delivers securities to a stock broker every weekday morning just before the stock market opens at 10 A.M. As shown on the diagram, the bank is located at First and *A* Streets, and the stock broker at Third and D Streets. The messenger wants to keep the distance that he travels at the minimum, but vary his route to forestall robbery attempts. How many different routes can he take?

b. If there is a second bank messenger delivering securities to an office at Fourth and E Streets, how many routes can he take?

6.2 PERMUTATIONS

In Chapter 1 in the discussion of sets and how they are defined, we were careful to point out that the order in which the elements of a set are listed is not important. For example, the set $\{a,b,c,d\}$ is the same set as $\{b,a,d,c\}$. In the next chapter we are going to study concepts that require knowing how many different orders there are for listing the elements of a given set.

Each listing in a different order of the same set of elements (each element different from the others) is called a **permutation**. For example, there are six different orders for listing the elements of the set $\{a,b,c\}$: *abc, acb, bac, bca, cab,* and *cba*. These are the six **permutations** of the elements of $\{a,b,c\}$. When a set has only three or four elements, the number of its permutations can be obtained by listing the permutations and counting them. However, if there are more than four elements in the set, the list becomes too long for this method to be practical. We shall, therefore, use the fundamental principle of counting to develop a formula for computing the number of permutations possible for the elements of any finite set.

This formula is the first of several that we shall need for the study of probability. We shall develop them using elementary procedures to emphasize the **structure** of the counting patterns involved. It is essential to understand this structure thoroughly to use the formulas correctly—to know which formula applies in a given situation.

Let us visualize the braces denoting a set as a container in which there are the same number of element holders as there are elements in the set. The number of permutations of the elements of the set is the number of different ways that we can arrange the elements in the element holders. Let us picture the element holders arranged in a line inside the set containers:

$$\{\_\_, \_\_, \_\_, \ldots, \_\_\}$$

We shall perform a sequence of operations—choosing elements to be placed in each element holder—starting with the first and continuing until each holder has an element in it. From the fundamental principle of counting, the number of ways this can be done is the product of the numbers of ways to perform each operation in the sequence; that is, the number of ways to perform the first operation times the number of ways to perform the second times the number of ways to perform the third, and so on.

Suppose the set contains five elements. When the element is selected for the first element holder, there are five elements to choose from—the holder can be filled in five different ways. When the element is selected for the second holder, there are only four elements to choose from—one has already been placed in the first holder. The selection can be made in four different ways. For the third holder there are three elements to choose from; for the fourth, two; and when we get to the fifth holder, there is exactly one element remaining. The total number of ways in which the complete selection process can be made is $5 \cdot 4 \cdot 3 \cdot 2 \cdot 1 = 120$. Thus, there are 120 permutations for a set of five elements. We shall use the symbol $P(5,5)$ to designate the number of permutations of a complete set of five elements. The reason for this particular form for the symbol will become evident in a succeeding section.

Note that for a set with three elements we can use the same pattern for the computation:

$$P(3,3) = 3 \cdot 2 \cdot 1 = 6$$

This is in agreement with our previous result.

The formula can be generalized to apply to a set with n elements. There are n choices for the first selection, $n-1$ for the second, $n-2$ for the third, and so on until there is exactly one element left for the nth selection. In symbols

$$P(n,n) = n(n-1)(n-2)\ \ldots\ (1) \qquad (6.2.1)$$

6.2.1 Factorials

Products of the form used in Equation 6.2.1 have enough uses in mathematics to justify assigning them a special name and symbol. The product of a counting number times each succeeding smaller counting number down to one is called a **factorial.** $5 \cdot 4 \cdot 3 \cdot 2 \cdot 1$ is called **factorial five**, or **five factorial**. The product in Equation 6.2.1 is called **factorial** n, or n **factorial**. A factorial is symbolized with an exclamation point. 5! denotes factorial five. $n!$ denotes factorial n.

$$n! \ = \ n(n-1)(n-2)\ldots(1)$$

The symbol for a factorial is a deceptively simple one. When the factorial symbol is used, we must keep in mind the product that it denotes. There is an **implied grouping symbol** around a factorial. No arithmetic operations can be performed on the number itself. For example,

$2 \cdot 3!$ does **not** equal $6!$

Similarly, there is no simple way to combine the factors in the expression, $3!4!$. It means the product of $3!$ and $4!$; that is, $6 \cdot 24$.

The arithmetic operations that can be performed with factorials are useful and worth remembering. For example,

$$4! = 4(3!)$$
$$5! = 5(4!)$$
$$n! = n(n-1)!$$

Notice that the **factorial symbol applies only to the number immediately to its**

left. In the expression above we have $n(n-1)!$. The meaning is, "n times factorial $(n-1)$".

In expanding a factorial into its factors, we can stop at any time and express the remaining factors as a factorial. For example, $6! = 6 \cdot 5 \cdot 4 \cdot 3!$. This means that six factorial is equal to six times five times four times three factorial. This property is useful when working with the quotients of factorials. For example,

$$\frac{6!}{3!} = \frac{6 \cdot 5 \cdot 4 \cdot 3!}{3!} = 6 \cdot 5 \cdot 4.$$

Note that $(6!)/(3!)$ does **not** equal $2!$.

A consequence of this property of factorials is that the quotient obtained when a factorial is divided by a smaller factorial is **always** a whole number.

In a later section we shall encounter a formula that under some circumstances yields the symbol $0!$, that is, zero factorial. The definition of factorial n does not make sense for $n = 0$. We shall, however, wish to use $0!$, and it is most convenient for it to have the value, **one**. Therefore, we **arbitrarily define**

$$0! = 1$$

The following table gives the numerical values for factorials up to $10!$. As can be seen, they very quickly become large numbers.

$$0! = 1$$
$$1! = 1$$
$$2! = 2$$
$$3! = 6$$
$$4! = 24$$
$$5! = 120$$
$$6! = 720$$
$$7! = 5040$$
$$8! = 40{,}320$$
$$9! = 362{,}880$$
$$10! = 3{,}628{,}800$$

6.2.2 Applications of Permutations to Counting

The counting pattern represented by the number of permutations of the elements of a set can be used in a substantial number of situations. Whenever an ordering is involved and the same element cannot appear more than once in the sequence, the formula for the number of permutations can be used.

In Problem 10 of Exercise 6.1 we found the number of sets of officer designations possible for an executive committee of three members in which each committee member received an office. We can consider the three offices in order; and the result is, therefore, the number of permutations of three elements: $3!$ or 6.

It is now evident that in each of the situations mentioned at the start of Section 6.1 the arrangement described is essentially a permutation of the elements involved. In filling 10 positions with 10 people, the process is exactly that described in developing the formula for counting the number of permutations. There is a possibility for indecision about which of the positions to list first, and which to list second, and so on. The order in which the positions are listed has no effect on the result. Regardless of the order in which the positions are listed, each person will be assigned to each position exactly 9! times in the complete list of the permutations. To make this plausible let us assign a particular person, say Person B, to a particular position, say the sixth, and see what happens with the rest of the assignments. Note that when this particular assignment is made, we have removed one person and one position from the selection process. There are now nine people to assign to nine positions. These assignments can be made in 9! different ways. With each of them Person B is assigned to position six. Notice that a "different way" occurs whenever there are **any** differences in two separate listings. It does not require that there be differences in every position. We shall discuss the case of complete disagreement between two listings in Section 6.5.

In the psychologist's word-color associations it is assumed that there is a list of 10 words and a list of 10 colors. Each color is to be associated with one and only one word. For example, there are 10 colors to choose from for the first word, nine for the second, and eight for the third, and so on. This is the same pattern that we followed in developing the formula for counting permutations.

When the 10 men and women are being paired by the sociologist, imagine that the women are lined up in a row. One by one men are chosen to be paired with the women in the line. There are 10 choices for the first pairing, nine for the second, and so on. The same pattern is followed as that for developing the formula for counting permutations.

Let us consider now a situation such as described in Problem 11 of Exercise 6.1. Here we are to make an ordering of three elements, but there is a set of five elements to choose from. Using the procedures that we have used before, we see there are five choices for the first office. When that choice has been made, there are four choices for the second office and, finally, three choices for the third office. Thus, the total number of possibilities is

$$\text{Number of possibilities} = 5 \cdot 4 \cdot 3 = 60$$

The procedure that we follow here is the same as that followed to obtain the number of permutations of a set of five elements. However, we stop after we have selected three elements. The number of possibilities in such a situation is called the number of **permutations** of **five elements, taken three at a time**. It is designated with the symbol, $P(5,3)$. Thus,

$$P(5,3) = 5 \cdot 4 \cdot 3 = 60 \tag{6.2.3}$$

Suppose we are to make a selection of three officers from a group of 10 committee members. We will have a choice of 10 people for the first office, nine for the second, and eight for the third. That is,

$$P(10,3) = 10 \cdot 9 \cdot 8 = 720 \tag{6.2.4}$$

Note that when three elements are selected from the set, there are $(10 - 3)$, or seven, which are not selected. We can take the expression used to compute

$P(10,3)$ and multiply it and divide it by the same number without changing its value. That is, we can write

$$P(10,3) = \frac{10 \cdot 9 \cdot 8 \cdot 7!}{7!}$$

From the definition of the factorial,

$$10 \cdot 9 \cdot 8 \cdot 7! = 10!$$

Further, since $7 = (10 - 3)$, $7! = (10 - 3)!$. Therefore, we can write the formula for the number of permutations of 10 things taken three at a time as

$$P(10,3) = \frac{10!}{(10-3)!} \tag{6.2.5}$$

You may wonder why the expression for $P(10,3)$ is manipulated in this fashion, particularly since the resulting expression is more complicated than the original. Such manipulations are common practice in mathematics, and they are done so that a neat and compact **general** formula can be written. When Equation 6.2.4 is generalized to obtain a formula for the number of permutations of n things taken r at a time, we have

$$P(n,r) = n(n-1)(n-2)\ldots(n-r+1)$$

This is a cumbersome expression, and the factor $(n-r+1)$ may be confusing. Using the manipulations that lead to Equation 6.2.5, we have

$$P(n,r) = \frac{n!}{(n-r)!} \tag{6.2.6}$$

The form of Equation 6.2.6 proves to be a more convenient expression to use in further calculations and developments.

6.2.3 Circular Permutations

In some situations involving ordering there is no distinguishable starting place, and only **adjacencies** are of interest. The seating of people around a circular table is an example. Let us consider the number of seating arrangements possible for six people around a circular table. There are $P(6,6) = 6!$ different orders in which these people can be placed in a line. Six of these orders can be represented as follows:

abcdef
bcdefa
cdefab
defabc
efabcd
fabcde

Note that if we form each of these lines into a circle, we cannot tell one circle from another. The diagram

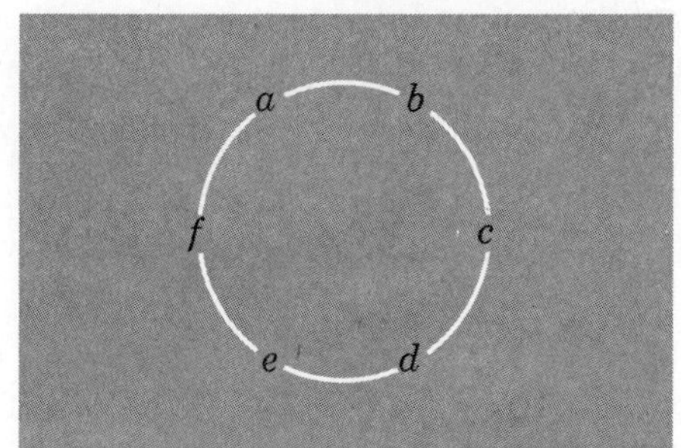

can be used as the representation for any of them. Although each starts with a different letter, they all have the same order of elements as we proceed clockwise around the circle. Thus, the 6! different orders for arranging the people in a line can be grouped in sets of six arrangements each. Each of these groups form the same circular arrangement. The number of distinguishable ways in which the people can be arranged in a circle is, therefore, (6!)/6, or 5!.

Permutations of this type are called **circular permutations**. In general, the number of circular permutations for a set of n elements is

$$CP(n,n) = (n-1)! \tag{6.2.7}$$

If r elements are to be selected from a set of n and arranged in a circle, we have

$$CP(n,r) = \frac{n!}{r(n-r)!} \tag{6.2.8}$$

Note that this formula can also be written as

$$CP(n,r) = \frac{1}{r}P(n,r)$$

$P(n,r)$ gives the number of arrangements in a line for r elements selected from a set of n. The $1/r$ takes care of the fact that the linear arrangements in $P(n,r)$ can be separated into groups of r arrangements that are the same when placed in a circle.

EXERCISE 6.2

1. Compute the numerical value for each of the following:
   a. $P(4,2)$ b. $P(6,2)$
   c. $P(5,4)$ d. $P(3,3)$
   e. $CP(7,4)$

2. a. List the permutations of the letters a, b, c, and d.
   b. How many different 4-digit numbers can be formed from the digits 1, 2, 3, and 4, if each digit is used only once?

3. Consider a circular permutation that can be "turned over"; that is, the order of the elements **reversed**, without a distinction being made. How

will this affect the number of possibilities? Compute the number of distinguishable orders in which three different keys can be placed on a key ring.

4. A person is asked to list in the order of his preference the five motion pictures nominated for an Oscar. How many different listings are possible?

5. An art appreciation class is shown 10 paintings. Each student is asked to select the five he likes best and list them in the order of his preference. How many different listings are possible?

6. A \$1500 and a \$1000 dollar scholarship are being awarded in each of the following subject areas: business, social science, and art. Seven students are competing for the business scholarship; nine, for the social science scholarship; and six, for the art scholarship. In how many different ways can the awards be made?

7. A company has four jobs open, one in each of four different departments. Nine people apply for the jobs. In how many different ways can four of the applicants be chosen to fill the jobs?

8. In a history quiz there is a list of five events, and a list of eight dates. A date is to be selected from the list and matched with each event.
   a. If each date can be used with only one event, in how many ways can the match-ups be made?
   b. If each date can be used as often as desired, how many sets of match-ups are possible?

9. Ships often form signals with colored flags hoisted so that they are in a vertical row. If a ship has eight different flags and a signal is formed using three flags at a time, how many different messages are possible?

10. Ten contestants are selected for the final judging in a beauty contest. If there are five different awards to be made and each contestant can receive no more than one award, how many different final results are possible?

11. a. Five salesmen are being assigned to five different districts. In how many ways can this assignment be made?
   b. If three of the districts are on the east coast and two on the west coast, how many different assignments can be made?

12. A theater troupe consists of four actors and three actresses. One play in their repertoire has parts for two men and two women. In how many different ways can they cast the play?

13. A company has just completed the training of four salesmen and four service technicians. These are to be assigned with one salesman and one service technician going to each of four district offices. In how many ways can this be done?

14. If the company of Problem 13 has six district offices to which the men can be sent,
   a. How many different ways can the assignment be made if each office that gets a salesman also gets a service technician?

b. How many different ways can the assignment be made if the only restriction is that all the salesmen will be sent to different offices and all the technicians will be sent to different offices?

6.3 COMBINATIONS

In Chapter 1 we found that a set with n elements has 2^n subsets ($2^n - 1$ of these have elements in them.) We are now concerned with how many subsets there are with a specific number of elements. For example, if a set contains five elements, it has 32 subsets, 31 of which are not empty: How many of these subsets have four elements? How many have three elements? and so on.

In this context, that is, when the **number** of **elements** in a subset is the matter of principal interest, the subsets are called **combinations**. We use the expression, "combinations of n things taken r at a time," to denote the different sets of r elements that can be formed using the elements of a set containing n elements. It is assumed, of course, that $n \geq r$.

A few examples of combinations are:

The number of different committees with three members each that can be formed from a group of five people is the number of combinations of five things taken three at a time.

A person has eight new records. There is time to play just four of them before dinner. The number of different selections of the four records to be played is the number of combinations of eight things taken four at a time.

An advertising man is given a list of 10 products to be featured in display ads for the next month. Each ad has room to display five products. The number of different selections of products is the number of combinations of 10 things taken five at a time.

A psychologist has eight people available to assist with experiments. An experiment requires six assistants. The number of different groups of assistants that he can choose is the number of combinations of eight things taken six at a time.

We shall use the symbol,* $C(n,r)$, to designate the number of combinations of n things taken r at a time.

We wish now to be able to compute the numerical value for $C(n,r)$. Note that if we take each of these subsets and arrange the r elements in all the orders possible, we have the permutations of n things taken r at a time. Since we form the $C(n,r)$ subsets **and** then arrange them in order, the total number of ways

*The symbol $\binom{n}{r}$ is frequently used to designate the number of combinations of n things taken r at a time. This symbol is also called a **binomial coefficient** because it occurs in the expansion of a binomial raised to the nth power. We shall not explore the reasons why the same formula occurs in two apparently unrelated situations.

this can be done is the product of the number of subsets (combinations) times the number of possible orderings. In the previous section this was found to be $r!$. Thus, we have

$$C(n,r) \cdot (r!) = P(n,r)$$

In the previous section we found that

$$P(n,r) = \frac{n!}{(n-r)!}$$

Therefore, by substitution,

$$C(n,r) \cdot (r!) = \frac{n!}{(n-r)!}$$

and by rearranging, we have

$$C(n,r) = \frac{n!}{r!(n-r)!} \tag{6.3.1}$$

Equation 6.3.1 provides a formula for computing the number of combinations of n things taken r at a time.

Example 6.3.1

Let us consider the number of combinations of a set of five elements taken three at a time. Equation 6.3.1 says that

$$C(5,3) = \frac{5!}{3!(5-3)!} = \frac{5!}{3!\cdot 2!} = \frac{120}{6 \cdot 2} = 10$$

Let us consider the set $\{a,b,c,d,e\}$. The three-element subsets (combinations) are

$\{a,b,c\}$ $\{a,d,e\}$
$\{a,b,d\}$ $\{b,c,d\}$
$\{a,b,e\}$ $\{b,c,e\}$
$\{a,c,d\}$ $\{b,d,e\}$
$\{a,c,e\}$ $\{c,d,e\}$

We see that there are 10 combinations, as predicted. Each of these combinations can be arranged in $3! = 6$ different orders, as we have seen. Thus, the number of permutations of five things taken three at a time is $6 \cdot 10 = 60$. This agrees with our findings in the previous section.

Let us verify that the formula for the number of combinations (Equation 6.3.1) is consistent with the total number of subsets for a set. We can use a set of five elements as a test case. This set has a total of $2^5 = 32$ subsets, including the set itself and the null set.

The entire set is obtained by taking the combinations of five elements five at a time.

$$C(5,5) = \frac{5!}{5!(5-5)!} = \frac{5!}{5!0!} = 1$$

Recall that we define $0!$ equal to one.

Continuing with subsets of four, three, two and one elements, we have

$$C(5,4) = \frac{5!}{4!(5-4)!} = \frac{5!}{4!1!} = 5$$

$$C(5,3) = \frac{5!}{3!(5-3)!} = \frac{5!}{3!2!} = 10$$

$$C(5,2) = \frac{5!}{2!(5-2)!} = \frac{5!}{2!3!} = 10$$

$$C(5,1) = \frac{5!}{1!(5-1)!} = \frac{5!}{1!4!} = 5$$

The null set can be considered as the number of combinations of five things taken none at a time. That is, the number of ways in which we can obtain a subset with zero elements. There is just one way to do this: leave all the elements out. Our formula still holds, and we have

$$C(5,0) = \frac{5!}{0!(5-0)!} = \frac{5!}{0!5!} = 1$$

The total number of subsets is the sum of the number for each number of elements, or $1 + 5 + 10 + 10 + 5 + 1 = 32$, as we found before.

6.3.1 A Simple Problem in Sampling

Sampling is a technique used when a person is dealing with a large number of things that are **ostensibly** the same, but that have **measurable differences** among them. Sociologists and psychologists study the activities and characteristics of people. They seek knowledge and understanding that will be applicable to all people, but it is completely impractical for them to study all people individually. Accordingly, they select for study groups of people, or **samples**, that they hope and expect are **representative** of an entire population. In business, also, samples of customers, processes, policies, products, and so on are selected to be studied in various ways. Methods for obtaining truly representative samples can be quite complex, and a study of them is beyond the scope of this text. We shall confine our attention here to some of the computations that are made in analyzing the significance of tests and observations made on samples.

For example, it is of interest to a manufacturing company to know how many of the items produced on one of its assembly lines have been assembled properly and how many are defective. When the quantity of items produced is too large to permit testing each one individually, a **sample** of the products is selected to be tested. A complete evaluation of the results of the testing is a problem in statistics. We direct ourselves here to some of the computations upon which these evaluations are based.

One of the necessary computations is to count the ways in which a specified number of defective products appear in a sample, based on a known number of defective products in the total production.

Example 6.3.2

Let us consider a situation in which a company produces 100 of a certain type of item per day. We shall assume that 10 of these items are defective. Note that this figure is not known in an actual situation. Part of the method of satistical inference is to figure out what to expect under different assumptions. Actual experience is then compared with the computed results. Statistical procedures provide ways to determine which assumption is most likely to be correct. In performing the statistical analysis the answers to the following types of questions are needed: In how many different ways can a sample be selected from the day's production? In how many of these samples will there be exactly one defective item? and so on. We can use the formula for the number of combinations to provide answers to these questions.

First, the number of ways to select 10 items from a group of 100 is the number of combinations of 100 things taken 10 at a time, or

$$\text{Number of samples possible} = C(100, 10) = \frac{100!}{90!\,10!}$$

$$= \frac{100 \cdot 99 \cdot 98 \cdot 97 \cdot 96 \cdot 95 \cdot 94 \cdot 93 \cdot 92 \cdot 91}{10 \cdot 9 \cdot 8 \cdot 7 \cdot 6 \cdot 5 \cdot 4 \cdot 3 \cdot 2 \cdot 1}$$

$$= \text{approximately } 17.3 \text{ trillion } (17.3 \times 10^{12})$$

In computing the number of samples that have exactly one defective item, we use the information that there are 10 defective items and $100 - 10 = 90$ good items in the total group. The sample is to contain one defective item and nine good ones. The one defective item must be selected from the 10 defective items, and the nine good ones must come from the 90 good ones in the total. For mathematical purposes we assume that we have the day's production separated into two groups, the good items and the defective ones. To make this plausible imagine that the defective items are a different color from the good ones. Then, in choosing the nine good ones, you will make your selection from the 90 good ones in the total group; and in selecting the one defective one, the selection will be made from among the 10 defective ones in the group. Recall that the order in which the selection is made is not a factor in selecting combinations. The number of ways in which the selection can be made is the product of the number of ways to select the good ones and the number of ways to select the defective items.

$$\text{Number of samples with 1 defective} = \frac{90!}{81!\,9!} \cdot \frac{10!}{9!\,1!}$$

$$= \text{approximately } 7.05 \text{ trillion } (7.05 \times 10^{12})$$

Using the same considerations, we can find the number of ways in which zero, two, or three defective items can be selected in the sample.

Number of samples with zero defective items

$= C(90,10) \cdot C(10,0)$

$= \dfrac{90!}{80!10!} \cdot \dfrac{10!}{10!0!}$

= approximately 5.7 trillion (5.7×10^{12})

Number of samples with two defective items

$= C(90,8) \cdot C(10,2)$

$= \dfrac{90!}{82!8!} \cdot \dfrac{10!}{8!2!}$

= approximately 3.5 trillion (3.5×10^{12})

Number of samples with three defective items

$= C(90,7) \cdot C(10,3)$

$= \dfrac{90!}{83!7!} \cdot \dfrac{10!}{7!3!}$

= approximately 8.93 billion (8.93×10^{9})

It is clear that these numbers are too large to have much meaning for us. In the next chapter we shall see how to combine the numbers to obtain some meaningful and useful quantities. The computations in this chapter are the first step.

6.3.2 Sampling with Replacement

In the example of the preceding section it is assumed that as each item in the sample is selected, it is removed from the set of items being sampled. Thus, on each succeeding selection there is one less item to choose from. There are situations in which succeeding selections are made from the entire set. This process is called **sampling with replacement**. After each item of the sample has been selected, it is returned to the set prior to the selection of the next item in the sample. The possibility exists, of course, that the same item will be selected more than once for the sample.

The counting pattern involved in determining the number of different possibilities for selection is the same as that studied in Section 6.1. The number of possibilities for each step in the sequence of selection is the total number of elements in the set being sampled.

If the set has 10 elements, for example, and a sample of three elements is to be made **with replacement**, the number of choices for each of the three elements in the sample is the same – 10. The total number of possible sample selections is, therefore,

$$SR(10,3) = 10 \cdot 10 \cdot 10 = 10^3$$

$SR(10,3)$ represents the number of different samples with replacement of three items from a set of 10.

This formula can be generalized to give the number of different samples with replacement of r items from a set of n by noting that for each of the r selections there are n items to choose from. Thus,

$$SR(n,r) = n^r \qquad (6.3.2)$$

6.3.3 Selection of a Counting Pattern

The most difficult problem in counting the number of ways to make selections, and so forth, is to determine which counting pattern applies to a given situation. The student should try to imagine himself in the position of actually making the selections. Often a simple diagram of the selection procedure is useful and instructive. The following basic guidelines indicate the distinguishing characteristics of each counting pattern:

1. If selections are made in which each choice is removed from further consideration—**sampling without replacement**:
   a. If there is some means of distinguishing one selection from another—the order of choice can be determined—the counting pattern is a **permutation.**
   b. If the selections constitute forming **subsets** of a set—there is no way of distinguishing one element chosen from another—the counting pattern is a **combination.**

2. If selections are made from an entire set each time—**sampling with replacement**—the total number of elements is raised to the power of the number of selections made.

3. In all cases where there is more than one choice or selection: if the selections are **all** to be made—one selection is made **and** a second **and** a third, and so on—the total number of possibilities is the **product** of the numbers for each individual selection. If the choices are **alternatives**—one choice **or** a second **or** a third, for example—the total number of possibilities is the **sum** of the numbers for the individual selections.

In the next section we shall investigate a few more complex situations and develop rules to be applied to determine the appropriate counting patterns.

EXERCISE 6.3

1. Compute the numerical value for the following:
   a. $C(6,4)$ b. $C(8,3)$
   c. $C(8,5)$ d. $C(6,2)$
   e. $C(10,7)$

2. a. List the possible 2-element subsets that can be selected from the set $\{1,2,3,4,5\}$.
   b. How many 2-digit numbers can be formed from the elements of each subset in part *a*.
   c. How many permutations are possible if two elements are selected from a set of five elements.

3. A bowl contains 10 marbles. Three marbles are to be selected from the bowl.
   a. How many different choices are there for the first selection? The second? The third?
   b. How many total possibilities are there for the selections of (*a*)?
   c. In how many of the possibilities for (*b*) will the same three marbles be selected?

4. A psychologist has a set of 12 questions from which he selects five for use with each subject in an experiment. How many different selections are possible for the five questions?

5. A merchant has a display window at the front of his store. He has 10 products that he wishes to feature in his displays. If the window has room for only six of them, in how many different ways can he select the products to be displayed?

6. A man selects a sample of each of three different grades of carpet for inspection. There are eight colors in each of the grades. How many different color combinations are possible in his sample?

7. Ten men compete in a pentathlon (a five-part track-and-field event). How many different sets of winners are possible for the individual parts of the event?

8. A service foreman has nine men in his repair crew. If the men are sent out on jobs in teams of four, how many different teams can be chosen?

9. During the summer months the repair men of Problem 8 get many more calls. The service foreman divides the men into three crews of three men each, one crew to answer the first call, another for the second call, and the third for the third call. In how many ways can he form the crews?

10. A company is conducting a sales contest in which three salesmen will be selected for a special bonus. In the last week of the contest it becomes clear that eight salesmen have a chance to win. A press agent for the company wants to prepare advance publicity for release as soon as the contest is ended. How many different stories must be prepared to cover all the possible sets of winners?

11. An instructor wants to prepare a set of exam questions so that he can give each of his 50 students a different 5-problem exam. (Exams are "different" if no more than 4 of the 5 exam questions are the same.) How many problems must he prepare? (**Hint:** Use trial and error to find your solution.)

12. A woman has a dozen eggs of which three are spoiled. She selects three eggs to make a cake.
   a. In how many ways can she make her selection?
   b. In how many of the selections will there be no bad eggs?
   c. In how many selections will there be exactly one bad egg?
   d. In how many selections will there be two bad eggs?
   e. In how many selections will she get all three bad eggs?
   f. Does the sum of the answers to b, c, d, and e equal the answer to a?

13. Among the 12 questions of Problem 4, there are four that deal with dreams. In what percentage of the possible sets of five questions is there at least one question that deals with dreams?

6.4 COMPLEX COUNTING PATTERNS

In this section we shall consider a few examples of more complicated situations in which different counting patterns must be combined to find the number of possibilities. The fundamental principle of counting is the foundation for all of the counting. It must be kept in mind, however, that this principle applies in situations where there is a sequence and a number of possibilities for each part of the sequence. The total number of possibilities is then found by **multiplying** the numbers in each of the parts together. The key consideration is that a possibility occurs for the first part of the sequence **and** for the second part **and** for the third part, and so on.

There is another type of situation in which the occurrence of some of the possibilities in one or more of the parts makes it necessary to use a different counting pattern than used for the others. In such a situation one sequence of possibilities **or** another will occur. It is necessary to separate the sequence into two (or more) alternate sequences and compute the number of possibilities in each. The total number is found by **adding** the numbers in each of the alternative sequences. Example 6.4.2 contains a problem that must be solved in this way.

Example 6.4.1

Let us consider a situation in which different parts of the sequence involve different counting patterns. Suppose that a delegation to a convention has 10 members. The delegation must choose a chairman, a vice-chairman, and three other members to act as "Principal Delegates" at a policy meeting. How many different choices for these positions are possible?

The selection of the two officers, the chairman and vice-chairman, implies an **ordered** selection. There is a difference between being selected as chairman or as vice-chairman. The counting pattern is, therefore, to count permutations. There is no distinction made among the three Principal Delegates, and the different possibilities are combinations.

Since there are 10 delegates in all, the number of possible choices for the two officers is $P(10,2)$, the number of permutations of 10 things taken two at a time. After the officers are selected, there are eight delegates left from which to choose the three Principal Delegates. The number of possible choices is $C(8,3)$, the number of combinations of eight things taken three at a time.

The selection process consists of a sequence of selections. The total number of possibilities, therefore, is the product of the numbers for each part. Thus,

$$\text{Number of possibilities} = P(10,2)\cdot C(8,3)$$

$$= \frac{10!}{8!}\cdot\frac{8!}{5!3!}$$

$$= \frac{10\cdot 9\cdot 8\cdot 7\cdot 6}{3\cdot 2\cdot 1} = 5040$$

Example 6.4.2

Let us consider now a situation in which a selection process must be separated into alternative selections for counting. Suppose that as an inducement to obtain new members a book club offers a choice of five free books from a list of 10 books when a new member joins the club. The list of 10 books, however, contains two that are a set and must be chosen together. Choosing this set counts as the selection of two of the free books. How many different choices are possible?

The individual selections possible are combinations, since there is no order inherent in the selection process. A given set of books chosen has the same members regardless of the order in which they are selected. A complication arises, however, in handling the selection of the two-book set. When this set is chosen, the new member has only four choices instead of five—the two-book set and three others.

The number of possibilities is counted by separating the selection into two alternative sequences, one in which the two-book set is chosen, and the other in which it is not. If the two-book set is chosen, three other books may be selected from a list of eight. This selection can be made in $C(8,3)$ ways. If the two-book set is not chosen, five books are selected from the remaining eight. This can be done in $C(8,5)$ ways. The total number of selections possible is the **sum** of the numbers for each of the alternatives.

$$C(8,3) = \frac{8!}{5!3!} \qquad C(8,5) = \frac{8!}{3!5!}$$

$$= \frac{8\cdot 7\cdot 6}{3\cdot 2\cdot 1} \qquad = \frac{8\cdot 7\cdot 6}{3\cdot 2\cdot 1}$$

$$= 56 \qquad = 56$$

$$\text{Total number of selections} = 56 + 56 = 112$$

Example 6.4.3

In the previous example the "standard" counting pattern was complicated by requiring that one possible choice in a sequence be essentially a double choice. Let us consider now a situation in which the choice of one element from a set precludes the choice of another. Let us suppose that there is a group of eight applicants for four jobs with a company. Two of the applicants, however, are husband and wife, and company policy does not permit hiring two employees from the same family unit. How many different selections are possible for the new employees?

There are three alternative sequences in this situation: in one, the husband is hired, but not the wife; in the second, the wife is hired, but not the husband; and in the third, neither is hired. Since these are alternative sequences, we must compute the number of possibilities for each and then add these numbers to obtain the total number of possibilities.

If the husband is hired, three other new employees are selected from the six applicants remaining. The wife is excluded from consideration. The number of possibilities is $C(6,3)$. The order in which the other employees are selected is of no consequence in this situation.

If the wife is hired, the other three new employees are selected from the six applicants not including the husband. There are $C(6,3)$ ways in which the selection can be made.

If neither the husband nor wife is hired, all four new employees are selected from the other six applicants. There are $C(6,4)$ ways in which this selection can be made.

Thus, the total number of different selections possible is given by

$$\begin{aligned}\text{Number of possibilities} &= C(6,3) + C(6,3) + C(6,4)\\ &= \frac{6!}{3!3!} + \frac{6!}{3!3!} + \frac{6!}{4!2!}\\ &= 20 + 20 + 15\\ &= 55\end{aligned}$$

Notice that in each of these examples it is necessary to analyze the situation and break it into its component parts before proceeding with the computations. There are no mathematical formulas for making the analysis. The "mathematical approach" of recording the analysis in a diagram, or other orderly manner, does aid in making sure that the analysis is complete and accurate.

Example 6.4.4

Three men and three women participate in a panel discussion. Suppose that the moderator wishes to have them sit so that no two men and no two women are seated next to each other. How many different seating arrangements are possible?

If a woman is seated first, the lineup is W-M-W-M-W-M. It is possible as an alternative, of course, for a man to be seated first. In this case the lineup is M-W-M-W-M-W. These are the only two possible arrangements consistent with the requirements that neither men nor women be seated next to each other, and they are **alternative** arrangements.

The placing of the women and the men must be considered separately. The moderator's specification immediately separates them into two groups. The fact that we are dealing with individual people with separate identities makes the order in which they are seated of importance. The number of orders in which the women can be seated is $P(3,3)$; and the number of seating orders for the men is the same.

Thus, the total number of possible seating arrangements is the sum of the number of possibilities if a woman is seated first and the number if a man is seated first.

$$\begin{aligned}\text{Number of possibilities} &= P(3,3)\cdot P(3,3) + P(3,3)\cdot P(3,3)\\ &= (3!)(3!) + (3!)(3!)\\ &= 36 + 36 = 72\end{aligned}$$

Note that any specification that provides **where** the two groups are to be seated has the same number of possibilities. If all the men and all the women are to be seated together, the two possible arrangements of the groups are W-W-W-M-M-M and M-M-M-W-W-W. The total number of possibilities for this situation is computed in the same way as above.

Suppose the moderator specifies that the women are to be seated together, but no additional restriction is made for seating the men. This specification divides the panelists into four separate entities: the group of women plus the three men. These four entities can be arranged in $P(4,4)$ ways. The group of women can be arranged in $P(3,3)$ ways as before. Thus, the total number of possibilities is given by

$$\begin{aligned}\text{Number of possibilities} &= P(4,4)\cdot P(3,3)\\ &= 24\cdot 6 = 144\end{aligned}$$

The foregoing situation can be analyzed in other ways. For example, we can list the four possible locations for the group of women: W-W-W-M-M-M, M-W-W-W-M-M, M-M-W-W-W-M, and M-M-M-W-W-W. For each of these **alternative** arrangements there are $P(3,3)$ orders for seating the women and $P(3,3)$ orders for seating the men. The total number of possibilities is, therefore, $4\cdot P(3,3)\cdot P(3,3)$, or 144 as computed above.

If there are no restrictions at all placed on where the panelists are to sit, there are six people to arrange in order. There are $P(6,6)$ ways this can be done.

Example 6.4.5

There are eight employees in a particular department of a small company. According to the company's policies up to three of these employees can be on vacation at the same time. How many different vacation schedules are possible for the first week in August?

The number of employees who are on vacation during this particular week may be three, **or** two, **or** one, **or** none. If there are three, the number of different possibilities for the individuals who are on vacation is $C(8,3)$. If there are two on vacation, the number of possibilities is $C(8,2)$. If there is one, the number of possibilities is $C(8,1)$; and if none, there is just one possibility. The total number of different possibilities is the **sum** of the number for each of the alternatives. Thus,

$$\begin{aligned}\text{Total possibilities} &= C(8,3) + C(8,2) + C(8,1) + 1\\ &= \frac{8!}{5!3!} + \frac{8!}{6!2!} + \frac{8!}{7!1!} + 1\\ &= 56 + 28 + 8 + 1 = 93\end{aligned}$$

In each of the examples in this section we have analyzed a situation to determine the applicable counting patterns. Such analysis is an essential part of using the formulas for counting. In many cases there are different ways to approach the problem and different ways to solve it. The same basic counting patterns can be applied in many ways in a variety of contexts. It is not possible to list situations and an appropriate counting formula to be used by mere substitution of numbers. The counting formulas can be applied only if the essential features of their structure are thoroughly understood.

EXERCISE 6.4

1. A committee of three employees is to be elected as a negotiating team in a contract dispute. Eight employees are nominated, and two of them are from the same department. It is decided that no more than one member from any department should be on the team. How many different committee selections are possible?

2. A psychologist has seven sets of questions from which he selects four to make up an aptitude test. Two of the question sets, however, are related in such a way that one set should not be used without the other. How many different selections are possible?

3. A company has 10 vice-presidents. A president, an executive vice-president, and an executive board of five members are to be selected from this group. How many possibilities are there for the results?

4. The situation in the preceding problem is altered somewhat by further considerations. There are three senior vice-presidents from whom the president and executive vice-president are selected. The third one of

these is certain to be placed on the executive board. How many possibilities are there?

5. In a company bowling tournament three members of the Quality Control Department and three members of the Service Department are competing for the six top awards.
   a. In how many different orders can they finish?
   b. In how many of these will the Service Department team win the top three prizes?
   c. In how many will one team finish in consecutive places (first, second, third; third, fourth, fifth, etc.)?
   d. In how many will both teams finish in consecutive order?

6. Refer to Example 6.4.1. In determining the number of possibilities for the selection of the delegation, we assumed that the chairman and vice-chairman were chosen first. Demonstrate the fact that the order of making the selections is not important by computing
   a. The number of possibilities if the chairman is selected first, then the three Principal Delegates, then the vice-chairman.
   b. The number of possibilities if the three Principal Delegates are chosen first, and then the chairman and vice-chairman. In all computations the selections are to be made from a group of 10 delegates.

7. A company that markets eight product lines selects from two to four of them to feature in its weekly advertisements in a newspaper. How many different possibilities are there?

8. If two of the product lines of Problem 7 are such that they are always selected together, how many possible selections will there be?

9. If two of the product lines of Problem 7 are such that they are never selected together, how many possibilities are there?

10. A crew of nine field workers is making a survey of customer brand preference for canned soup. Three large supermarkets, A, B, and C, in a city are selected as operating locations for the survey. The nine workers are divided into three teams of three workers each, one team to work in each market. How many different ways are there to form the teams?

11. A delegation, consisting of a chairman and two members, is to be selected from a group of seven people. There are two members of the group whose feelings toward each other are such that neither will serve on the delegation if the other is chairman, although they would agree to serve together as the regular members of the delegation if selected. How many different selections of the delegation are possible?

12. A bowl contains 10 marbles: three red, three yellow, two blue, and two green.
   a. Two marbles are selected from the bowl. How many possibilities are there for this selection? In how many of these will there be either a blue or a green marble?
   b. Three marbles are selected. In how many of the possible selections will there be at least one yellow marble?

6.5 PARTITIONS

The partitioning of sets was discussed briefly in Chapter 1. In that discussion the principal interest was directed toward partitions determined by the definition of the subsets to be formed. For example, a group of people can be partitioned into subsets on the basis of their age, their occupation, their income, and so on. We turn our attention now to partitions made on the basis of the **number** of **elements** in each subset, or cell. Groups of items are often divided up into smaller groups with specified numbers of elements. Our interest in such partitions centers on counting the number of ways that the partitioning can be done.

The concept of partitioning is essentially another way of viewing some of the ideas that we have discussed in preceding sections. Many of the examples in this chapter can be discussed from the point of view of the partitioning of a set. In Example 6.4.1 we computed the number of ways to select a chairman, vice-chairman, and three principal delegates from a delegation with 10 members. The selection process can be viewed as a partitioning of a set of 10 elements into two cells with one element each (chairman and vice-chairman), one cell with three elements (principal delegates), and one cell with five elements (unselected delegates). Any sampling problem in which there is sampling without replacement can be viewed as a partitioning of the set being sampled. The different approaches to the same basic problem are presented for two principal reasons. First, one point of view seems "more natural" and, thus, easier to apply than another in some situations and, second, studying a subject from different viewpoints provides a more complete understanding of it.

There are several general considerations for the partitioning process that we should discuss before considering the different types of partitions in detail. First, since the cells of a partition are **subsets**, the principal counting pattern that we shall use is **combinations**. Notice, however, that in special situations it may be necessary to place the elements of one or more cells of the partition in a particular order. The counting pattern must include the number of permutations possible for those particular cells.

Second, the cells themselves may or may not be identifiably different. That is, it may or may not be possible to set up a specific order for the selection of the cells. If an order can be specified for the selection of the cells, the partition is an **ordered partition**. If there is no identifiable order for the selection, the partition is an **unordered partition**.

6.5.1 Ordered Partitions

The distinguishing characteristic that makes a partition in terms of the number of elements in the cells into an **ordered** partition is that the **cells** are **identifiably different** in some way. In Example 6.4.1 the cells of the partition can be identified by the name of the office for the two one-element cells, and by the number of elements in the other two cells.

In computing the number of possibilities for an ordered partition we consider the cells in sequence. As we shall see, it does not matter in which order they are considered, so long as they are considered in **some** particular order. Let us consider the situation described in Problem 10 of Exercise 6.4. A nine-member crew is partitioned into three, three-member teams. Each team is to visit a particular supermarket in a city. The teams can be identified by the market they are to visit. If the markets are designated as A, B, and C, we have Team A, Team B, and Team C. Thus, the partition is **ordered.**

We select the team for each of the markets in sequence. For Market A there are $C(9,3)$ possible selections. After the selections have been made for Market A, there are six workers left unassigned. We choose from among them to obtain the survey team for Market B. There are $C(6,3)$ possibilities. There are three workers left unassigned, so these will go to Market C. The total number of possibilities is

$$\begin{aligned}\text{Total number of partitions} &= C(9,3)\cdot C(6,3)\cdot 1\\ &= \frac{9!}{3!6!}\cdot\frac{6!}{3!3!}\\ &= \frac{9!}{3!3!3!}\\ &= 1680\end{aligned}$$

Let us now consider a case in which we divide nine people into three groups, one with four members, one with three members, and one with two members. This partition is automatically an ordered partition because we can distinguish the cells from each other by the number of elements that each contains. To compute the number of ways the partitioning can be done, we start with one of the cells, let us say the four-element cell. The number of ways that this cell can be formed is $C(9,4)$. There are now five people left from which to form the other two cells. If we choose the three-element cell next, the number of possibilities is $C(5,3)$. The two people remaining make up the two-element cell. The total number of possibilities is given by

$$\begin{aligned}n(OP) &= C(9,4)\cdot C(5,3)\\ &= \frac{9!}{5!4!}\cdot\frac{5!}{3!2!}\\ &= \frac{9!}{4!3!2!}\\ &= 1260\end{aligned}$$

The symbol, $n(OP)$, is used here to denote the number of different ordered partitions of a given cell makeup.

If we change the order of selection and choose the three-element cell first, followed by the two-element and the four-element cells, we have

$$\begin{aligned}n(OP) &= C(9,3)\cdot C(6,2)\\ &= \frac{9!}{6!3!}\cdot\frac{6!}{4!2!}\end{aligned}$$

$$= \frac{9!}{3!4!2!}$$

$$= 1260$$

This is the same result as we obtained before.

Let us look carefully at the expressions used in the foregoing computations. Note that in terms of factorials the number of different ordered partitions is given by

$$n(OP) = \frac{9!}{4!3!2!}$$

In the formula the numerator, $9!$, is obtained from the number of elements in the set being partitioned. The three factors in the denominator, $4!$, $3!$, and $2!$, are obtained from the number of elements in each of the cells.

Similarly, in computing the number of different survey teams possible for the market study we have the expression,

$$\frac{9!}{3!3!3!}$$

The numerator is obtained from the number of members of the crew being partitioned. The three $3!$'s in the denominator come from the number of elements in the cells. This suggests that we can write a general formula for the number of **ordered** partitions that can be formed from a set. If a set has n elements, the formula for the number of ways in which it can be partitioned into cells containing $a, b, c, \ldots, k$ elements is

$$n(OP) = \frac{n!}{a!b!c!\ldots k!} \tag{6.5.1}$$

In this formula the number of elements in each cell may or may not be the same. The numbers, $a, b, c, \ldots, k$, represent the numbers of elements in **all** of the partitions. Thus, if there are to be two cells with the same numbers of elements, that number must appear as a factorial twice in the denominator. For example, we wish to form an **ordered** partition of a set with 20 elements into two cells of three elements each, three cells with four elements each, and one cell with two elements. The number of different partitions of this type is

$$n(OP) = \frac{20!}{3!3!4!4!4!2!}$$

6.5.2 Unordered Partitions

If a set is partitioned in such a way that there is no way to distinguish one cell from another, the partition is said to be **unordered**. For example, if a set of six elements is merely to be divided into three subsets of two elements each, there is no way of distinguishing one cell from another. If the elements of the set are a, b, c, d, e, and f, **one** such partitioning yields the subsets, $\{a,b\}$, $\{c,d\}$, and $\{e,f\}$. That is, we have selected a and b as the elements of the first subset, c and d as the elements of the second subset, and e and f as the elements of the third subset. However, the same partitioning would have resulted if we had selected c and d as the first subset, and a and b as the second subset. In fact, there are $P(3,3)$ or $3!$, different orders in which we could have selected the same subsets. Thus,

$$n(OP) = (3!)n(UP)$$

where $n(OP)$ represents the number of ordered partitions, and $n(UP)$ represents the number of unordered partitions.

In the previous section we found that the number of ordered partitions of a six-element set into three two-element subsets is given by

$$n(OP) = \frac{6!}{2!2!2!} = 90$$

Thus, we have that the number of unordered partitions of a six-element set into three two-element subsets is

$$n(UP) = \frac{1}{3!} \cdot \frac{6!}{2!2!2!} = 15$$

In this example all of the cells of the partition have the same number of elements and, thus, are indistinguishable. It is necessary in an unordered partition that all of the cells have the same number of elements. Whenever the cells have different numbers of elements, we have a means of telling them apart. They become an ordered partition.

Let us consider now a case in which some of the cells have one number of elements and others a different number. We assume that there is no means of identifying the cells except by the number of elements. We have a partial ordering of the partition. The cells having the same number of elements form an unordered "subpartition," and the subpartitions with different numbers of elements are ordered. To compute the number of ways such a partitioning can be made, the cells are first grouped according to the number of elements they contain. The **denominator** of the counting formula is then written considering each of the groups as an unordered partition. The numerator is the factorial of the total number of elements in the partitioned set. If there are k cells in a group with the same number of elements, the factor, $k!$, $(P(k,k))$, must be included in the denominator to take care of the different orders in which those cells can be formed.

For example, let us form an unordered partition of a set of 10 elements into two three-element subsets and one four-element subset. The number of ordered partitions is

$$n(OP) = \frac{10!}{3!3!4!} = 4200$$

The two three-element cells cannot be distinguished, however, and there will be $P(2,2)$ of the ordered partitions with the same two three-element cells. Thus,

$$n(UP) = \frac{1}{2!}n(OP) = \frac{10!}{2!3!3!4!} = 2100$$

In the previous section we found the number of ways a set with 20 elements can be formed into an ordered partition with two cells of three elements each, three cells of four elements each, and one cell of two elements to be

$$n(OP) = \frac{20!}{3!3!4!4!4!2!}$$

If we wish the same cell arrangement in an unordered partition, we find that there are $P(2,2)$ of the ordered partitions with the same three-element cells and $P(3,3)$ with the same four-element cells. Therefore,

$$n(UP) = \frac{1}{2!} \cdot \frac{1}{3!} \cdot \frac{20!}{3!3!4!4!4!2!}$$

6.5.3 Ordered Listings—Agreements and Disagreements

In our study of probability there are situations in which we are interested in the number of agreements and disagreements between two ordered listings of the same set of elements. An example of such a listing familiar to students occurs in a type of quiz sometimes given. There is a list of questions or topics, and a set with the same number of words or phrases. The words or phrases are to be matched to the questions. The student forms an ordered listing of the set of words or phrases, hoping that his order for the responses matches that of the teacher. Our interest is in counting the number of different ways in which there can be a specified number of disagreements between the two listings. That is, in a quiz with 10 questions, in how many different ways can a student "miss four"—have four elements in his list disagree with the teacher's list?

Another example of comparing listings is to have two people rate a list of items in the order of their preference for them. Those items that appear in the same position on both lists constitute the "agreements"; and those that appear in different positions, the "disagreements."

The computation for the number of ways in which two lists can have a specific number of disagreements consists of two considerations, the number of ways to "choose" the positions on the lists in which there is disagreement, and the number of ways in which those particular items can disagree. For example, suppose there are five items in a list, and we are interested in the number of ways in which two lists can have three disagreements. We must first determine how many ways there are to specify which three items disagree, and then find the number of ways for disagreement among these items to occur.

We have already studied the way to determine the number of ways to specify the items that disagree. It is the number of combinations of n things taken r at a time, where there are to be r disagreements in a list of n items. Determining the number of ways in which the disagreements can occur is slightly more complicated. We shall demonstrate the procedure for three and for four disagreements, and present a formula without proof that applies to larger numbers of disagreements.

First, let us note that there is no way in which there can be only one disagreement between two lists. If the lists each contain only one element, there can be but one order in which to list them. If there is more than one element in the lists, the element in the "wrong" place is occupying the place of another element. That element must also be in the "wrong" place.

Disagreement between two elements can occur in just one way—the elements are interchanged.

In our further considerations we assume that one of the listings is designated as the "standard." Other listings are compared to it. We can consider a single listing and its disagreements with a "standard" order.

To find the number of ways in which there can be three disagreements, let us designate the three elements as "1," "2," and "3" and assume that the standard order for these elements is their ordinary numerical order. The question is now, "How many ways can we list these three numbers so that no number appears in its normal numerical order?" That is, 1 cannot come first; 2 cannot come second; and 3 cannot be third. Let us use a diagram showing the possible listings:

1st   2nd   3rd

In the first position we can have either 2 or 3:

1st	2nd	3rd
2		
3		

If 2 is in the first position, we have 1 or 3 as possibilities for the second position. If we choose 1, however, that leaves 3 for the third position—its normal order. Therefore, we must choose 3 for the second position and 1 for the third. Similarly, if 3 is in the first position, 1 must be the second, and 2, third. We have, therefore, as the only possibilities for complete disagreement with normal order:

1st	2nd	3rd
2	3	1
3	1	2

There are exactly two possibilities.

Let us now construct a diagram showing the ways to have four disagreements in order. We designate the elements as the numbers 1, 2, 3, and 4. The standard order is their normal numerical sequence. Figure 6.5.1 shows a tree diagram with the possibilities for placing the "wrong" number in the first position. When 2 is in the first position, any of the other numbers can be placed

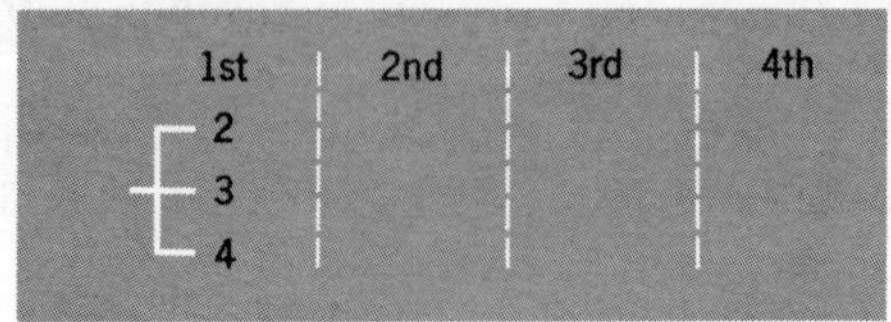

Figure 6.5.1

in the second position. When 3 is in the first position, 1 and 4 are the only possibilities for the second spot. When 4 is in the first position, only 1 and 3 can come second. These possibilities are shown in Figure 6.5.2.

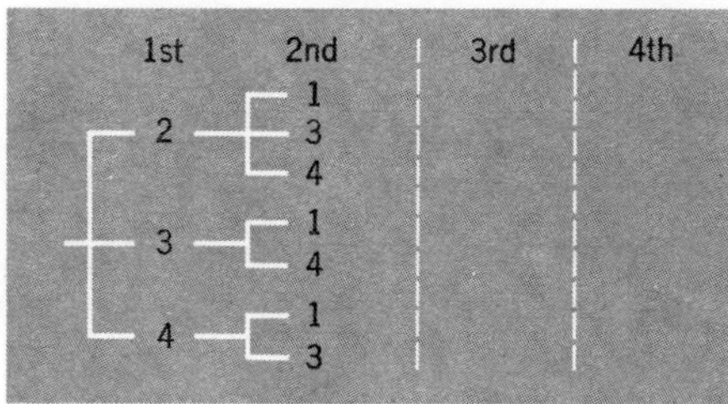

Figure 6.5.2

The remaining positions are filled with the unused elements, keeping in mind that no element can appear in its proper numerical position. Figure 6.5.3 shows all of the possible sequences, each with all of the elements in an incorrect position.

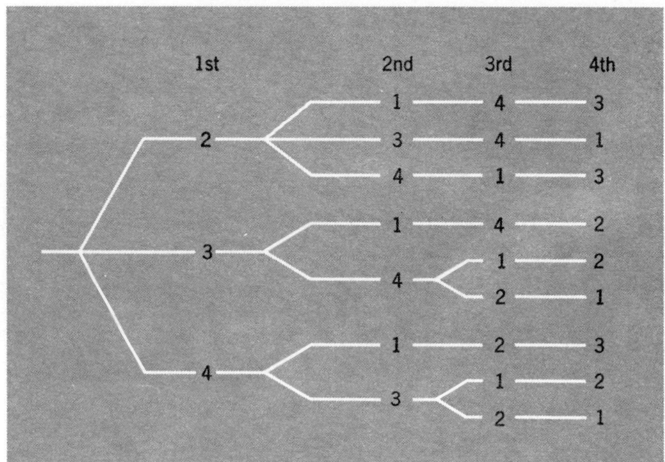

Figure 6.5.3

As can be seen in Figure 6.5.3, there are nine possible sequences in which each of the four elements is out of its normal position.

The preceding discussion illustrates the considerations involved in determining the number of ways in which two lists of the same elements of a set can disagree. Tree diagrams can be used to find the number of possibilities for disagreements between longer lists, but the diagrams are quite large and cumbersome. The following formula is stated without derivation. It was formulated by inductive reasoning, and a proof by deductive reasoning was then worked out to establish its validity. The process is too lengthy to include here.

We let $D(r,r)$ represent the number of ways there can be r disagreements between two lists of r elements—that is, for the two lists to be in complete disagreement.

$$D(r,r) = r!\left(1 - \frac{1}{1!} + \frac{1}{2!} - \frac{1}{3!} + \ldots + (-1)^r \frac{1}{r!}\right) \tag{6.5.2}$$

The factor, $(-1)^r$, is the mathematical way to handle the changes in algebraic sign of successive terms. When the factorial in a term is an even number, (-1) raised to that power is $(+1)$; when the factorial is an odd number, (-1) to that power is (-1).

Equation 6.5.2 is the general formula that applies to all values of r. Note that for $r = 1$, the formula gives $D(1,1) = 1(1 - (1/1!)) = 0$, in agreement with our previous reasoning. For larger values of r, however, the first two terms always add to zero. They can be omitted without affecting the result. We have

$$D(r,r) = r!\left(\frac{1}{2!} - \frac{1}{3!} + \frac{1}{4!} + \ldots + (-1)^r \frac{1}{r!}\right) \tag{6.5.3}$$

In the compact symbology of algebra the formula becomes

$$D(r,r) = r! \sum_{k=0}^{r} (-1)^k \frac{1}{k!} \tag{6.5.4}$$

Note that for $r = 2$, Equation 6.5.3 becomes

$$D(2,2) = 2!\left(\frac{1}{2!}\right) = 1$$

When $r = 3$, we have

$$D(3,3) = 3!\left(\frac{1}{2!} - \frac{1}{3!}\right)$$
$$= 6\left(\frac{1}{2} - \frac{1}{6}\right) = 2$$

For $r = 4$,

$$D(4,4) = 4!\left(\frac{1}{2!} - \frac{1}{3!} + \frac{1}{4!}\right)$$
$$= 24\left(\frac{1}{2} - \frac{1}{6} + \frac{1}{24}\right)$$
$$= 24\left(\frac{12}{24} - \frac{4}{24} + \frac{1}{24}\right) = 9$$

In each case the formula agrees with our previous results.

Returning now to the consideration of computing the number of ways there can be a specified number of disagreements between the listings of the same elements, we let $D(n,r)$ represent the number of ways to have r disagreements between two lists of n elements, where $n \geq r$. We must first compute the number of possible choices for the r elements that disagree. Then we compute the number of ways in which the disagreement can occur. The value of $D(n,r)$ is the product of these two numbers. As we have learned earlier, r items can be selected from a list of n elements in $C(n,r)$ ways. We can use Equation 6.5.2 or 6.5.3 to find the number of ways in which there can be disagreement among the r items. Therefore

$$D(n,r) = C(n,r) \cdot D(r,r) \tag{6.5.5}$$

Example 6.5.1

Consider the orderings (permutations) of the first five letters of the alphabet: a, b, c, d, e. In how many of these will all of the letters appear in the "proper" position? The "proper" position for a letter is the position that it occupies in the alphabet: first for a, second for b, and so on. In how many will just one letter be out of place? Two letters? Three letters? Four letters? All five letters?

There are $5!$ different orders in which the five letters can be listed. There is just one of these in which they are all in their proper order. In none of the permutations will there be just one letter out of place. As we have seen, $D(1,1) = 0$.

When two letters are out of place, three must be in their proper places. There is only one way for two letters to be in the wrong places—they must be interchanged. $D(2,2) = 1$. There are $C(5,2)$ ways to select the letters that are not in their proper order. We have therefore,

$$\begin{aligned} D(5,2) &= C(5,2)\cdot D(2,2) \\ &= \frac{5!}{3!2!}(1) \\ &= 10 \end{aligned}$$

where $D(5,2)$ represents the number of ways in which there can be two disagreements or "errors" in a list of five elements.

To illustrate, the list of the 10 sets of letters, each with two letters out of their "proper" order is shown below. Each out-of-place letter is called a "disagreement." The out-of-place letters are underlined.

$$\begin{array}{ll} \underline{b}\ \underline{a}\ c\ d\ e & a\ \underline{d}\ c\ \underline{b}\ e \\ \underline{c}\ b\ \underline{a}\ d\ e & a\ \underline{e}\ c\ d\ \underline{b} \\ \underline{d}\ b\ c\ \underline{a}\ e & a\ b\ \underline{d}\ \underline{c}\ e \\ \underline{e}\ b\ c\ d\ \underline{a} & a\ b\ \underline{e}\ d\ \underline{c} \\ a\ \underline{c}\ \underline{b}\ d\ e & a\ b\ c\ \underline{e}\ \underline{d} \end{array}$$

When we have three letters out of order, these can be selected in $C(5,3)$ ways. There are $D(3,3)$ ways for three letters to be in the "incorrect" position. Thus

$$D(5,3) = C(5,3)\cdot D(3,3) = \frac{5!}{3!2!}(2) = 20.$$

When four letters are out of place, the four can be selected in $C(5,4)$ ways. The number of ways in which the four letters can be incorrectly placed is given by $D(4,4)$. As we have seen, $D(4,4) = 9$. That is, there are nine possible sequences in which each of the four elements is out of its normal position. In our example these could be any set of four out of the five letters. As an illustration let us assume that the out-of-place letters are to be a, b, c, and d, with e as the "correct" one. The nine sequences in which this occurs are listed below in the order of their appearance in the diagram shown in Figure 6.5.3.

$b\ a\ d\ c\ e$	$c\ a\ d\ b\ e$	$d\ a\ b\ c\ e$
$b\ c\ d\ a\ e$	$c\ d\ a\ b\ e$	$d\ c\ a\ b\ e$
$b\ d\ a\ c\ e$	$c\ d\ b\ a\ e$	$d\ c\ b\ a\ e$

The total number of possibilities for four "disagreements" with a sequence of five elements is

$$\begin{aligned} D(5,4) &= C(5,4) \cdot D(4,4) \\ &= \frac{5!}{4!} \cdot (9) = 45 \end{aligned}$$

The last situation is the number of listings in which all five letters are out of place. When $r = 5$ we have, using Equation 6.5.3

$$\begin{aligned} D(5,5) &= 5!\left(\frac{1}{2!} - \frac{1}{3!} + \frac{1}{4!} - \frac{1}{5!}\right) \\ &= 120\left(\frac{1}{2} - \frac{1}{6} + \frac{1}{24} - \frac{1}{120}\right) \\ &= 60 - 20 + 5 - 1 \\ &= 44 \end{aligned}$$

In summary,

For zero disagreements	$D(5,0) = 1$	
one disagreement	$D(5,1) = 0$	
two disagreements	$D(5,2) = C(5,2) \cdot D(2,2) =$	10
three disagreements	$D(5,3) = C(5,3) \cdot D(3,3) =$	20
four disagreements	$D(5,4) = C(5,4) \cdot D(4,4) =$	45
five disagreements	$D(5,5) = 44$	
Total possibilities	$= 120$	

We see that the sum of the numbers of possibilities for the six different situations is 120, $(P,(5,5))$, as it should be.

We restate the general counting rule. When a set of n elements is to be listed in order and compared with a list in a particular order (called the "proper" or "standard" order), the number of ways in which there can be r disagreements in the lists is given by

$$D(n,r) = C(n,r) \cdot D(r,r)$$

$D(r,r)$ can be computed using either Equation 6.5.2 or 6.5.4.

Note that the formula can also be written as

$$D(n,r) = C(n,n-r)\cdot D(r,r) \qquad (6.5.6)$$

In this form $C(n,n-r)$ represents the number of ways to select the $n-r$ items that are in their "proper" places. This number is the same as the number of ways to select those in the "wrong" places. The set of elements in the proper order and the set of disagreements comprise an ordered partition of the complete set of elements listed. It is implied in these considerations that $C(n,r) = C(n,n-r)$, which is, indeed, true.

EXERCISE 6.5

1. Find the number of ordered partitions of a set containing eight elements, in which there are
   a. Two cells of 4 elements each.
   b. Four cells of 2 elements each.
   c. Two cells of 3 elements and one of 2 elements.
   d. Two cells with 2 elements and one with 4 elements.

2. Find the number of unordered partitions possible for a set with eight elements in which there are
   a. Two cells of 4 elements each.
   b. Four cells of 2 elements each.

3. Consider a set with 12 elements. Find the number of ways to partition it
   a. Forming cells with 3 elements, 4 elements, and 5 elements respectively.
   b. Forming two cells with 3 elements and one with six elements, assuming the 3-element cells are ordered.
   c. Forming two 4-element and two 2-element cells, assuming the partition is unordered.

4. Show that $C(n,r) = C(n,n-r)$.

5. Compute the following:
   a. $D(6,4)$
   b. $D(6,2)$
   c. $D(8,6)$
   d. $D(8,5)$

6. Consider the set $\{a,b,c,d\}$. Find the number of ways to list the elements in which there are 0, 1, 2, 3, and 4 letters out of their normal alphabetical order. Verify that these account for all of the permutations of the four letters.

7. a. A group of eight people meet to play golf. They have arranged for two "tee" times, 9 and 9:15 A.M., at the golf course. In how many ways can the two foursomes be selected?
   b. Later the same eight people meet to play cards. In how many ways can they be separated into two groups of four for bridge?

8. a. Ten men on an army base are grouped into two duty sections of five men each. One duty section is to work during the day, and the other is to stand by for emergencies at night. How many different assignments to sections are possible?
   b. The same 10 men play basketball on off-duty days. How many different pairs of teams can be formed?

9. Twelve executives of a company arrange a weekend conference. They are to be divided into four three-man groups. One group is to meet with the controller of the company to discuss budget preparation; one group is to discuss a proposed pension plan; and the other two groups are to discuss topics of their own choosing. How many different groupings are possible?

10. A group of nine people are to be formed into three committees of three members each. How many different ways can this be done?

11. If there are two people in the group of Problem 10 who insist on being on the same committee, how many possibilities are there?

12. If two of the people in the group in Problem 10 are to be placed on different committees, how many possibilities are there?

13. In an examination a list of six answers is to be matched with a list of six questions. In how many ways can a person give three wrong answers? Four wrong answers?

14. In an experiment on aesthetic sense two people are shown eight paintings and asked to rate them in order of their "liking." The people are said to have similar "taste" if their lists show six or more agreements in order of preference. Considering all of the possible sets of preference orders, what percentage meet the "similar-taste" criterion?

15. A high school athletic league has six teams. Local sportswriters predict the order of finish of the six teams for each major sport. A writer is considered an "expert" if he correctly predicts the winner and the correct position for at least three of the other teams.
    a. How many different predictions are possible?
    b. How many of these qualify as "expert" predictions?

7

probability

7.0 INTRODUCTION

If a person holds a cubical block in his fingers, stretches out his arm, and releases the block, he can predict with certainty that the block will fall to the floor. He can use an algebraic formula to compute precisely, for example, the length of time that it takes the block to reach the floor, and the speed that it has when it strikes. But when the block comes to rest, which of its six faces will be up? This question cannot be answered with certainty. Any one of the six faces could be up, and we have no formulas to predict which one it will be.

The falling of the block is governed by physical laws and is called a **deterministic** process. These are processes in which the results of performing certain operations or "experiments" are always the same if the conditions of the experiment are the same.

The process by which the block eventually comes to rest depends on factors that cannot be determined precisely, and this prevents our being able to predict the result in advance. Such processes are called **random**, or **stochastic** processes. The best that we can do in these cases is to state the **relative likelihood** of the different possible results. These relative likelihoods are called **probabilities**.

There are two general types of stochastic processes for which the numerical values of probabilities are determined and used. In one type, the nature of the phenomena makes it possible to determine everything that can possibly happen. The relative likelihood of these possibilities can be found by analysis and deduction. The probabilities associated with such phenomena are called **analytical probabilities**.

In the other type of stochastic process the relative likelihoods of different occurrences are found by statistical inference. Most of the actions and behavior of people fall into this category. Policy-producing studies in business, behavioral research in psychology and sociology, for example, use statistical methods to infer relationships and correlations among the various actions of people, and to compute the relative likelihood of different actions in a given set of circumstances. The probabilities obtained from such studies are called **statistical probabilities.**

In general the theory of probability is the study of the methods used in analyzing and describing stochastic, or random, phenomena. There are two principal branches in this study. In one branch the possible occurrences are regarded as discrete, separate events. In the other the occurrences form a continuum. The falling block is an example in which the results are discrete events. There are six faces to the block; and the block comes to rest with exactly one of these up. There are six separate and distinct outcomes possible.

The type of situation in which the occurrences form a continuum is exemplified by a traffic study in which a person measures the time interval between successive automobiles at a point along a street. The elapsed time between automobiles can be any positive number, and the set of real numbers is a continuum. (There is an uncountable quantity of numbers between any two numbers, no matter how nearly equal they are.) The study of probabilities in this latter type of situation requires calculus. We shall confine our study to the basic properties of probability and to the methods for computing analytical probabilities in several different situations.

7.1 SIMPLE PROBABILITY

In the introductory comments we used several terms without indicating their meaning. Such terms as "experiment," "event," and "outcome" are assigned specific meanings in the study of probability. These definitions are sometimes slightly different from those in general usage. In our discussions we shall present, as they are needed, definitions for the terms that are to have special meanings.

The term "probability" is an undefined term. We consider the **probability** of an event to be a **numerical** response to the question, "How likely is the occurrence of the event?" Thus, probability is a **measure** of the likelihood of an event. However, the meaning of "likelihood" is left to our intuition. A better understanding of the concept of probability is gained from a knowledge and understanding of its properties.

7.1.1 Terms and Definitions

Probability is used as a measure of the likelihood that a particular occurrence will take place, or happen under specified circumstances. The situations in which the things are supposed to happen are all lumped together under the name "experiment." Any operation, or activity, in which we are concerned with what happens, but cannot predict with certainty what will happen, is called an **experiment**. In formal language we have the definition:

Definition 7.1.1

> **Experiment:** Any well-defined procedure, operation, or activity, in which the result, or outcome, is observed or measured, but **cannot** be **predicted** with **certainty** is called an **experiment**.

Note that the definition uses the term "well-defined." This means that the experiment must be described carefully enough so that exactly the same experiment can be repeated as often as desired. However, we shall frequently refer to a single performance of an experiment as simply "an experiment."

The term **outcome** is used for the **individual possibilities** that can result from an experiment.

Definition 7.1.2

> **Outcome:** The result of a single performance of an experiment is called its **outcome**.

The outcome of an experiment is a very specific result. For example, in an experiment consisting of rolling a pair of dice it is necessary to identify one die from the other. One outcome possible for this experiment is that die 1 shows "2," and die 2 shows "5." This is a different outcome from die 1 showing "5," and die 2 showing "2," although for most purposes the consequence is the same for both. We can describe as an **event** for this experiment the rolling of a "5" and a "2." This event occurs for either of the two outcomes above.

Any **single performance** of an experiment has **exactly one outcome**. The set of all the outcomes possible is called the **sample space**.

Definition 7.1.3

> **Sample Space:** The set of **all** the outcomes possible for an experiment is called the **sample space** of the experiment.

The symbol S is used to designate the sample space. Note that it is a **set**. O's with subscripts are used to designate outcomes—the elements of the sample space.

The term sample "space" is commonly used in probability theory instead of sample "set" to convey the idea that the set referred to is the universal set for the discussion. There is no implication of any concept of region or area involved. "Space" is another example of a word used in mathematics to have a special meaning different from the one it has in normal usage.

One of the early difficulties in the study of probability is to understand clearly the concepts of "outcomes" and "events" and the differences between them. In strictly mathematical terms we have the following definition for "event."

Definition 7.1.4

> **Event:** Any **subset** of the sample space for an experiment is called an **event**.

This definition is simple, but its implications are not. Much of the difficulty stems from the way in which events are described. The description, or "name" given to an event is generally based on the interests of those observing the "experiment." For example, if the "experiment" is a baseball game, the outcome is the final score of the game—Cardinals, 3 and Mets, 0. This outcome could "result" in many different events. "The Cardinals won." "The Mets lost." "Gibson pitched another shutout." In certain types of betting pools the event would be "The total number of runs scored is three."

In making use of the concepts of probability, we set up an experiment and describe, or define, the events that we are interested in. These events include as elements all outcomes of the experiment that produce the result described by the event. If the event contains only one outcome, it is called a **simple event**. If it contains more than one outcome, it is a **compound event**. An event is said to **occur** if the outcome of an individual performance of an experiment is an element of the event. We note an important difference between **outcomes** and **events**. Outcomes are the **individual** results of the experiments. That is, no single performance of the experiment can result in more than one **outcome.** Events, on the other hand, can occur in **one** or **more** ways; and more than one event can result from a single performance of an experiment. The "compound" part of a "compound event" implies that the event can occur in more than one way, not that it has more than one part.

Example 7.1.1

Three new salesmen in a company are assigned to district offices —one to New York, one to Chicago, and one to Los Angeles. The salesmen are named Byrd, Foxx, and Hunter. The process of making the assignments is the **experiment**. The **outcomes** are shown in the accompanying table, and designated O_1, O_2, O_3, O_4, O_5, and O_6.

Outcomes

Office	O_1	O_2	O_3	O_4	O_5	O_6
New York	Byrd	Byrd	Foxx	Foxx	Hunter	Hunter
Chicago	Foxx	Hunter	Byrd	Hunter	Byrd	Foxx
Los Angeles	Hunter	Foxx	Hunter	Byrd	Foxx	Byrd

The sample space, S, is the set $\{O_1, O_2, O_3, O_4, O_5, O_6\}$. We can specify **events** in a great variety of ways. As **simple events** we have

Event: Byrd goes to New York, and Hunter goes to Chicago. $\{O_2\}$

Event: Foxx goes to New York, Byrd to Chicago, and Hunter to Los Angeles. $\{O_3\}$

There are six simple events possible, each one a description of one of the outcomes; but these events can be **described**, or **named**, in a variety of ways.

The **compound events** possible include, but are not restricted to:

Event: Foxx goes to New York. $\{O_3, O_4\}$

Event: Byrd goes to Los Angeles. $\{O_4, O_6\}$

Event: Hunter does not go to New York. $\{O_1, O_2, O_3, O_4\}$

Recall from Chapter 1 that a set with n elements has 2^n subsets. One of these is the set itself, and one is the null set. In the experiment of the example above, the event corresponding to the set S itself is that the assignments are made; that is, the experiment is performed and the outcome observed. The event corresponding to the null set is that the assignments are not made. Since the assignments are made, the null set corresponds to something that cannot happen—an impossibility. We have $2^6 - 2 = 62$ other possible different events. These can be described in many ways, and the same event can be designated in more than one way. In the above example, as we have seen, some of the compound sets can be described in simpler language than can the simple events.

7.1.2 Basic Properties of Probability

The first goal in the use of probability theory is to associate with each event that **can** occur as the result of the experiment a **number**, which is the **measure** of the relative likelihood that this event **will** occur. This number is called the **probability** of the **event**. We shall symbolize events with capital E's and their probability with $p(\quad)$. Thus, $p(E_2)$ represents the probability that event E_2 will occur. The numbers from zero to one, inclusive, have been arbitrarily chosen as the numbers to be used as probabilities. If we use the symbology of algebra, this statement becomes

$$0 \leq p(E) \leq 1$$

We shall customarily refer to the number assigned to the **measure** of the probability of an event as "the probability of the event." The less likely an event is to occur, the smaller the probability of that event, and vice versa. The combination of this concept with the first property of probability leads to the second property of probability: if an event is **impossible**, its probability is **zero**; and if the event is **certain to happen**, its probability is **one**.

The third basic property provides a way to study combinations of events. If two events are **mutually exclusive**, that is, they both do not occur as the result of any one outcome of an experiment, the probability that **either** will occur is the **sum** of their individual probabilities.

The computations that we make in our study of probabilities are based on the algebra of sets. We, therefore, will find useful a statement of the basic properties of probability in set language. If we have an experiment whose possible outcomes comprise the sample space, S, and we consider the events, $E_1, E_2, E_3, \ldots, E_n$ that are the subsets of S,

Probability Property 1 $\quad 0 \leq p(E_k) \leq 1 \qquad k = 1,2,3,\ldots,n$

Probability Property 2 $\quad p(S) = 1 \qquad p(\emptyset) = 0$

Probability Property 3 $\quad$ If $E_j \cap E_k = \emptyset$, $p(E_j \cup E_k) = p(E_j) + p(E_k)$

In these properties $p(S)$ is the probability that **any** one of the possible outcomes occurs; that is, the event, S, is a certainty. $\emptyset$ is the subset of S that does not include any of the **possible** outcomes. This is the set language for an impossibility. $E_j \cap E_k$ is the set-language way of specifying the set of outcomes that are common to both E_j and E_k. The statement $E_j \cap E_k = \emptyset$ says there are no outcomes common to E_j and E_k; that is, E_j and E_k are mutually exclusive events. $E_j \cup E_k$ is the set language way of specifying that **either** E_j or E_k will occur. (Since $E_j \cap E_k = \emptyset$, both cannot occur in the same experiment.)

Observe that the properties of probability are statements of its "mathematical nature." The concept that behaves mathematically according to these properties is what mathematicians call probability. The more general, intuitive concepts of probability come from a person's own experience. The body of probability theory that has been developed are the ramifications of a concept with the three basic properties listed above.

We can draw one conclusion from these properties immediately. Consider an event E_k and its **complement**, E'_k. (Recall that we regard events as **subsets.**) We can associate the complement of an event with the **nonoccurrence** of that event. Since the union of any subset with its complement is equal to the parent set, $E_j \cup E'_j = S$. Thus, $p(E_j \cup E'_j) = 1$; and $p(E_j) + p(E'_j) = 1$. Rearranging, we have

$$p(E_j) = 1 - p(E'_j) \tag{7.1.1}$$

Equation 7.1.1 is a very useful relationship. It is often much easier to determine the probability that an event will **not** happen than the probability that it will. Equation 7.1.1 enables us to evaluate the probability of an event by finding the probability that it will not happen.

7.1.3 Equally Probable Outcomes

We shall discuss primarily experiments with equally probable outcomes. That is, each of the possible outcomes is just as likely to occur as any of the others. If the same experiment is repeated several times, the outcomes occur in a "random" order. There is no bias for or against the occurrence of any of them. This is the type of situation that exists in a "fair" lottery or a "random" choice. For example, a quantity of differently numbered balls are placed in a container and thoroughly mixed. If someone reaches into the container and selects one of the balls at random, any one number has the same likelihood of being selected as any other. Or, if a deck of cards is thoroughly shuffled and a card is selected without looking, any one of the cards has an equal likelihood of being selected.

We can use the basic properties of probability to compute the probability of one of a set of equally probable outcomes. For example, suppose that two balls, one red and one blue, which are identical otherwise, are placed in an urn. A person is blindfolded and asked to select one of the balls. What is the probability that the red one is chosen?

The two possible outcomes of this "experiment" are: O_1, the red ball is chosen; and O_2, the blue ball is chosen. Let us call the event that the red ball is chosen, R; and the event that the blue ball is chosen, B. That is, let $R = \{O_1\}$, and $B = \{O_2\}$. These are mutually exclusive events, and so the probability that **either** occurs is the sum of their individual probabilities. If we call the probability of R, $p(R)$, and the probability of B, $p(B)$, we have

$$p(R \cup B) = p(R) + p(B)$$

In this case $R \cup B = S$, the sample space. Therefore

$$p(R \cup B) = p(S) = 1 \tag{7.1.2}$$

We have decided that the probability of choosing the red ball is the same as the probability of choosing the blue one. That is,

$$p(R) = p(B)$$

If we substitute $p(R)$ for $p(B)$ in Equation 7.1.2, we have

$$\begin{aligned} p(R) + p(R) &= 1 \\ 2p(R) &= 1 \\ p(R) &= \frac{1}{2} \end{aligned}$$

Note: We use this cumbersome procedure here to obtain a result that we could have obtained quickly by intuition to illustrate the processs that we need for later computations. Observe that we were careful to distinguish between the outcomes, O_1 and O_2, and the simple events occurring as a result of these outcomes. Events are **sets**, and outcomes are the **elements** in these sets. In simple situations these distinctions are not needed for understanding. They will prove very useful, however, in our later studies.

We can extend the process used in the foregoing example to compute probabilities for experiments in which there are more than two outcomes.

Example 7.1.2

Recall the experiment of dropping the cubical block discussed at the beginning of this section. Let us number the faces of the block, 1, 2, 3, 4, 5, and 6. Let us designate the outcome that the face numbered "1" is up when the block comes to rest as O_1; the outcome when "2" is up as O_2, and so on. The sample space consists of the set of outcomes, $\{O_1, O_2, O_3, O_4, O_5, O_6\}$. We designate the simple events containing each of the outcomes as E_1, E_2, E_3, E_4, E_5, and E_6. We assume that there is equal probability for the block to land with any of its faces up and that it will not land on edge. Thus

$$p(E_1) = p(E_2) = p(E_3) = p(E_4) = p(E_5) = p(E_6) \tag{7.1.3}$$

Since two faces cannot be up at the same time, the events are mutually exclusive. The union of these events is the sample space. Therefore

$$p(E_1) + p(E_2) + p(E_3) + p(E_4) + p(E_5) + p(E_6) = 1 \tag{7.1.4}$$

If we wish to know the probability that any of the faces, say 3, lands up, we can substitute from Equation 7.1.3 into Equation 7.1.4 to obtain

$$p(E_3) + p(E_3) + p(E_3) + p(E_3) + p(E_3) + p(E_3) = 1$$

or

$$6p(E_3) = 1$$

$$p(E_3) = \frac{1}{6}$$

The procedure in Example 7.1.2 can be extended to experiments with any number of **equally probable** outcomes. In an experiment with n equally probable outcomes, the probability of each of the n **simple events** (those containing only one outcome) is given by

$$p(E_j) = \frac{1}{n} \tag{7.1.5}$$

where E_j represents any one of these events.

7.1.4 The Urn Model

There are many situations in various fields of study that can be simulated with an urn model. In an urn model one or more urns are set up containing various numbers of differently colored balls. Experiments are then devised consisting of various procedures for selecting one or more balls from the urns. The selection process is designed to simulate the operation of the phenomenon being studied.

We shall occasionally use urn models to illustrate the properties of probability. We can use an urn to illustrate the relationship among outcomes, events, and the sample space. Let us suppose we have an urn containing 10 balls. The experiment consists of reaching into the urn and selecting one of the balls at random.

Implicit in the use of an urn is that each of the balls has an individual identity even though several of the balls may have the same color. (It is assumed that all of the balls are of the same size and weight, so that they cannot be identified by "feel" during the selection process.) Thus, although the balls are ostensibly identical, we must take account of their separate identities. Selection of each ball produces a different outcome. With an urn containing 10 balls, there are exactly 10 outcomes to an experiment in which one ball is selected. If three of the balls are red, three are blue, and four are yellow, we can designate the outcomes as R_1, R_2, R_3, B_1, B_2, B_3, Y_1, Y_2, Y_3, and Y_4.

The sample space is the set containing the 10 outcomes. That is

$$S = \{R_1, R_2, R_3, B_1, B_2, B_3, Y_1, Y_2, Y_3, Y_4\}$$

We can consider the event of drawing a blue ball from the urn. This event "occurs" for any one of the three outcomes, B_1, B_2, or B_3. Thus, if E_B represents the event, a blue ball is drawn,

$$E_B = \{B_1, B_2, B_3\}$$

It is assumed in working with an urn model that any of the balls has an equal chance of being drawn. Thus, all 10 outcomes are equally probable. From our earlier considerations (Equation 7.1.5) the probability of a simple event containing just one of the outcomes is 1/10. From set algebra we know that

$$E_B = \{B_1\} \cup \{B_2\} \cup \{B_3\}$$

We know from the nature of the urn model selection process that these three simple events must be mutually exclusive. Thus

$$p(E_B) = \frac{1}{10} + \frac{1}{10} + \frac{1}{10}$$

$$= \frac{3}{10}$$

We note that the probability of drawing a blue ball is equal to that fraction of the balls in the urn that are blue.

EXERCISE 7.1

1. An experiment consists of selecting from among three people, Hilton, Milton, and Tilton, a manager and an assistant manager for a branch bank. Describe the sample space for the experiment.

2. An experiment consists of noting the last digit in the license number of the automobiles passing a particular corner in a city. Describe the sample space for the experiment.

3. An experiment consists of selecting three people from an office staff of 10 to attend a convention. Describe the elements (outcomes) of the sample space. How many elements (outcomes) will there be in the sample space?

4. An experiment consists of selecting from a production lot of 20 a set of five items for test.
   a. Describe the elements (outcomes) in the sample space.
   b. How many elements (outcomes) are there in the sample space?

5. Four of the items in the lot of 20 in Exercise 4 are defective.
   a. Assume we are interested in the number of defective parts in the sample. List some of the events that will give this information.
   b. How many elements (outcomes) are there in the event, "exactly one item is defective."

6. An experiment consists of selecting a committee of three people from a group of five. The five are designated A, B, C, D, and E.
   a. List the possible outcomes for this experiment.
   b. List the outcomes in the event, "A is selected."
   c. List the outcomes in the event, "B and D are selected."
   d. List the outcomes in the event, "C is not selected."

7. An organization is raffling a television set. They sell 1000 tickets. What is your probability of winning if you buy one ticket? Three tickets?

8. A merchant devises a guessing contest as a sales promotion. With each purchase the customer is entitled to enter a guess of how many beans there are in a jar. The person who guesses closest to the actual number wins a prize. The merchant places a sign on the jar stating (truthfully) that there are more than 5000, but no more than 15,000 beans in the jar. What is the probability of guessing the correct number in a single guess?

9. An urn contains 12 balls, three white, three blue, and six yellow.
   a. What is the probability of drawing a yellow ball?
   b. Use Equation 7.1.1 to find the probability of drawing either a blue or a yellow ball.

10. An urn contains 12 balls, four white, five yellow, and three green.
   a. What is the probability of drawing a yellow ball?
   b. What is the probability of drawing a ball that is not white?
   c. What is the probability of drawing a ball that is either white or green?

11. An urn contains 20 balls, four red, five white, five green, and six yellow. Two balls are selected at random and drawn from the urn.
   a. Describe the sample space.
   b. List the different **events** that are possible with respect to the color of the balls drawn.

12. Referring to the situation described in Problem 11,
   a. Use the methods of Chapter 6 to compute the number of different outcomes possible.
   b. In the text it was stated that if there are n outcomes for an experiment, there are $2^n - 2$ events possible. Contrast this number with the number of events listed for Exercise 11*b*. Explain the difference.

13. Two coins, a nickel and a dime, are flipped simultaneously. Assume that neither coin lands on edge.
   a. List the elements in the sample space.
   b. What is the probability that both coins land "heads"?

14. Two quarters are flipped simultaneously. Assume that neither lands on edge. What is the probability of getting one "heads" and one "tails"?

7.2 COMPOUND EVENTS

A compound event is an event that can occur as a result of more than one outcome to an experiment. In set language, a compound event is a subset of the sample space that contains more than one element (outcome).

Example 7.2.1

A new chief accountant is being selected for Workwell, Inc. The candidates for the job are Frank Able, Donald Baker, Mary Carter,

Betty Doerr, and Clyde Eaton. The sample space for the "experiment" of making the selection is the set containing five individual people as elements (outcomes).

Let us consider as events the following:

$E_1 =$ A woman is selected.

$E_2 =$ A man is selected.

Each of these events is a compound event. E_1 occurs if either Mary Carter or Betty Doerr is selected. That is,

$E_1 = \{$Mary Carter, Betty Doerr$\}$

E_2 occurs if Frank Able, Donald Baker, or Clyde Eaton is selected. That is,

$E_2 = \{$Frank Able, Donald Baker, Clyde Eaton$\}$

In accordance with our definitions in the previous section, these two events are **mutually exclusive**. Suppose the office politicians at Workwell, Inc. cannot dig up any information that would indicate that any of these five candidates has any greater or less chance of being selected. The five outcomes are equally probable.

What is the probability that the new chief accountant will be a woman?

If we let A represent the outcome, Able gets the job; B, that Baker gets it; C, that Carter is selected; D, that Doerr is chosen; and E, that Eaton is; we have, using Equation 7.1.5

$$p(A) = p(B) = p(C) = p(D) = p(E) = \frac{1}{5}$$

Since

$$E_1 = \{C\} \cup \{D\}$$

$$\begin{aligned} p(E_1) &= p(C) + p(D) \\ &= \frac{1}{5} + \frac{1}{5} \\ &= \frac{2}{5} \end{aligned}$$

Thus, the probability, $p(E_1)$, that a woman will be selected is 2/5.

Note that there are two outcomes in E_1 and there are five in S. Therefore

$$p(E_1) = \frac{2}{5} = \frac{n(E_1)}{n(S)}$$

where $n(E_1)$ represents the number of elements in E_1, and $n(S)$ represents the number of elements in S.

Example 7.2.2

Suppose we have a group of 10 people and three are to be selected to attend a convention. (See Problem 3 of Exercise 7.1.) Joe is one of the 10 people, and he wishes to compute his chances of being selected. The outcomes of the experiment are sets of three people each, the three who have been selected. We recall that there are $C(10,3) = 120$ different sets of three people each that can be chosen from a set of 10. Thus, there are 120 different outcomes possible – $n(S) = 120$.

Let J denote the event of Joe's selection. If Joe is selected, two others are selected with him. These two are selected from the other nine people in the group. There are $C(9,2) = 36$ ways that the selections can be made. That is, Joe is one of those selected in 36 of the outcomes. Thus, the event, J, contains 36 elements.

Since each of the outcomes is equally probable, the probability that any one of them will happen is 1/120 from Equation 7.1.5. The 36 outcomes in which Joe is selected are all different; and the event of his selection, J, is the union of 36 simple events, each containing one of these outcomes as its only element. Thus, using the third basic property of probability, the probability that Joe is selected, $p(J)$ is

$$p(J) = 36 \cdot \frac{1}{120}$$

$$= \frac{3}{10}$$

We note in this example, as in Example 7.2.1, the probability of an event is the number of outcomes "contained" in that event divided by the total number of outcomes in the sample space. In symbols, for Example 7.2.2

$$p(J) = \frac{n(J)}{n(S)}$$

We can state as a general result that in an experiment with **equally probable outcomes** the probability of any event, E, is given by

$$p(E) = \frac{n(E)}{n(S)} \qquad (7.2.1)$$

where $n(E)$ is the number of outcomes in the event, E, and $n(S)$ is the number of outcomes in the sample space, S.

(The proof of Equation 7.2.1 in this general form follows the procedure we have used in Examples 7.2.1 and 7.2.2. It is time-consuming and, therefore, will be omitted.)

We can paraphrase Equation 7.2.1 as, "In an experiment with **equally probable** outcomes, the probability of a given result, or event, is the number of ways the event can happen, divided by the total number of things that can happen." This is the fundamental relationship for the probability of "random" events, events that happen purely by chance. Notice that "the total number of things that can happen" means the total number of outcomes, not the total number of events.

It may seem that we used an unnecessarily complicated method for finding the probability that Joe would be chosen to attend the convention in Example 7.2.2. After all, there are 10 people to choose from and three are to be selected. Joe has three chances out of 10 to be selected, and this gives a probability of .3.

The concepts of probability are deeply rooted in intuition, and it is actually a measure of the validity of the formula that it gives the result we feel intuitively is the correct one. In simple situations, such as the cases in our two examples thus far, our intuition is quite adequate to determine the probabilities. As the situations become more complex, however, we find that the mathematical formulations derived from the basic intuitive concepts and extended into more general terms become necessary in computing probabilities.

Let us consider a more complex situation.

Example 7.2.3

A factory is preparing a shipment of 20 items. An inspector for the customer arrives at the factory to inspect the items before they are shipped. He does not have time to inspect all 20 of the items, so he tells the factory that he will select a sample of four items to be tested. If any of the four is defective, he will reject the entire shipment. What is the probability that he will accept the shipment if there are three defective items in the lot of 20?

The experiment is the selection of a set of four items out of the set of 20. All of the items have an equal likelihood of being selected. There are $C(20,4)$ ways to choose the four items; and the experiment, therefore, has $C(20,4)$ different outcomes. That is,

$$n(S) = C(20,4) = 4845$$

where $n(S)$ is the number of outcomes in the sample space, S.

From the point of view of the selection of good and defective items, the events are

E_1 = selection of 4 good items and 0 defective

E_2 = selection of 3 good items and 1 defective

E_3 = selection of 2 good items and 2 defective

E_4 = selection of 1 good item and 3 defective

By their nature these events are mutually exclusive. It is not possible for a sample to have exactly one defective and exactly two defectives, for example.

The only one of these events that will result in the acceptance of the shipment is E_1. The number of outcomes which result in E_1 is

$$n(E_1) = C(17,4) = 2380$$

(If there are three defective items in the lot, there are 17 good ones. The sample of four must be selected from these 17 if there are to be no defective items in the sample.)

The probability that the shipment will be accepted is, from Equation 7.2.1,

$$p(E_1) = \frac{n(E_1)}{n(S)} = \frac{2380}{4845}$$

$$= \text{approximately } .49$$

Example 7.2.4

Let us consider the situation in Example 7.2.3 with the modification that the inspector will accept the shipment if no more than one item in his sample is defective. In other words, he will accept the shipment if either one or none of the sample items is defective.

Referring to the events defined in Example 7.2.3, he will accept the shipment if either event E_1 or event E_2 occurs. We can define the event of acceptance, which we symbolize A, as

$$A = E_1 \cup E_2$$

Events E_1 and E_2 are mutually exclusive; neither contains any of the outcomes of the other. Therefore, from the basic properties of probability

$$p(A) = p(E_1) + p(E_2)$$

$p(E_1)$ was computed in Example 7.2.3 to be .49. The number of outcomes in E_2 is given by

$$n(E_2) = C(17,3) \cdot C(3,1)$$

$$= 680 \cdot 3$$

$$= 2040$$

Therefore

$$p(E_2) = \frac{2040}{4845}$$

$$= \text{approximately } .42$$

The probability of acceptance of the shipment is

$$p(A) = p(E_1) + p(E_2)$$

$$= .49 + .42 = .91$$

Recall that the sum of $p(A)$ and $p(A')$ is 1, where A', the complement of A, is the event, the shipment is not accepted; that is, it is rejected.

Thus, we have

$$p(A') = 1 - p(A)$$

$$= 1 - .91$$

$$= .09$$

as the probability the shipment will be rejected.

In summary, we have discussed only situations in which all the outcomes are equally likely. Although this may appear to be a severe restriction, it applies in all cases where there is random selection or choice. That is, unless there is some sort of bias or "loading" present, random (stochastic) processes are considered to involve "experiments" with equally likely outcomes. When events are considered that can happen in more than one way (can result from more than one outcome), the general relationship giving the probability of such events was given in Equation 7.2.1:

$$p(E) = \frac{n(E)}{n(S)}$$

In this relationship, $p(E)$ is the probability of the event; $n(E)$, the number of ways it can happen (the number of outcomes producing the event); and $n(S)$, the total number of outcomes possible.

Remember that we have discussed only one event at a time, although the event can happen as the result of more than one outcome to an experiment. In succeeding sections we shall discuss the relationships among the probabilities for more than one event. In particular, in the next section we shall consider situations in which different events are produced by the same outcome.

EXERCISE 7.2

1. Each year the Selective Service conducts a lottery to determine the order in which young men become eligible for the draft. Each of the dates for the year is written on a piece of paper and placed in a capsule. The capsules are placed in a drum and thoroughly mixed. They are drawn out one by one, and the order in which they are drawn is recorded. The 19-year-olds born on the date drawn first are called up first, and so on. In a year with 365 days what is the probability that the first day drawn is in December?

2. At a meeting of a labor organization a secret ballot is conducted on whether or not to strike. The vote is 497 for, 290 against, with 13 abstentions. After the meeting a reporter selects a delegate at random for an interview. What is the probability that this delegate voted against the strike?

3. An urn contains 10 balls, three red, three blue, and four yellow. Two balls are selected from the urn at random.
   a. What is the probability they are both red?
   b. What is the probability that one is red and the other is yellow?

4. An urn contains 10 balls, five red and five blue.
   a. Two balls are selected at random from the urn. What is the probability that one is red and one is blue?
   b. Four balls are selected at random. What is the probability that two are red and two are blue?

c. Six balls are selected at random. What is the probability that three are red and three are blue?
d. Eight balls are selected at random. What is the probability that four are red and four are blue?

5. A class is given a true-false test with four questions. If a student guesses at each of the answers,
a. What is the probability he will guess four answers correctly?
b. What is the probability he will get three correct?

6. Three students are to be chosen by lot from 10 members of the student council to attend a convention. Bill and Randy, who are members of the council, usually take opposite sides in the discussions of the council.
a. What is the probability that both will be selected to attend the convention?
b. What is the probability that Bill will be chosen and that Randy will not?

7. Refer to Section 6.3, Example 6.3.2, relating to the selection of samples for testing from a company's production line.
a. Compute the approximate probability that the selected sample will have no defective items, one defective item, two defective items, and three defective items.
b. What is the most likely number of defectives; that is, what number of defective items has the highest probability?
c. What percentage of the **sample** is the number of defectives in the "most likely sample" found in (*b*)? Compare this with the percentage of defectives assumed to be in the entire lot.
d. What is the probability that the number of defectives in the sample is different from that in the "most likely sample"? If the number of items in the sample—the sample size—is increased, how do you expect this probability will change?

8. Recall the woman with the dozen eggs, three of which are bad. If she selects three of them to make a cake, what is the probability that **at least** one of the eggs is bad?

9. A psychologist is conducting an experiment on thought transference. He has five cards marked *a*, *b*, *c*, *d*, and *e*. He arranges the cards in an arbitrary sequence and places them so that one subject can see the cards and the other cannot. The subject who can see the cards concentrates on their order, and the other subject writes down the order that he thinks the cards are in. What is the probability that he will get three or more cards in their proper place purely by chance?

10. A merchant sets up a guessing contest as described in Problem 8 of Exercise 7.1.
a. A person enters five different guesses. What is the probability that one of them is the correct number?
b. What information is needed to compute the probability that one of the guesses wins the prize?

11. Referring to Example 7.2.1, suppose that the office politicians determine that the probability that the new chief accountant will be a man is twice as great as the probability that it will be a woman. If all three men are still equally likely to be chosen, and both women have equal probability of being selected, what is the probability that Mary Carter will be selected? Frank Able? (**Hint:** Consider two sets of outcomes (events). Then compute the probabilities of the elements of the events.)

12. Two urns each contain 10 balls, three red, three blue, and four yellow. One ball is drawn at random from each urn.
    a. What is the probability they both are red?
    b. What is the probability that one is red and the other is yellow?

7.3 CONDITIONAL PROBABILITY

The general idea of a "random" process is that one outcome of an experiment is just as likely to occur as any other. This contrasts with experiments in which there is a bias favoring some outcomes over others (e.g., loaded dice). We shall study primarily experiments with equally probable outcomes, although in a later section we discuss the effects of biasing experiments.

A consequence of equally probable outcomes and the properties of probability is that the probabilities of **events** are determined by the relative frequency of their occurrence. This relative frequency is found in one of two ways: by computations, using the counting techniques that we learned in the previous chapter, or from statistical studies.

For example, suppose that the enrollment records for a college show a total of 5000 students, 3000 men and 2000 women. If a student is selected at random, the probability the student selected is a man is

$$p(M) = \frac{3000}{5000} = .6$$

Note that the selection of a man is an **event**. The **outcome** of the experiment is that a specific student is selected. The possible outcomes—the sample space—comprise a list of the entire student body of the college. The event, the student is a man, contains all of the men students of the college. From the records—statistics—there are 3000 outcomes in this event.

If we have a group of 10 students, six men and four women, and we select two of them at random, the sample space will consist of all of the possible pairs of students, $C(10,2)$ in all. The number of these which consist of two women is $C(4,2)$. Thus, the probability that two women are chosen is

$$p(WW) = \frac{C(4,2)}{C(10,2)} = \frac{6}{45} = \frac{2}{15}$$

One of the principal uses of probability is as an aid in predicting what will happen under certain circumstances. Since we like our predictions to be as accurate as possible, we make use of all the information that we can get. In an experiment in which many different events are possible, knowledge that one event occurs can affect the probability of another event. For example, let us suppose that 1500 of the 5000 students in the college mentioned above have part-time jobs. 1200 of the jobholders are men and 300 are women. The overall statistics say that if a student is chosen at random, the probability he has a job is

$$p(J) = \frac{1500}{5000} = .3$$

But suppose that the student chosen is a woman. With this additional information the statistics say

$$p(J \mid W) = \frac{300}{2000} = .15$$

$p(J \mid W)$ is translated as "the probability that the student has a job **given that the student is a woman**." In the terminology of probability $p(J \mid W)$ is the **conditional probability** of "J" given that "W" occurs.

The traditional language of probability can cause confusion to students unless the special meanings for the words are clearly understood. In the preceding discussion the phrase, "the probability of J given that W occurs," was used. In the normal usage of language we might infer that an event called W occurs **followed by** an event called J. No such time sequence is implied. There is a single experiment with a single outcome. The "events" are descriptions of different aspects of that outcome. The concepts are most clearly expressed in terms of sets. Conditional probability is the fraction of the outcomes (elements) in one event (set), which are also elements of another set. In the example we are discussing, $p(J \mid W)$, the probability that a student has a job, given the student is a woman, is the same as the fraction of the women students who have jobs. The difficult language is necessary to discuss the concepts of probability **in general**, instead of in terms of a specific example. This is one of the great powers of mathematics—the same concepts can be applied to a myriad of specific examples. If we discussed only examples, it would be necessary to redevelop all the concepts involved for each new application.

In computing $p(J \mid W)$ above, as soon as we had the information that the student chosen was a woman, we used only the statistics relating to the women students. It is useful, however, to have relationships available for finding $p(J \mid W)$ in terms of the other probabilities that relate to the situation. Since these relationships cannot be derived directly from the basic properties of probability, we form a new **definition**.

Definition 7.3.1

Conditional Probability: The **conditional probability** of an event, E_2, given that the event E_1 occurs, is the probability that both E_2 and E_1 occur divided by the probability that E_1 occurs. (It is assumed that the probability of E_1 is greater than zero.) In symbols,

$$p(E_2 \mid E_1) = \frac{p(E_2 \cap E_1)}{p(E_1)} \quad \text{if} \quad p(E_1) > 0 \qquad (7.3.1)$$

In this definition $p(E_2|E_1)$ is the conditional probability of E_2 given E_1 occurs, and $p(E_2 \cap E_1)$ is the probability that both E_2 and E_1 occur. Using the symbols of the previous example, this is written

$$p(J|W) = \frac{p(J \cap W)}{p(W)}$$

Here $p(J \cap W)$ is the probability that a student selected at random is a woman **and** has a job. From the statistics there are 300 women with jobs at the college. Thus

$$p(J \cap W) = \frac{300}{5000} = .06$$

From the other data we have

$$p(W) = \frac{2000}{5000} = .4$$

The formula for conditional probability gives

$$p(J|W) = \frac{p(J \cap W)}{p(W)} = \frac{.06}{.4} = .15$$

which agrees with our computation using the statistics about women students.

Notice two important items at this stage:

First, we have not previously encountered the symbols $p(E_1 \cap E_2)$ or $p(J \cap W)$. If we recall that events are **sets of outcomes**, the **intersection** of two sets E_1 and E_2, $(E_1 \cap E_2)$, denotes the set of outcomes that will result in the occurrence of **both** E_1 and E_2. If E_1 and E_2 happen to be mutually exclusive events, their intersection is the null set. That is, $E_1 \cap E_2 = \emptyset$. In this case $p(E_1 \cap E_2) = p(\emptyset) = 0$; and $p(E_2|E_1) = 0$. In words, if E_2 and E_1 are mutually exclusive events, E_2 cannot occur if E_1 does. (The probability of E_2, given that E_1 occurs, is zero.)

In the example of the college students we introduced three "events" in connection with the experiment of selecting a student at random: M, the stusent is a man; W, the student is a woman; and J, the student has a job. Thus, $J \cap W$ represents the event that the student is a woman and has a job.

Our second observation is that there is a difference between $p(E_2|E_1)$ and $p(E_1|E_2)$. We have used $p(E_2|E_1)$ in our previous discussions. It is the probability that event E_2 occurs given that E_1 occurs. $p(E_1|E_2)$ is the probability that E_1 occurs given that E_2 occurs, an entirely different situation.

Relating this distinction to our example of the college students, we note the difference between $p(J|W)$ and $p(W|J)$. We have already computed $p(J|W)$, the probability that the student has a job given that she is a woman. $p(W|J)$ is the probability that the student is a woman, given that "he" has a job; that is, the fraction of students with jobs who are women. If we use our definition,

$$p(W|J) = \frac{p(W \cap J)}{p(J)}$$

From set algebra $W \cap J = J \cap W$. This is still the set of women students who have jobs. But note that the denominator is now $p(J)$, the probability that a student has a job. From our previous statistics the number of working students is 1500, and $p(J) = .3$. Thus

$$p(W \mid J) = \frac{p(W \cap J)}{p(J)} = \frac{.06}{.3} = .2$$

The concept of conditional probability has the effect of restricting the sample space of an experiment to the outcomes of the "given" event. We are saying in effect that we **know** that one event has occurred. This event is no longer a possibility; it is a certainty. The outcome of the experiment must be one of those contained in that event. If we use a Venn diagram to depict the outcomes of events of an experiment we have a diagram such as Figure 7.3.1.

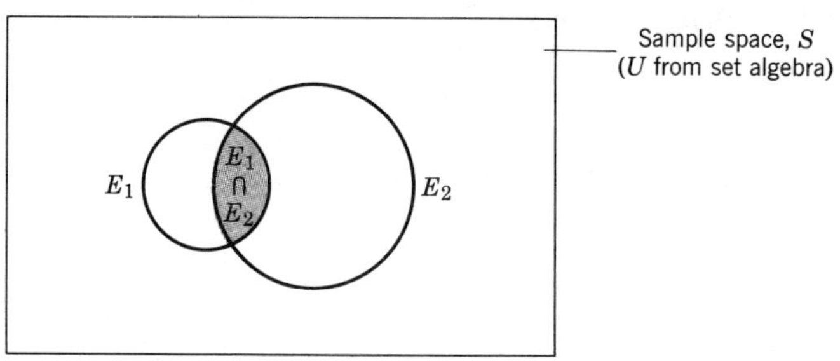

Figure 7.3.1

The total sample space is the area S. E_1 and E_2 are shown, and their intersection is shaded. If we let the size of the circle be proportional to the probability of an event, in this diagram E_2 has a larger probability than E_1. $p(E_2 \mid E_1)$ is the size of $E_1 \cap E_2$ **relative to** E_1. $p(E_1 \mid E_2)$ is the size of $E_1 \cap E_2$ **relative to** E_2. It can be seen that $E_1 \cap E_2$ is a larger fraction of E_1 than it is of E_2. Thus, $p(E_2 \mid E_1)$ should be greater than $p(E_1 \mid E_2)$. Suppose that $p(E_2) = .4$, $p(E_1) = .2$, and $p(E_1 \cap E_2) = .1$. Then

$$p(E_2 \mid E_1) = \frac{.1}{.2} = .5$$

$$p(E_1 \mid E_2) = \frac{.1}{.4} = .25$$

Example 7.3.1

A large oil company ran a series of public-service-type ads in a newspaper. In an attempt to measure the effect of these ads on the sales of the company's product, a survey was conducted of 500 people in the city served by the newspaper. The people were asked first if they regularly read the newspaper. 280 said they did; 220 said they did not. Then the people were asked if they used the brand of gasoline sold by the company. Of the 280 who read the paper, 42 said they did; 238 said they did not. Of the 220 who did not read the paper, 33 said they used the gasoline; 187 did not. A survey crew chief made spot checks of the data taken. If he selects at random one of the persons interviewed, what is the probability the person buys the company's gasoline? If the person reads the newspaper with the ads, what is the probability that he buys the gasoline? If the person does not read the newspaper, what is the probability he buys the gasoline?

Let E_1 be the event, the person reads the newspaper.

Then E_1' is the event, the person does not read the newspaper.

Let G be the event, the person buys the gasoline.

We want to find $p(G)$, $p(G \mid E_1)$, and $p(G \mid E_1')$. Since there is a total of $42+33=75$ gasoline buyers,

$$p(G) = \frac{75}{500} = .15$$

$$p(E_1) = \frac{280}{500} = .56$$

$$p(E_1') = \frac{220}{500} = .44$$

$$p(E_1 \cap G) = \frac{42}{500} = .084$$

$$p(E_1' \cap G) = \frac{33}{500} = .066$$

Using these values we compute $p(G \mid E_1)$ and $p(G \mid E_1')$:

$$\begin{aligned} p(G \mid E_1) &= \frac{p(E_1 \cap G)}{p(E_1)} \\ &= \frac{.084}{.56} \\ &= .15 \end{aligned}$$

$$\begin{aligned} p(G \mid E_1') &= \frac{p(E_1' \cap G)}{p(E_1')} \\ &= \frac{.066}{.44} \\ &= .15 \end{aligned}$$

Thus, according to this survey, the ads did not influence the sales of gasoline. The probability that the person interviewed uses the gasoline is the same whether he reads the newspaper or not.

Example 7.3.2

An inspector is preparing to examine a group of 10 items by selecting two to be tested in order. If the first one passes the test, he will accept the lot without testing the second. If the first one fails, he will test the second. Suppose that he will accept the remaining items if the second passes, but reject all of them if it fails. There are three defective items in the group. If the first one is defective, what is the probability that the second will also fail?

There are seven good items and three defective ones. In selecting the items the following sequences are possible: GG, GD, DG, DD, where G represents the selection of a good item and D, a defective one. Since there are seven good items to choose from, the first possibility, GG, can be done in $7 \cdot 6 = 42$ ways. The second sequence, GD, can be made in $7 \cdot 3 = 21$ ways, and so on. The following is a summary of the total number of ways the choices can be made:

$$\begin{array}{ll} GG: & 7 \cdot 6 = 42 \\ GD: & 7 \cdot 3 = 21 \\ DG: & 3 \cdot 7 = 21 \\ DD: & 3 \cdot 2 = 6 \\ \textit{Total} & \quad\;\; = 90 \end{array}$$

We are interested in $p(D_2 | D_1)$, the probability that the second item is defective, given that the first one is. From Definition 7.3.1,

$$p(D_2 | D_1) = \frac{p(D_2 \cap D_1)}{p(D_1)}$$

D_1, the event of getting a defective item first, occurs in sequence DG or DD, a total of $21 + 6 = 27$ ways. Thus

$$p(D_1) = \frac{27}{90} = .3$$

$D_2 \cap D_1$ can occur in six ways. Thus

$$p(D_2 \cap D_1) = \frac{6}{90} = .067$$

Therefore

$$p(D_2 | D_1) = \frac{p(D_2 \cap D_1)}{p(D_1)} = \frac{.067}{.3}$$
$$= .22$$

In some cases it is possible to obtain the value for a conditional probability by direct observation or in some way different from a computation such as those illustrated above. In such cases we can use the conditional probability in other computations. We observe that since

$$p(E_2 | E_1) = \frac{p(E_1 \cap E_2)}{p(E_1)}$$

we can rearrange this to obtain

$$p(E_1 \cap E_2) = p(E_1) \cdot p(E_2 | E_1) \tag{7.3.2}$$

This states that the probability that both E_1 and E_2 occur is the product of the probability that E_1 occurs and the probability that E_2 occurs given that E_1 occurs.

Example 7.3.3

An urn contains 10 balls: four white, three red, and three yellow. Two balls are drawn in succession (without replacement). The first ball drawn is yellow. What is the probability the second is yellow also?

A knowledge of what is in the urn after the first draw—nine balls, of which two are yellow—enables us to compute the probability the second ball is yellow as 2/9. That is, we can compute $p(Y_2 | Y_1)$ directly from our knowledge of the experiment. We know also that the probability of drawing a yellow ball on the first draw, $p(Y_1)$ is 3/10. We can compute the probability of drawing two yellow balls, $p(Y_1 \cap Y_2)$, using Equation 7.3.2:

$$p(2Y) = p(Y_1 \cap Y_2) = p(Y_1) \cdot p(Y_2 | Y_1)$$

$$= \frac{3}{10} \cdot \frac{2}{9} = \frac{1}{15}$$

In this case we can also obtain the same result using the methods developed earlier. We can compute the probability of drawing two yellow balls by considering the outcomes of the experiment as pairs of balls. The number of outcomes in which both are yellow is $C(3,2)$, or 3. The total number of outcomes in the sample space is $C(10,2)$, or 45. Thus, the probability that both balls are yellow is

$$p(2Y) = \frac{3}{45} = \frac{1}{15}$$

This agrees with the result obtained above by considering drawing the two balls one at a time.

Example 7.3.4

A candidate for Congress assesses his prospects for winning the primary election in his district. The surveys show that the probability that he will win is .5. Since a majority of the voters in his district are in his party, he figures that he has a probability of .6 of winning in the general election, if he wins the primary. What is the probability he will win both the primary and general elections?

Let P represent the event of winning the primary and G, the general election.

The information he has is

$$p(P) = .5$$

$$p(G | P) = .6$$

Thus, the probability he wins both elections is

$$p(P \cap G) = p(P) \cdot p(G | P)$$

$$= (.5)(.6) = .3$$

Equation 7.3.2 expresses an important relationship in probability theory. It provides a way of computing the probability for various combinations of events. In set language these combinations are the intersections of the individual events. The probability that two events, A and B, will both occur, $p(A \cap B)$, is sometimes called the **joint probability** of A and B. Note that Equation 7.3.2 can be arranged in two ways:

$$p(A \cap B) = p(A) \cdot p(B|A)$$

or

$$p(A \cap B) = p(B) \cdot p(A|B)$$

By successive computations the joint probability for any number of events can be obtained. For example, the joint probability for three events, A, B, and C is given by

$$p(A \cap B \cap C) = p(A \cap B) \cdot p(C|A \cap B)$$

or

$$p(A \cap B \cap C) = p(A \cap C) \cdot p(B|A \cap C)$$

or

$$p(A \cap B \cap C) = p(A) \cdot p(B \cap C|A)$$

and so on.

The concepts of conditional probability are used also in analyzing the results of surveys as discussed in Section 1.4. The results of such surveys are usually presented as percentages. These percentages, when changed to their equivalent fractions, are the probabilities that a typical member of the group surveyed will have the characteristics specified.

Example 7.3.5

A survey is taken to obtain information about people's sources of news. 100 people are interviewed with the following results:

52 regularly read a newspaper.
39 regularly watch TV news.
28 regularly read a news magazine.
17 read a newspaper and watch TV news.
12 read both a newspaper and a news magazine.
11 watch TV news and read a news magazine.
7 do all three.

As we have seen earlier, the simplest way to analyze this information is to display it in a Venn diagram.

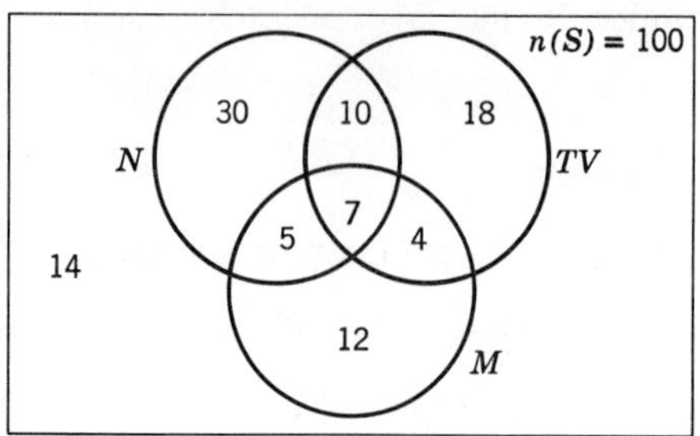

We assume that the 100 interviewees are representative of the general public. (One of the most difficult parts of making public opinion surveys is the selection of the sample to be interviewed so that it is properly representative.) From the information displayed in the diagram we can draw many conclusions. Here are a few examples:

The probability that a person who reads a newspaper also watches TV news, $p(TV|N)$ is 17/52.

The probability that a person who watches TV news does not read either a newspaper or news magazine, $p((N \cap M)'|TV)$ is 18/39.

The probability that a person who does not read a newspaper does either watch TV news or read a news magazine, $p((TV \cup M)|N')$, is 34/48.

EXERCISE 7.3

1. In an experiment with equiprobable outcomes, $p(E_1) = .2$ and $p(E_1 \cap E_2) = .08$. Find $p(E_2|E_1)$

2. In an experiment with equiprobable outcomes $p(E_1|E_2) = .5$ and $p(E_1 \cap E_2) = .2$. What is the probability of E_2?

3. The probability of an event A is .8. If A happens, the probability that event B will happen also is .6. What is the probability that both A and B happen?

4. The sample space of an experiment with equiprobable outcomes is $S = \{O_1, O_2, O_3, O_4, O_5, O_6\}$.
   $E_1 = \{O_1, O_4\}$ $\quad E_2 = \{O_2, O_4, O_6\}$ $\quad E_1 \cap E_3 = \{O_1\}$
   a. Find $p(E_2|E_1)$.
   b. Find $p(E_1|E_2)$.
   c. If $p(E_1|E_3) = \frac{1}{4}$, what is $p(E_3)$?
   d. If $p(E_2|E_3) = \frac{1}{2}$, what is $p(E_2 \cap E_3)$?

5. An urn contains seven balls, four red and three white. The balls are removed one by one by random selection. What is the probability that the order in which the colors are drawn is red, white, red, white, red, white, red?

6. An urn contains eight balls, four red and four white. An experiment consisting of drawing two balls from the urn is performed.
   a. Compute the probability that both balls are red assuming they are both drawn at the same time.
   b. The experiment is repeated, drawing one ball at a time. When the first ball is drawn, its color is noted. Use conditional probability to compute the probability both balls are red. Compare the result with the result of part *a*.

7. A survey of study programs of first-year college students showed that

75% take a course in English, 15% take a foreign language, and 6% take both. A first-year student is selected at random.

a. What is the probability that he is taking a foreign language?
b. If it is known that the student is taking English, what is the probability that he is taking a foreign langauge?
c. If it is known the student is taking French, what is the probability he is taking English?

8. A study of the special orders received for new automobiles at an assembly plant shows that in one model line 70% of the orders called for power brakes, 50% called for power steering, and 40% called for both.

a. If a customer orders power steering, what is the probability that he also wants power brakes?
b. If he orders power brakes, what is the probability that he wants power steering also?

9. Nine new employees of a large bank are assigned to work in different branches under a procedure in which there is equal likelihood of assignment to any branch. Three are to be assigned to the downtown branch, three to the westside branch, and three to the northside branch. One of the new employees, Jones, wants to work at the westside branch, and another employee, Smith, wants to work downtown.

a. What is the probability that both will receive the assignment they wish?
b. Jones has a friend in the personnel office, who tells him he is being assigned to the westside branch. In the light of this information, what is the probability that Smith will get the assignment that he wants?

10. A voter survey in a state yielded the following responses to the question, "Do you think the governor is doing a good job?

	Yes	No
Democrat	150	100
Republican	60	90
Independent	60	40

If this survey is representative of the entire population of the state

a. What is the probability that a voter thinks the governor is doing a good job?
b. If the voter is a Republican, what is the probability that he thinks the governor is doing a poor job?

11. A survey of airline passengers showed that 70% flew in tourist and 30% in first class. In the tourist section 60% of the passengers favored separate sections for smokers and nonsmokers. In first class 50% favored the separate section. What is the probability a passenger selected at random favored the separation?

12. In a suburb of a large city a survey was made of the number of automobiles owned by families of different income ranges. The results shown are the numbers of families in each category.

	Number of Cars		
Income	0	1	2 or more
Less than $10,000	1	20	4
$10,000–$20,000	3	25	17
More than $20,000	2	8	20

Assume that these results are typical of the community.

a. What is the probability that a family does not own an automobile?
b. What is the probability the family has two or more automobiles if the income is less than $10,000 per year?
c. If the family owns one automobile, what is the probability the income is more than $20,000?

13. A company gives an aptitude test to applicants for technical jobs. Their experience is that the probability an applicant can pass the test is .6. If an applicant passes the test, the probability is .8 that he can do the work satisfactorily. If he does not pass the test, the probability he can do the work is .4.

a. What is the probability that an applicant can both pass the test and do the work?
b. What is the probability that an applicant can do the work without regard to the test?

14. One hundred men who thought they might be color blind were given a test in which they viewed for a few seconds a card on which the outline of a letter was printed in a pattern of various colored dots. Card 1 had an *A* printed in red and green dots. Card 2 had a *B* in yellow and blue dots; and Card 3 had a *C* in orange and brown dots.

38 men correctly identified the *A*.
75 men correctly identified the *B*.
47 men correctly identified the *C*.
30 men correctly identified the *A* and the *B*.
18 men correctly identified the *A* and the *C*.
26 men correctly identified the *B* and the *C*.
12 men correctly identified all three.

If one of the men is selected at random,

a. What is the probability a man could identify the *C*, given that he could identify the *A*?
b. If he properly identified the *B*, what is the probability that he could identify both of the others?
c. If he identified the *A*, what is the probability that he could identify the *C*, but not the *B*?
d. If he identified the *A* and the *B*, what is the probability that he could not identify the *C*?

7.4
INDEPENDENT EVENTS AND UNION OF EVENTS

We have been studying situations in which the occurrence of one event influences the probability of the occurrence of another. We turn our attention now to situations in which there is no such influence. When the probability of an event is not affected by the occurrence or nonoccurrence of another, the two events are said to be **independent**.

In the previous section we used the symbol $p(E_2|E_1)$ to represent the probability of E_2 given that E_1 occurs. It follows from the meaning of independence, that **if E_1 and E_2 are independent,**

$$p(E_2|E_1) = p(E_2) \qquad (7.4.1)$$

Equation 7.4.1 is an algebraic way of saying that the probability of E_2 is not affected by the occurrence of E_1. It is useful as a statement of the meaning of independence. Another statement, which can be obtained as a consequence of Equation 7.4.1, proves to be more useful in making computations involving probabilities. Recalling Definition 7.3.1,

$$p(E_2|E_1) = \frac{p(E_2 \cap E_1)}{p(E_1)}$$

we can replace $p(E_2|E_1)$ by $p(E_2)$ if E_1 and E_2 are independent. Thus, we have

$$p(E_2) = \frac{p(E_2 \cap E_1)}{p(E_1)}$$

The equation can be rearranged as

$$p(E_2 \cap E_1) = p(E_2) \cdot p(E_1) \qquad (7.4.2)$$

This equation states that the probability that **both** E_1 and E_2 occur is the product of their individual probabilities, when E_1 and E_2 are independent. This equation is used as a definition of independence.

Definition 7.4.1

Independent Events: Two events, E_1 and E_2, are said to be **independent** if, and only if,

$$p(E_1 \cap E_2) = p(E_1) \cdot p(E_2)$$

This definition can be extended to include any number of events. We say that the set of events, $E_1, E_2, E_3, \ldots, E_n$, are all independent of each other if, and only if,

$$p(E_1 \cap E_2 \cap E_3 \cap \ldots \cap E_n) = p(E_1) \cdot p(E_2) \cdot p(E_3) \cdot \ldots \cdot p(E_n) \qquad (7.4.3)$$

Example 7.4.1

Suppose we have an urn containing two balls, one red and one white. An experiment consists of selecting a ball, noting its color and **returning** it to the urn. The urn is then shaken; a second ball is selected and its color noted. For each selection the probability of choosing a red ball is the same — 1/2.

The sample space is the set, $\{RR, RW, WR, WW\}$, with the first letter of each pair representing the color of the first ball selected and the second, the second selection.

If R_1 represents the event that the first selection is a red ball,

$$R_1 = \{RR, RW\}$$

There are two outcomes in R_1 and four in S. Therefore

$$p(R_1) = \frac{2}{4} = \frac{1}{2}$$

Let R_2 represent the event that the second selection is a red ball. Then

$$R_2 = \{RR, WR\}$$

As before, there are two outcomes in R_2 and four in S. Thus

$$p(R_2) = \frac{2}{4} = \frac{1}{2}$$

The event, $R_1 \cap R_2$, that both selections are red consists of the single outcome, $\{RR\}$. Its probability is

$$p(R_1 \cap R_2) = \frac{1}{4}$$

We observe that

$$p(R_1) \cdot p(R_2) = \frac{1}{2} \cdot \frac{1}{2} = \frac{1}{4} = p(R_1 \cap R_2)$$

Thus, Equation 7.4.2 is satisfied, and the two events are independent. We expect intuitively from the nature of the experiment that this is true. The chance of drawing a red ball as the second selection is not affected by what happened on the first draw, since the contents of the urn are exactly the same for both draws.

In previous sections we have discussed **mutually exclusive events**. Since it is possible to interpret the word "independent" as "having no connection with," we may be tempted to associate independent events with mutually exclusive events. However, these are two very different concepts and are, in fact, antithetical.

Consider two events, E_1 and E_2, whose probabilities $p(E_1)$ and $p(E_2)$ are **not** zero. Let us consider two cases:

Case 1 E_1 and E_2 are mutually exclusive.

From our definition of mutually exclusive, these events cannot occur together. That is, $E_1 \cap E_2 = \emptyset$, and $p(E_1 \cap E_2) = 0$.

For E_1 and E_2 to be independent

$$p(E_1) \cdot p(E_2) = p(E_1 \cap E_2)$$

But we have said neither $p(E_1)$ nor $p(E_2)$ is zero and $p(E_1 \cap E_2) = 0$. Therefore E_1 and E_2 **cannot be independent** if they are mutually exclusive.

Case 2 E_1 and E_2 are independent.

The definition of independence says

$$p(E_1) \cdot p(E_2) = p(E_1 \cap E_2)$$

Since neither $p(E_1)$ nor $p(E_2)$ is zero, $p(E_1 \cap E_2) \neq 0$. Therefore, $E_1 \cap E_2 \neq \emptyset$, and they are **not mutually exclusive**.

Example 7.4.2

Let us extend our comments about the experiment of Example 7.4.1 in which we make two successive selections from an urn. In choosing the first ball we can select either a red one or a white one. The event of choosing a red ball and that of choosing a white one are certainly mutually exclusive. They are not independent since if the event R_1 occurs, the event W_1 cannot.

However, if we make **two** selections, the choice of the second ball is independent of the choice of the first, as we have seen. And the events of choosing a red ball first and white ball second are not mutually exclusive.

Before proceeding let us summarize the concepts about the special types of events that we have studied thus far. Since events are sets, we can illustrate them with Venn diagrams. Figure 7.4.1 shows the general case of two events. As we have seen, the probability that E_1 given E_2 occurs, $p(E_1|E_2)$, is given by the ratio of the area of $E_1 \cap E_2$ to the area of E_2; that is, $[p(E_1 \cap E_2)]/p(E_2)$.

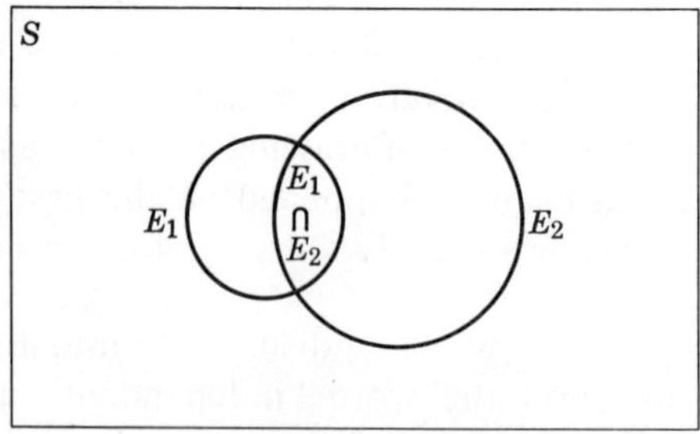

Figure 7.4.1

In the **special case** of **independent events**, this ratio is exactly the same as the ratio of the area of E_1 to the entire sample space. Figure 7.4.2 shows one possibility for this situation. The areas of the events have been arranged so that $p(E_2) = 1/2$, $p(E_1) = 1/4$, and $p(E_1 \cap E_2) = 1/8$. Further, the area of $E_1 \cap E_2$ is one-fourth the area of E_2. That is, $p(E_1|E_2) = 1/4 = p(E_1)$. Also, the area of $E_1 \cap E_2$ is one-half the area of E_1, so that $p(E_2|E_1) = 1/2 = p(E_2)$. Thus, the events E_1 and E_2 are independent.

In a statistical study independence of two events, A and B, means that the

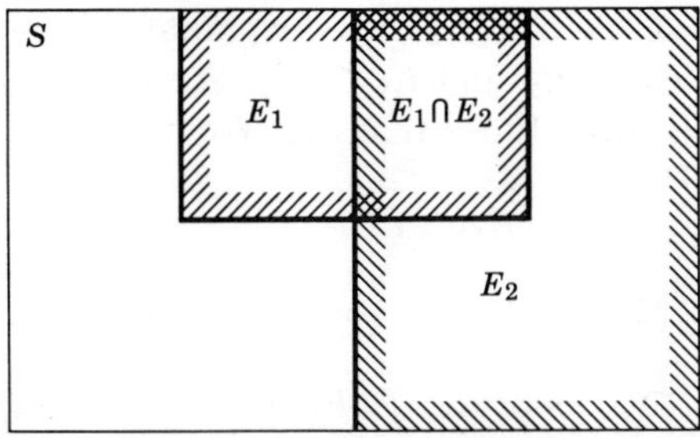

Figure 7.4.2

probability that one individual has characteristic A is the same whether or not he has characteristic B. In the example at the beginning of Section 7.3 the probability that a student has a job is .15 if the student is a woman $(p(J \mid W))$; but it is .4 if the student is a man $(p(J \mid M))$. Thus, the probability of having a job is not independent of sex. In Example 7.3.1, however, the probability of using the brand of gasoline turned out to be independent of whether a person reads a newspaper.

The situation for mutually exclusive events is shown in Figure 7.4.3. The sets E_1 and E_2 are disjoint and, thus, mutually exclusive.

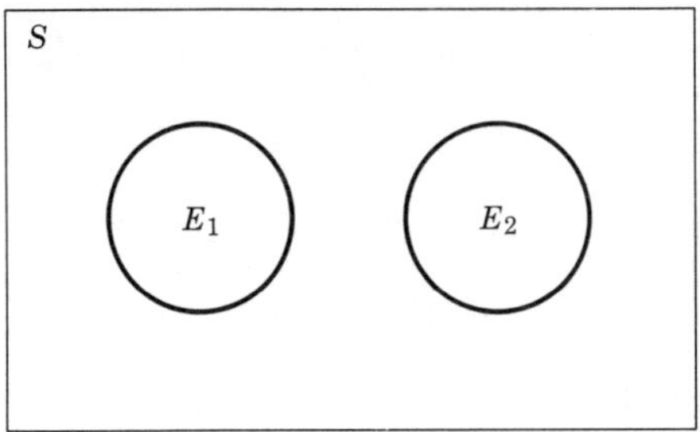

Figure 7.4.3

7.4.1 Probability of the Union of Events

The third basic property of probability says that for two **mutually exclusive** events, the probability that **either** will occur (the probability of the union of the events) is the sum of their individual probabilities. That is, when $E_1 \cap E_2 = \emptyset$,

$$p(E_1 \cup E_2) = p(E_1) + p(E_2)$$

We wish now to investigate the situation in which the events are not mutually exclusive. What is the probability that **either** of two events will occur if we put no restrictions on the events except that they be in the same sample space?

We shall assume that the events E_1 and E_2 are not mutually exclusive; that is, $E_1 \cap E_2 \neq \emptyset$. Therefore, one or more of the outcomes in the sample space belong both to E_1 and E_2. Let us separate the event E_1 into two mutually exclusive sets, $E_1 \cap E_2'$ and $E_1 \cap E_2$. We are separating the **outcomes in** E_1 that **are not in** E_2 from those **that are**. We have, therefore,

$$(E_1 \cap E_2') \cup (E_1 \cap E_2) = E_1 \tag{7.4.4}$$

Since $(E_1 \cap E_2')$ and $(E_1 \cap E_2)$ are mutually exclusive,

$$p[(E_1 \cap E_2') \cup (E_1 \cap E_2)] = p(E_1 \cap E_2') + p(E_1 \cap E_2)$$

Using Equation 7.4.4 to replace $(E_1 \cap E_2') \cup (E_1 \cap E_2)$ with its equal, E_1, we have

$$p(E_1) = p(E_1 \cap E_2') + p(E_1 \cap E_2)$$

This equation can be rearranged as

$$p(E_1 \cap E_2') = p(E_1) - p(E_1 \cap E_2) \tag{7.4.5}$$

In words this equation says that the probability that E_1 **will** occur but E_2 **will not** is the probability of E_1 minus the probability that both E_1 and E_2 will occur.

Now let us consider $E_1 \cup E_2$. $E_1 \cup E_2$ can be separated into the two mutually exclusive events, $(E_1 \cap E_2')$ and E_2, that is, into the outcomes of E_2 and those of E_1 that are not in E_2. From set algebra

$$E_1 \cup E_2 = (E_1 \cap E_2') \cup E_2 \tag{7.4.6}$$

Since $(E_1 \cap E_2')$ and E_2 are mutually exclusive,

$$p[(E_1 \cap E_2') \cup E_2)] = p(E_1 \cap E_2') + p(E_2)$$

Substituting from Equation 7.4.6, we have

$$p(E_1 \cup E_2) = p(E_1 \cap E_2') + p(E_2)$$

We can now replace $p(E_1 \cap E_2')$ with its equal from Equation 7.4.5 to obtain

$$p(E_1 \cup E_2) = p(E_1) - p(E_1 \cap E_2) + p(E_2)$$

or if we rearrange,

$$p(E_1 \cup E_2) = p(E_1) + p(E_2) - p(E_1 \cap E_2) \tag{7.4.7}$$

Equation 7.4.7 is the relationship that we seek. It says in words that, for two events in the same sample space, the probability that either or both of two events will occur is the **sum** of their **individual probabilities minus** the **probability** that **both will** occur.

This result is directly analogous to the result that we obtained in Chapter 1 for the number of elements in the union of two sets. The analogy should not be surprising, since events are sets of outcomes, and their probabilities are directly related to the number of outcomes (elements) that they contain.

To help clarify the meaning of Equation 7.4.7, let us illustrate its derivation with Venn diagrams. Figure 7.4.4 shows E_1 and E_2, with E_1 separated into

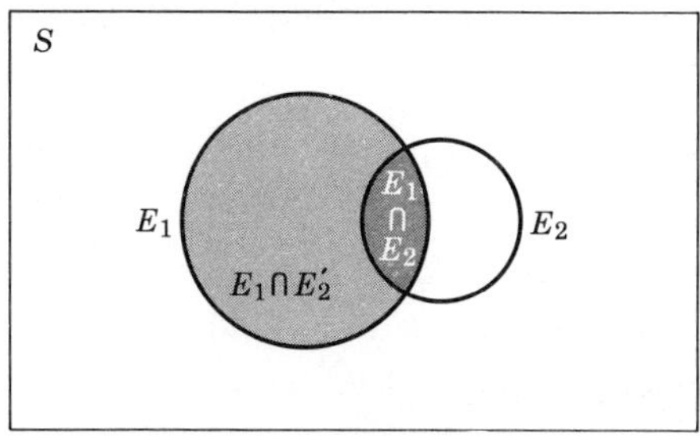

Figure 7.4.4

the disjoint sets $E_1 \cap E_2'$ and $E_1 \cap E_2$. Thus, using the diagrams in the manner illustrated in Section 7.3, we have

$$p(E_1) = p(E_1 \cap E_2') + p(E_1 \cap E_2)$$

or

$$p(E_1 \cap E_2') = p(E_1) - p(E_1 \cap E_2)$$

Figure 7.4.5 shows $E_1 \cup E_2$ as the union of the disjoint sets, $E_1 \cap E_2'$ and E_2.

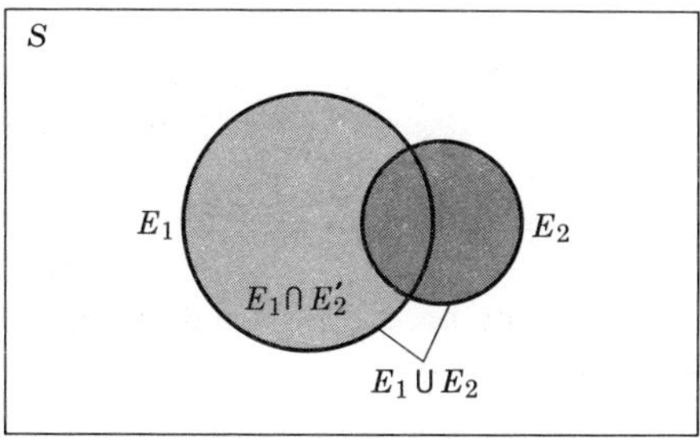

Figure 7.4.5

Thus, if we use the diagram as before,

$$p(E_1 \cup E_2) = p(E_1 \cap E_2') + p(E_2)$$

If we substitute for $p(E_1 \cap E_2')$ as before,

$$p(E_1 \cup E_2) = p(E_1) + p(E_2) - p(E_1 \cap E_2)$$

We can make two observations:

i. If E_1 and E_2 are **mutually exclusive**, $E_1 \cap E_2 = \emptyset$. Thus, Equation 7.4.7 becomes

$$p(E_1 \cup E_2) = p(E_1) + p(E_2) - p(\emptyset)$$
$$= p(E_1) + p(E_2)$$

This is, of course, the third basic property of probability.

ii. If E_1 and E_2 are **independent**, $p(E_1 \cap E_2) = p(E_1) \cdot p(E_2)$. In this case Equation 7.4.7 becomes

$$p(E_1 \cup E_2) = p(E_1) + p(E_2) - p(E_1) \cdot p(E_2) \tag{7.4.8}$$

Example 7.4.3

If a family has three children, what is the probability that either the oldest or the youngest child is a boy? Assume the probability a child is a boy is 1/2.

The sample space listing the children in order of their birth, is the set,

$$\{BBB, BBG, BGB, BGG, GBB, GBG, GGB, GGG\}$$

If we call E_1 the event, the first child is a boy,

$$E_1 = \{BBB, BBG, BGB, BGG\}$$

If we call E_2 the event, the third child is a boy,

$$E_2 = \{BBB, BGB, GBB, GGB\}$$

Since the sample space consists of eight equally probable outcomes, each one has a probability of 1/8.

Therefore

$$p(E_1) = \frac{4}{8} = \frac{1}{2}$$

$$p(E_2) = \frac{4}{8} = \frac{1}{2}$$

The event $E_1 \cap E_2$ is the set, $\{BBB, BGB\}$ and

$$p(E_1 \cap E_2) = \frac{2}{8} = \frac{1}{4}$$

We have, thus, the probability that either the oldest or youngest child (or both) is a boy is

$$p(E_1 \cup E_2) = \frac{1}{2} + \frac{1}{2} - \frac{1}{4} = \frac{3}{4}$$

Example 7.4.4

A nationwide survey on the reactions of people to various measures proposed to control inflation brought these responses:

Favor price controls	68%
Favor wage controls	57%
Favor both wage and price controls	34%

If we assume that this survey was made on a representative sample, what is the probability that a person selected at random favors either price or wage controls?

The meaning of the 68% is that 68 out of every 100 people favored price controls. Therefore, the probability a person favors price controls is .68. Similarly, the probability that a person favors wage controls is .57; and the probability that a person favors both is .34. Using Equation 7.4.7 we have, where P represents the event, favors price controls, and W represents the event, favors wage controls,

$$p(P \cup W) = p(P) + p(W) - p(P \cap W)$$

$$= .68 + .57 - .34 = .91$$

In the example above we observe that the two events, P and W, are not independent. $p(P)\cdot p(W) = (.68)(.57) = .39$. This is not the same as the observed probability, $p(P \cap W)$. In surveys it is often of interest to determine whether the responses to different questions are independent or are related in some way. In fact, many surveys are made for the specific purpose of determining such relationships, called **correlations**. The amount by which $p(A \cap B)$ differs from $p(A)\cdot p(B)$ is used as a measure of the correlation between the events A and B.

In Example 7.4.4 if we wish to determine the probability that a person favors **neither** price nor wage controls, we must find $p((P \cup W)')$. The event that a person favors neither type of control is symbolized $P' \cap W'$, where P' represents the event, does not favor price controls, and W' represents the event, does not favor wage controls. We recall from our previous work with sets that one of De Morgan's laws says

$$P' \cap W' = (P \cup W)'$$

Therefore

$$\begin{aligned} p(P' \cap W') &= p((P \cup W)') \\ &= 1 - p(P \cup W) \end{aligned}$$

In this case the probability that a person favors neither type of controls is

$$\begin{aligned} p(P' \cap W') &= 1 - .91 \\ &= .09 \end{aligned}$$

If two events are independent, their complements are also independent. We can find the probability that neither of two independent events occurs by finding the product of the separate probabilities that the events do not occur.

If E_1 and E_2 are two independent events, E'_1 and E'_2 are also independent. Therefore, the probability that neither event occurs, $p(E'_1 \cap E'_2)$ is given by

$$p(E'_1 \cap E'_2) = p(E'_1)\cdot p(E'_2) \tag{7.4.9}$$

Since $p(E'_1) = 1 - p(E_1)$ and $p(E'_2) = 1 - p(E_2)$,

$$\begin{aligned} p(E'_1 \cap E'_2) &= (1 - p(E_1))\cdot(1 - p(E_2)) \\ &= 1 - p(E_1) - p(E_2) + p(E_1)\cdot p(E_2) \\ &= 1 - [p(E_1) + p(E_2) - p(E_1)\cdot p(E_2)] \\ &= 1 - p(E_1 \cup E_2) \end{aligned} \tag{7.4.10}$$

Equation 7.4.10 is true for **any** two events. Equation 7.4.9 holds when E_1 and E_2 are **independent** events.

EXERCISE 7.4

1. Given that two events, E_1 and E_2, are independent, $p(E_1) = .3$ and $p(E_2) = .4$.
   a. What is the probability that both E_1 and E_2 occur?

b. What is the probability that either E_1 or E_2, or both, occur?
c. What is the probability that neither occurs?

2. Given two mutually exclusive events E_1 and E_2. The probability of E_1 is .2, and the probability of E_2 is .4.
a. What is the probability that either E_1 or E_2 occurs?
b. What is the probability that they both occur?

3. The respective probabilities of two independent events, E_1 and E_2, are $p(E_1) = .5$ and $p(E_2) = .4$. What is the probability that either E_1 or E_2, but not both, will occur?

4. An urn contains 20 balls: five are red, seven are white, and eight are yellow.
a. A ball is selected at random. What is the probability that it is either red or white?
b. A ball is selected, and its color noted. It is replaced in the urn, and a second selection is made at random. What is the probability that the first ball is yellow and the second one is white?
c. The experiment of (*b*) is repeated. What is the probability that either (or both) of the balls is yellow?
d. A ball is selected and its color noted. A second ball is selected without replacing the first. What is the probability that both balls are red?

5. A pair of dice is rolled once. What is the probability of rolling either a "7" or "11"?

6. A card is selected at random from a standard deck of 52 playing cards.
a. What is the probability that it is either an ace or a face card?
b. What is the probability that it is either an ace or a heart?

7. An experiment has 10 outcomes, O_1 to O_{10}.

$$E_1 = \{O_1, O_3, O_7, O_9\}$$

$$E_2 = \{O_3, O_4, O_6, O_7, O_8\}$$

a. What is $p(E_2|E_1)$?
b. What is $p(E_1|E_2)$?
c. What is $p(E_1 \cap E_2)$?
d. What is $p(E_1 \cup E_2)$?
e. Are the events independent?

8. Two urns each contain 10 balls. In the first there are five red, two white, and three yellow. In the second there are two red, four white, and four green. An urn is selected at random and a ball drawn from it.
a. What is the probability the ball is red?
b. What is the probability the ball is green?

9. In the situation of Problem 8 above, a ball is selected at random from the first urn and placed in the second. The second urn is shaken and a ball drawn at random. What is the probability this ball is red?

10. A man is applying for a position as trainee for the astronaut program. He must pass both an aptitude examination and a physical examination. From the information that he has been able to obtain about exams, he figures the probability of passing the aptitude test is 0.7, and the probability of passing the physical exam is 0.6. If these two examinations are independent,

a. What is the probability he will pass both?
b. At least one?
c. Neither?

11. An electric grill has two principal sources of failure, the heating element and the power cord, and the occurrence of one type of failure is independent of the other. The probability that the power cord fails within the first year is .6, and the probability that the heating element fails within the first year is .7.

a. What is the probability that the grill will work for a year without any failures?
b. What is the probability that both the cord and the heating element will fail?

12. In organizing an election campaign the chairman of a political action group asked the members to support the campaign either by working at campaign headquarters or by a financial contribution, or both. Twenty-five percent expressed willingness to work, 68% were willing to contribute money, and 15% were willing to do both. If a member of the group were selected at random, what is the probability that he would be willing either to work or to contribute money, but not both?

13. A man provides financing for three new businesses. In one of the three the probability that the business will survive its critical first year is .7. In the second the probability of survival for the first year is .5; and in the third, .4. If the three businesses are completely independent, what is the probability that at least one of the three will survive the first year?

14. A company has a policy of giving its employees the day off on their birthdays. Two employees are chosen at random to be interviewed about their reaction to the policy. For that year

a. What is the probability that both their birthdays fall on Monday?
b. What is the probability that the birthday of either of them falls on Monday?
c. Considering just one of the employees, what is the probability his birthday falls on a weekend?
d. What is the probability that both people's birthdays fall on a weekend?
e. What is the probability that either one's birthday falls on a weekend?

15. In a study of the effect that using the same brand name on different kinds of products has on product sales, the following data were obtained: A company that produces a wide range of food products uses the same brand name on margarine and coffee. Customers in supermarkets were asked whether or not they use this brand in these two products; 13.2% said that they use the margarine regularly, 7.3% said they use the coffee, and just under 1% said they use both. On the basis of this information what can be concluded about brand-name loyalty between these products?

7.5
REPEATED TRIALS

In our discussions of probability thus far we have dealt with experiments defined to consist of a single operation or activity. In those cases where the events of interest included sequences of outcomes, the experiment was carefully defined so that the sample space consisted of appropriate sequences of outcomes. In the previous section, for example, we discussed the probability that the oldest or youngest of the three children in a family is a boy (Example 7.4.3). We described an experiment as "having three children," and the sample space lists as outcomes all of the possible sequences of three children—for example, boy, boy, boy and boy, girl, boy.

It is often useful and more convenient to consider such experiments as a series of **trials**, each having its own set of outcomes. These trials are essentially experiments within experiments, and we shall use the term "trial" to make the distinction between a total experiment and "parts" of an experiment. There are two reasons for this new approach. First, there is no way to relate the results of one experiment directly to the results of another using the definitions and properties on which the theory of probability is based. Second, in these experiments the interest focuses on the result of the entire series of trials. The separation into parts is merely an aid to studying these results. Thus, when an experiment is regarded as a sequence of activities, we call each part of the sequence a **trial**, and consider the experiment as comprising a number of trials.

Each trial has its own set of outcomes, and we shall form events relating to the individual trials in a sequence. We shall then combine the trial events to form events for the entire experiment.

Example 7.5.1

Let us consider an urn containing two balls, one red and one white. We conduct an experiment consisting of three trials. In each trial a ball is selected at random from the urn, and its color noted; it is then replaced in the urn. Thus, the identical operation is performed in each trial.

Each trial has two possible outcomes: a red ball is drawn or a white ball is drawn. Let us use the symbol R_1 to denote that a red ball is drawn on the first trial; R_2 to denote a red ball is drawn on the second trial; and R_3 to denote a red ball is drawn on the third trial. Similarly, W_1, W_2, and W_3 denote a white ball drawn on the respective trials. The sample space for the three trials are, respectively,

$$S_1 = \{R_1, W_1\} \qquad S_2 = \{R_2, W_2\} \qquad S_3 = \{R_3, W_3\}$$

Let us define three events as follows:

$$E_1 = \{R_1\} \qquad E_2 = \{R_2\} \qquad E_3 = \{R_3\}$$

That is, E_1 is the event of drawing a red ball in the first trial; E_2 is the event of drawing a red ball in the second trial; and E_3 is the event of drawing a red ball in the third trial. As we have seen earlier,

$$p(E_1) = \frac{1}{2} \qquad p(E_2) = \frac{1}{2} \qquad p(E_3) = \frac{1}{2}$$

The conditions for each trial are identical, and so these events are independent. Therefore, the probability of drawing three red balls as the outcome of the **experiment** is

$$p(3R) = p(E_1) \cdot p(E_2) \cdot p(E_3)$$

$$= \frac{1}{2} \cdot \frac{1}{2} \cdot \frac{1}{2} = \frac{1}{8}$$

Using the methods of preceding sections, the sample space of the experiment is, with R representing the drawing of a red ball and W, the drawing of a white ball,

$$S = \{RRR, RRW, RWR, RWW, WRR, WRW, WWR, WWW\}$$

The event of drawing three red balls in succession is $\{RRR\}$. The eight outcomes in the sample space are equally probable, and so

$$p(RRR) = \frac{1}{8}$$

in agreement with the result above.

There is little to choose between the two approaches in Example 7.5.1 for the experiment considered there. In more complex experiments with many parts, the repeated-trials approach is substantially simpler.

The important fact that has been illustrated is that the probability of a **specific sequence** of outcomes of successive trials is the product of the individual probabilities of those outcomes in their respective trials. That is, if an experiment consists of a sequence of n trials and $E_1, E_2, E_3, \ldots, E_n$ are events formed from the outcomes of the respective trials, the probability of the sequence of events $E_1 \cap E_2 \cap E_3 \cap \ldots \cap E_n$ is given by

$$p(E_1 \cap E_2 \cap E_3 \cap \ldots \cap E_n) = p(E_1) \cdot p(E_2) \cdot p(E_3) \cdot \ldots \cdot p(E_n) \qquad (7.5.1)$$

In Example 7.5.1 we used an experiment of identical trials and a sequence of independent events. It is not necessary for the trials to be identical or related activities. However, the successive events must be independent for Equation 7.5.1 to hold. If they are not independent, the probabilities for the events following the first are replaced by the appropriate conditional probabilities. Further, it is not implied that the terms "sequence" or "successive" have any relation to time—that is, to trials performed one after another. The trials are **ordered**, in that they are identified individually; but the ordering is not necessarily related to time. Thus, in Example 7.5.1 the three selections from the same urn could just as well be three selections performed at any time from three different urns with the same contents.

Let us now consider an example in which the events for different trials are not independent.

Example 7.5.2

Consider an urn containing 10 balls, three red, three white, and four yellow. An experiment consists of three trials—selecting one ball from the urn **without replacing it**. What is the probability of selecting three red balls?

We consider each trial in sequence. For the first selection the probability of selecting a red ball is 3/10 – three of the 10 balls in the urn are red. That is, $p(R_1) = 3/10$.

To compute the probability of selecting three red balls in succession, we assume that **each** of the **selections is "successful."** Thus, for the second selection there are nine balls left in the urn, of which two are red. The probability of selecting a red ball in the second trial is, therefore, 2/9. That is $p(R_2|R_1) = 2/9$.

For the third trial there are eight balls in the urn, and one of them is red. Therefore, $p(R_3|R_1 \cap R_2)^* = 1/8$.

The event of selecting three red balls is symbolized $R_1 \cap R_2 \cap R_3$. Its probability is given by

$$\begin{aligned} p(R_1 \cap R_2 \cap R_3) &= p(R_1)\cdot p(R_2|R_1)\cdot p(R_3|R_1 \cap R_2) \\ &= \frac{3}{10}\cdot\frac{2}{9}\cdot\frac{1}{8} \\ &= \frac{1}{120} \end{aligned}$$

Using the approach from previous sections we call the outcome of selecting three red balls from the urn, *RRR*. The experiment is to select three balls at once from the urn. There is $C(3,3) = 1$ way to make this selection. The total number of selections of three balls from the 10 in the urn—the number of outcomes in the sample space—is $C(10,3) = 120$. Therefore,

$$p(RRR) = \frac{C(3,3)}{C(10,3)} = \frac{1}{120}$$

Let us now consider a few additional examples to illustrate how an experiment can be considered as a sequence of trials.

Example 7.5.3

An inspector is going to check a shipment of 12 items by selecting a sample of three items for test. If there are three defective items in the shipment, what is the probability that the three items selected for test will all be good?

With our previous methods we solve this problem by first

*The symbol $R_1 \cap R_2$ is used to symbolize the compound event R_1 **and** R_2 even though, technically speaking, we do not include outcomes from the second trial in R_1 nor outcomes from the first trial in R_2. This use of the symbol "$\cap$" to mean "and" is consistent with the use of "$R_1 \cup R_2$" to denote the compound event R_1 **or** R_2.

determining that there are $C(9,3)$ ways of choosing three items from the nine good ones. There are $C(12,3)$ ways of choosing three items from the shipment of 12. Thus, the probability of choosing three good ones is

$$p(3G) = \frac{C(9,3)}{C(12,3)}$$

$$= \frac{\frac{9!}{6!3!}}{\frac{12!}{9!3!}} = \frac{9!}{6!3!} \cdot \frac{9!3!}{12!}$$

$$= \frac{9 \cdot 8 \cdot 7}{12 \cdot 11 \cdot 10} = \frac{21}{55}$$

Let us now consider the selection of the three items as a sequence of three "trials."

Since there are nine good items in the 12, the probability that the first one selected is good is

$$p(G_1) = \frac{9}{12}$$

Assuming that the first item selected is good, there are now eight good ones in the 11 remaining. Thus

$$p(G_2) = \frac{8}{11}$$

Similarly, if we assume that the second one is good,

$$p(G_3) = \frac{7}{10}$$

Therefore, the probability that all three items are good is

$$p(3G) = \frac{9}{12} \cdot \frac{8}{11} \cdot \frac{7}{10} = \frac{21}{55}$$

Example 7.5.4

Let us repeat the experiment of Example 7.5.3 with the exception that we wish to find the probability that one of the three items is defective.

There are two alternative types of problems in a situation of this kind. If we wish the probability that a **specific** one of the three selections yields a defective item, we proceed as in the previous example, and for the particular item that is to be defective we use the probability that the item is defective in the place of the probability that the item is good.

If we want to know the probability that **any** of the three is defective, we must consider all of the sequences that produce this result. In this case these sequences are GGD, GDG, and DGG. We must compute the probability for each and, then, since they are mutually exclusive events, we must add the three probabilities to obtain the probability that we seek.

In the first of these sequences, GGD, we have, since nine of the 12 are good and three are defective,

$$p(GGD) = \frac{9}{12} \cdot \frac{8}{11} \cdot \frac{3}{10} = \frac{9}{55}$$

For the second sequence, GDG:

$$p(GDG) = \frac{9}{12} \cdot \frac{3}{11} \cdot \frac{8}{10} = \frac{9}{55}$$

For the third sequence, DGG:

$$p(DGG) = \frac{3}{12} \cdot \frac{9}{11} \cdot \frac{8}{10} = \frac{9}{55}$$

Therefore, the total probability that we choose two good items and one defective one is

$$p(2G1D) = \frac{9}{55} + \frac{9}{55} + \frac{9}{55} = \frac{27}{55}$$

Note that, although the order of the numbers in the numerators in the three sequences above changed position, their product is the same in each case. This agrees with our intuition that the three possible sequences are equally likely to occur.

Example 7.5.5

In a certain congressional district 50% of the voters are Caucasian; 25%, Negro; and 25%, Oriental. This means, of course, that if a voter in that district is selected at random, the probability that he is a Caucasian, $p(C)$, is 1/2; the probability that he is a Negro, $p(N)$, is 1/4; and the probability that he is an Oriental, $p(O)$, is 1/4. These data can be interpreted to say that two out of four voters in the district are Caucasian; one out of four is a Negro; and one out of four is Oriental. However, if we choose a group of four voters at random, what is the probability that we choose two Caucasians, one Negro, and one Oriental?

We can consider this as a sequence of four trials in which the outcomes are to include two Caucasians, one Negro, and one Oriental. The probability of choosing them in that order is

$$p(CCNO) = p(C) \cdot p(C) \cdot p(N) \cdot p(O)$$

$$= \left(\frac{1}{2}\right)\left(\frac{1}{2}\right)\left(\frac{1}{4}\right)\left(\frac{1}{4}\right) = \frac{1}{64}$$

There are $\frac{4!}{2!1!1!}$ orders in which the same number in each category can be selected. Thus

$$p(2C1N1O) = \frac{4!}{2!} \cdot \frac{1}{64} = \frac{24}{2} \cdot \frac{1}{64}$$

$$= \frac{3}{16}$$

The result in this example, besides illustrating the method of using repeated trials in the computation of probabilities, indicates one of the problems in taking small random examples and using them as representatives of large groups. With a sample of four the probability of obtaining the same ethnic distribution in the sample that exists in the district is only 3/16. This important aspect of studying samples instead of entire populations is considered in greater detail in the study of statistics.

As the foregoing example indicates, "multiple" independent events can be considered as a sequence by finding the product of the probabilities of the individual events in the sequence. Remember that the sequence is inherently an **ordered** process. That is, the result of a single sequence of trials gives the probability for the events to occur **in that order**. If the situation is such that the order is not important, then all of the possible orders for those events must be considered. The set of different orders in which the same results can occur is a set mutually exclusive events, and the probability that **any** of them occurs is the sum of the probabilities of the individual sequences. In the foregoing example the probability for each of the different orders of selection of the voters was the same, 1/64. Since there are (4!)/(2!), or 12, of them, we added the 12 (1/64)'s together to get the final result, 3/16.

7.5.1 Bernoulli Trials (Binomial Probabilities)

One important special type of experiment consists of repeated trials, each of which has exactly two outcomes. For example, in a manufacturing process there can be just two results for a given unit of production—it is acceptable or it is defective. When a potential customer considers a purchase, he either buys a given product, or he does not. In many tests related to psychological experiments a subject either "passes" the test, or he does not. Experiments of this general nature, in which the same trial is made repeatedly under "identical" conditions, and in which there are only two outcomes possible for each trial, are called **Bernoulli* trials**. (Because of the use of a theorem of algebra called the binomial theorem in the computation of their probabilities, Bernoulli trials are sometimes referred to as **binomial probabilities**.)

The two outcomes of a Bernoulli trial are called "success" and "failure," although there need be no concept of success or failure involved. The outcome of primary interest is usually called the "success."

To summarize, a sequence of identical operations, called trials, are Bernoulli trials if, and only if,

1. The outcome of any trial is independent of the preceding trial.
2. There are exactly two outcomes possible for each trial.
3. The probability of each outcome remains the same throughout the sequence.

*Named for James Bernoulli (1654–1705) who did some of the first work on the theory of probability.

4. There are a specified number of trials in the experiment.

We let p represent the probability of a "success" in any trial and q, the probability of a "failure." Then, since there are only two possible outcomes,

$$p + q = 1 \tag{7.5.2}$$

We shall also use two expressions equivalent to Equation 7.5.2:

$$p = 1 - q \qquad q = 1 - p$$

In developing the expressions used to compute the probabilities for Bernoulli trials, the number of trials to be made in a given experiment is usually symbolized by n. We are concerned with the probability of determining a given number of "successes," symbolized as k, in a sequence of n trials when the probability of a "success" in any one trial is p.

First, let us consider the sequence: success, failure, success. From our previous discussions we know that

$$\begin{aligned} p(SFS) &= p \cdot q \cdot p \\ &= p^2 q \end{aligned}$$

In this sequence there are two successes and one failure. The other sequences with two successes and one failure are (success, success, failure) and (failure, success, success). For these experiments

$$p(SSF) = p \cdot p \cdot q = p^2 q$$

$$p(FSS) = q \cdot p \cdot p = p^2 q$$

We see that all these sequences have the same probability. Note that in this experiment $n = 3$ (the number of trials) and $k = 2$ (the number of successes). The probability for **any** of the sequences is

$$p(SSF) = p^2 q = p^k q^{n-k}$$

The total probability for two successes and one failure in three trials is

$$\begin{aligned} p(2S1F) &= 3p^2 q \\ &= 3 \cdot p^k q^{n-k} \end{aligned}$$

The number of different sequences of results that give us two successes out of three trials (three) turns out to be given by $C(3,2)$. The $C(3,2)$ comes from counting the number of ways the two successes can be distributed through the sequence of three trials; that is, the number of ways we can "choose" the specific trials that are successes.

These considerations lead to the general equation giving the probability of obtaining k successes in a sequence of n trials if p is the probability for success in any one trial:

$$p(n,k) = C(n,k) \cdot p^k q^{n-k} \tag{7.5.3}$$

Since $q = 1 - p$, Equation 7.5.3 can be written also as

$$p(n,k) = C(n,k) \cdot p^k (1 - p)^{n-k} \tag{7.5.4}$$

Example 7.5.6

In a manufacturing process 10% of the items produced are defective. If a sample of 10 items is taken for testing, what is the probability that exactly one of them is defective?

Note that in previous examples we specified a specific number of items—a "batch" or "lot"—from which the sample is selected. A specific number of defective items are assumed to be in the "batch." We are now considering a process that is assumed to operate continuously with a **percentage** of the items defective. The first procedure above is "sampling without replacement"; the second, "sampling with replacement." There are situations in which each approach is appropriate.

Let us consider a good item to be a success and a defective item a failure. We seek the probability of nine successes in a sequence of 10 trials. Thus, $n=10$ and $k=9$. The probability an item is defective is 0.1 (10%); and as a result, the probability that an item is good is 0.9. That is, $p=0.9$ and $q=0.1$. Equation 7.5.3 gives

$$\begin{aligned} p(10,9) &= C(10,9)\cdot(0.9)^9(0.1)^1 \\ &= 10\cdot(.387)(0.1) \\ &= .387 \text{ (to nearest 3 decimals)} \end{aligned}$$

If we want the probability that **no more** than one item is defective, we have as possibilities (9 good, 1 defective) **or** (10 good, 0 defective). We computed the probability for (9 good, 1 defective) above. If we have 10 good and zero defective, Equation 7.5.3 gives

$$\begin{aligned} p(10,10) &= C(10,0)\cdot(0.9)^{10}(0.1)^0 \\ &= (1)(.348)(1) \\ &= .348 \text{ (to nearest 3 decimals)} \end{aligned}$$

The probability of (9 good, 1 defective) or (10 good, 0 defective) is given by

$$\begin{aligned} p(D \le 1) &= p(10,9) + p(10,10) \\ &= .387 + .348 = .735 \end{aligned}$$

Let us note one or two characteristics of the computations we have made, and also the form of Equations 7.5.3 and 7.5.4. Note first that in a sequence of n trials, there must be a probability for the **result** of **each trial** in our computation. That is, if we are considering the probability of three successes in five trials, there must be exactly two failures in order that there be exactly three successes. Thus, the probability computation includes the factors p^3q^2. The other factor in the computation, $C(5,3)$, "counts" the different orders in which these successes and failures can occur.

Whenever we are seeking a probability for other than one exact number of successes and failures, we must make more than one computation. The probability for each possible number of successes must be computed separately and the results **added** to get the desired probability. If we extend the situation to large numbers of trials, the computations can become quite lengthy. For example, if we wanted the probability of not more than 10 defective items out of a sample of 100 in the situation of Example 7.5.6, we require the following computations:

$$p(D \leq 10) = p(100,10) + p(100,9) + p(100,8) + p(100,7) + p(100,6) + p(100,5) + p(100,4) + p(100,3) + p(100,2) + p(100,1) + p(100,0)$$

where

$$p(100,10) = C(100,10)(0.9)^{90}(0.1)^{10}$$
$$p(100,9) = C(100,9)(0.9)^{91}(0.1)^{9}$$
$$p(100,8) = C(100,8)(0.9)^{92}(0.1)^{8} \qquad \text{and so on.}$$

The computations are obviously long and tedious. However, these probabilities are used frequently enough by those engaged in studies involving probabilities that tables of values for these quantities have been published. We shall restrict our examples to situations in which the computations are reasonably simple.

In the small samples that we have considered in our computations, the probability of finding the "expected" number of items in our examples has been relatively small. For example, in Example 7.5.6 we considered a process in which one out of 10 (10%) of the items produced are defective. Yet the probability of finding exactly one defective in a random sample of 10 is .387, less than an "even" chance. Earlier we considered a sample of four taken from a population 50% Caucasian, 25% Negro, and 25% Oriental. The probability of having that ethnic distribution in the sample was .1875, much less than an "even" chance. It is the nature of random processes that deviations from the expected or "average" values can be quite large in small samples. As the sample gets larger, the probability of finding the "expected" quantities increases. For example, in the production sample, as we select larger and larger samples, the probability that exactly 10% of the sample items are defective gets closer to one. When the sample includes all of the items produced, the percentage of defectives is certain to be 10%. In a sequence of trials that can in theory be continued indefinitely, the relative frequency of each outcome gets closer and closer to the probability of its occurrence. In flipping a coin, for example, if the probability of getting "heads" is 1/2, the number of heads actually obtained will approach one-half of the total number of flips. It is the nature of random processes, however, that we cannot expect it to be **exactly** one-half of the total. The deviations that can be expected and the interpretation of deviations of different amounts from the probable values are considered in detail in the study of statistics and statistical inference.

EXERCISE 7.5

1. In a sequence of five Bernoulli trials the probability of success in each trial is .6.

a. What is the probability of three successes in the sequence?
b. What is the probability of two successes?
c. What is the probability of five successes?
d. What is the probability of fewer than five successes? (**Hint:** Consider the alternative.)

2. In a sequence of five Bernoulli trials the probability of success in each trial is 1/2. Compute the probability for each result possible and show that the sum of these probabilities is one.

3. In a sequence of six Bernoulli trials the probability of success is .4 for each trial.
a. What is the probability of exactly three successes?
b. What is the probability of at least one success?

4. You are offered a choice of two games. In each there is a sequence of five plays. In one the probability of success on each trial is .7. To win, you must have at least four successes in the five trials. In the second the probability of success in each trial is .5. To win, you must have at least three successes in the five trials. In which game is your probability of winning greater and by how much?

5. A box contains eight light bulbs, of which four are good and four are not. A person selects three bulbs from the box.
a. What is the probability that all three are good?
b. What is the probability that two of the three are good?
c. What is the probability that at least one is good?

6. Three representatives each from three different ecology organizations (a total of 9) meet to discuss a pollution problem in their city. A committee of three is chosen at random to draft a petition to send to the city council asking for action on the problem. What is the probability that one member of each organization is chosen to draft the petition?

7. A legislative committee votes 9 to 6 to approve a controversial bill. A reporter interviews three of the committee members after the meeting.
a. What is the probability that all three legislators interviewed voted for the bill?
b. What is the probability that a majority of the three interviewees voted against the bill?

8. The voter registration in a precinct shows 60% Republican and 40% Democrat. An election worker chooses five voters at random to interview.
a. What is the probability that he chooses three Republicans and two Democrats?
b. What is the probability that he chooses three Democrats and two Republicans?
c. What is the probability that he chooses all Republicans?
d. What is the probability that he chooses all Democrats?

9. Brown eyes are a genetic characteristic. Suppose the genetic makeup of the parents in a family is such that the probability of their children having brown eyes is 3/4. If they have four children,

a. What is the probability that three of the four will have brown eyes?
b. What is the probability that they will all have brown eyes?
c. What is the probability that at least three will have brown eyes?

10. A line of people is waiting to be interviewed for employment. The probability that any given applicant will be hired is 0.6. If there are three jobs available, what is the probability that the fifth person in the line will be hired?

11. The probability that a student entering college will graduate is .3. Five students from the same neighborhood enter college at the same time.
a. What is the probability that they all will graduate?
b. What is the probability that a majority of them will graduate?

12. Twenty percent of the families in California move each year.
a. What is the probability that a family chosen at random will not move for five years?
b. What is the probability that a family chosen at random will move once in the next five years?

13. In an experiment on ESP one of two subjects selects a card at random from a special deck. Each card has **one** of **four** symbols printed on it. The first subject concentrates on the symbol for 10 seconds. At the end of the 10 seconds a signal sounds and the second subject writes down the symbol that he believes is on the card. The first subject selects another card and again concentrates on it for 10 seconds, and so on. The experiment consists of a sequence of five card selections and corresponding symbol recordings by the second subject. If there are no thought signals, and the second subject writes down symbols purely by guesswork,
a. What is the probability that he will guess all five correctly?
b. What is the probability that he will guess at least three of the five?

14. A box contains four items of which only one is good. A person selects an item from the box and tests it. He continues this process until he finds the good item.
a. What is the probability that he will not find the good item until all three defective ones have been selected?
b. What is the probability that he will find the good one on or before the second selection?

15. A manufacturing process produces items one after another automatically. Ten percent of the items are defective. An inspector takes an item off the production line and tests it. After the test is completed, he selects another and tests it, and so on.
a. What is the probability that the second item he selects will be defective?
b. What is the probability that the fifth item is the first defective one?
c. How many items must he test before the probability of finding a defective one is greater than 1/2?

16. In a critical situation where component failure can be very serious, if not fatal, (e.g., in a space vehicle) it is sometimes not possible to obtain components with adequate reliability. In such cases "back-up" parts are designed into the machine to perform a function in case of failure of the component that normally performs that function. If the probability a given part operates satisfactorily is .99,

a. What is the probability of failure if a single back-up part is provided?
b. How many back-up parts are needed to make the chance of failure "one in a million"?

7.6
STOCHASTIC PROCESSES; BAYES' THEOREM

In the previous section we studied a stochastic process consisting of a sequence of trials in which the same activity was repeated a number of times. The result in each stage of the sequence was independent of the result in the adjoining stage. We now turn our attention to stochastic processes in which the probabilities in one stage depend on the result of another stage. Again we caution the student that the "stages" in the process may or may not involve different activities. They may be only different aspects of the same activity that are studied independently. We wish to use the results of one stage of the process to make inferences about the results of another stage.

For example, let us consider a situation in which a company produces a complicated machine. In production some of the machines are assembled properly, and the machine operates satisfactorily. Others have minor defects and require adjustment before they operate satisfactorily. Still others contain defective components and require a major rework. The company has an inspection department that tests each machine, but the test is not perfect. It indicates that some of the satisfactory machines need adjustment. Those machines that need adjustment are sometimes passed without this fact being detected; and sometimes they are rejected completely and sent back for rework, although they do not need it. Some of the unsatisfactory machines are merely adjusted and then passed without replacement of the faulty component. We wish to use the results of the inspection tests to make inferences about the likelihood that a machine passing inspection will, in fact, operate satisfactorily, or about the likelihood of doing unnecessary rework.

Let us consider a situation in which a sociologist wishes to make inferences about a person on the basis of observations that he makes. For example, suppose a survey is made of a group of people, some of whom are in management positions, some in clerical positions, and some in sales positions. They are all asked how well they enjoy their work: do they enjoy it very much, are they neutral toward it, or do they actively dislike it? If we use the results of such a survey, what kind of predictions can be made about a person's occupation if it is known only how well he likes his work?

These situations are examples of many in which we want to make inferences about the first stages of a process from knowledge of the latter stages. The probabilities computed for this purpose are called **a posteriori probabilities**, probabilities obtained by generalizations from observations; as contrasted with **a priori probabilities**, probabilities deduced by reasoning—the kind we have studied up to now.

Before considering specific cases such as the examples cited above, let us construct a simple experiment to study how a posteriori probabilities can be computed.

Consider three urns, labeled A, B, and C. Urn A contains six red balls, five white balls, and four green. Urn B contains two red balls, three white balls, and five green ones. Urn C contains six red balls, 10 white balls, and four green ones. In a typical experiment a person is blindfolded. He then selects an urn at random (he does not know which urn he selects), draws out a ball, and notes its color. Given the color of the ball, what is the probability that the person selected Urn A, Urn B, or Urn C? In an experiment of selecting an urn at random, the probability of selecting any one of them is 1/3 — there is one way to make a specified selection out of three equally likely possibilities. How does noting the color of the ball drawn affect this probability?

We start our analysis by drawing a diagram of the process, as shown in Figure 7.6.1.

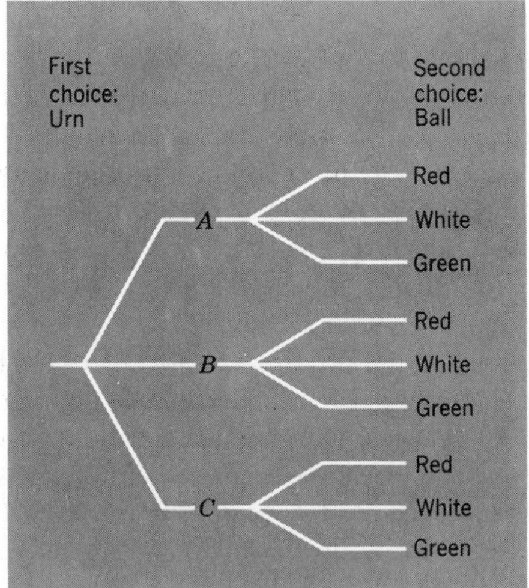

Figure 7.6.1

The outcomes for the experiment consist of pairs of results. The first element of the pair is the urn selected; and the second element is the ball selected—for example (A,R_2) and (C,G_3). (A,R_2) represents the selection of Urn A and then the choice of the second red ball in Urn A. We observe the result of the experiment as an **event**—the choice of a particular color ball. The outcomes in this event are all the pairs whose second element is a ball of that color. We note that there is a total of 45 balls in the three urns and, thus, 45 different outcomes possible. These outcomes, however, are **not all equally probable** because of the different number of balls in each of the urns. We cannot obtain the probability of drawing a red ball, for example, by noting there are a total of 14 red balls and forming the fraction, 14/45. We must compute the probabilities of the mutually exclusive events consisting of the choice of an urn and the choice of a ball. Thus, the probability of choosing a red ball, $p(R)$, is given by

$$p(R) = p(A \cap R) + p(B \cap R) + p(C \cap R) \tag{7.6.1}$$

where $p(A \cap R)$ is the probability of selecting Urn A and drawing a red ball from it; $p(B \cap R)$ is the probability of selecting Urn B and drawing a red ball from it; and $p(C \cap R)$ is the probability of selecting Urn C and drawing a red ball from it. As we have seen in Section 7.3,

$$\begin{aligned} p(A \cap R) &= p(A)\cdot p(R|A) \\ p(B \cap R) &= p(B)\cdot p(R|B) \\ p(C \cap R) &= p(C)\cdot p(R|C) \end{aligned} \tag{7.6.2}$$

Urn A contains 15 balls of which six are red. Urn B contains 10 balls of which two are red. Urn C contains 20 balls of which six are red. Thus

$$p(R|A) = \frac{6}{15} = \frac{2}{5}$$

$$p(R|B) = \frac{2}{10} = \frac{1}{5}$$

$$p(R|C) = \frac{6}{20} = \frac{3}{10}$$

The probability of choosing each urn is the same; and since there are three of them,

$$p(A) = p(B) = p(C) = \frac{1}{3}$$

We now have

$$p(A \cap R) = \frac{1}{3}\cdot\frac{2}{5} = \frac{2}{15}$$

$$p(B \cap R) = \frac{1}{3}\cdot\frac{1}{5} = \frac{1}{15}$$

$$p(C \cap R) = \frac{1}{3}\cdot\frac{3}{10} = \frac{1}{10}$$

and

$$\begin{aligned} p(R) &= \frac{2}{15} + \frac{1}{15} + \frac{1}{10} \\ &= \frac{9}{30} = \frac{3}{10} \end{aligned}$$

The question that we asked at the start of this example was: After observing that the ball drawn is red, what is the probability that it came from Urn A? If we use the symbology of Section 7.3, this probability is $p(A|R)$.

Recall from Section 7.3 the definition of conditional probability:

$$p(E_2|E_1) = \frac{p(E_2 \cap E_1)}{p(E_1)}$$

Using the symbols of this situation, we have

$$p(A|R) = \frac{p(A \cap R)}{p(R)} \tag{7.6.3}$$

Thus

$$p(A|R) = \frac{2/15}{3/10}$$

$$= \frac{4}{9}$$

This means that after we see that the ball drawn is red, the probability that Urn A was the urn selected is no longer 1/3, but has increased to 4/9.

In essence Equation 7.6.3 states that the probability that the red ball came from Urn A is determined by the relative contribution that the probability of drawing a red ball from Urn A makes to the total probability that we draw a red ball. We note that

$$p(B|R) = \frac{p(B \cap R)}{p(R)}$$

$$= \frac{1/15}{3/10} = \frac{2}{9}$$

and

$$p(C|R) = \frac{p(C \cap R)}{p(R)}$$

$$= \frac{1/10}{3/10} = \frac{1}{3}$$

Our results show that $p(A \cap R)$, the probability of selecting Urn A and then drawing a red ball from it, is larger than either $p(B \cap R)$ or $p(C \cap R)$. Thus, since the ball turned out to be red, the probability is greater that we selected Urn A in our first selection instead of Urn B or C.

The urn experiment described above is a model for a two-stage stochastic process. In our analysis we have made some inferences about what happened in the first stage of the process, when this is not known to us, based on the final outcome of the experiment, which is known. The generalized statement of the crucial relationship, Equation 7.6.3 is called **Bayes' theorem.** It is usually stated in the following form:

> Let $A_1, A_2, A_3, \ldots, A_k$ be a set of mutually exclusive events including **all** of the possible outcomes of the first stage of a two-stage stochastic process (i.e., let the A's be a partition of the sample space).
>
> Let $E_1, E_2, E_3, \ldots, E_n$ be a set of mutually exclusive events that include all of the possible outcomes of the second stage of the process (i.e., let the E's be a second partition of the sample space).
>
> Then,

$$p(A_i|E_j) = \frac{p(A_i \cap E_j)}{p(E_j)} \tag{7.6.4}$$

A_i is **any** of the events of the first stage and E_j is the **observed** event in the second stage.

This theorem is often stated with $p(E_j)$ replaced in the denominator by its equivalent,

$$p(E_j) = p(A_1) \cdot p(E_j | A_1) + p(A_2) \cdot p(E_j | A_2) + p(A_3) \cdot p(E_j | A_3) + \ldots + p(A_k) \cdot p(E_j | A_k)$$

Let us now apply this theorem to a specific example.

Example 7.6.1

Consider the situation of the sociologist in his study of how well people like their work. Suppose that a survey is made of the employees of a large insurance company. The information shown in the table of Figure 7.6.2 is obtained.

	People in Management	People in Clerical Work	People in Sales Work
Percent of number surveyed	20%	30%	50%
Enjoys work	50%	10%	30%
Neutral	30%	40%	50%
Dislikes work	20%	50%	20%

Figure 7.6.2

This "process" is diagrammed in Figure 7.6.3. The numbers on each branch of the diagram are the probabilities for the outcomes indicated.

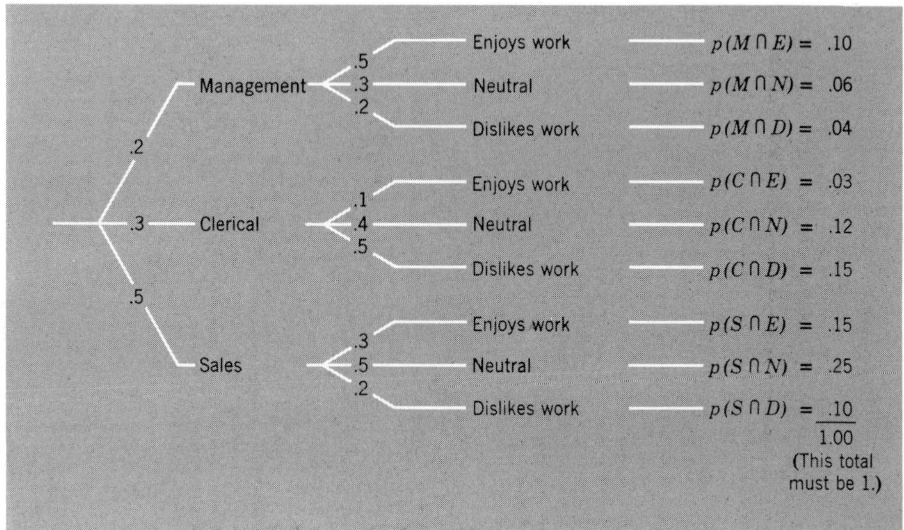

Figure 7.6.3

The ".2" on the first branch leading to "management" means that from **those surveyed** the probability a person selected at random is in management is .2 – that is, 20% of those surveyed are in management. The other probabilities on the diagram are obtained in similar manner from the data in the table of Figure 7.6.2.

The experiment consists of selecting an employee of the company at random and asking him how well he likes his work. Using his response we determine the probability that he is in management, in clerical work, or in sales work. We assume that the survey sample is representative of the entire company.

Let us first compute the probability of each response. The probability that an employee of the company enjoys his work, $p(E)$, is given by the sum of the probability that he is in management and likes his work, that he is in clerical work and likes his work, and that he is in sales work and likes his work. (We are assuming that **all** of the employees have been placed in one of these three categories.) In symbols,

$$p(E) = p(M \cap E) + p(C \cap E) + p(S \cap E)$$

$$p(M \cap E) = p(M) \cdot p(E|M) = (.2)(.5) = .10$$

$$p(C \cap E) = p(C) \cdot p(E|C) = (.3)(.1) = .03$$

$$p(S \cap E) = p(S) \cdot p(E|S) = (.5)(.3) = .15$$

In these computations the values for $p(M)$, $p(C)$, and $p(S)$ are given in the first line of the table in Figure 7.6.2, the percent of the number surveyed. The values for $p(E|M)$, $p(E|C)$, and $p(E|S)$ are obtained from the second line of the table in Figure 7.6.2, the percent in each job classification that enjoy their work. Thus

$$p(E) = .10 + .03 + .15 = .28$$

From Bayes' theorem, Equation 7.6.4,

$$p(M|E) = \frac{p(M \cap E)}{p(E)} = \frac{.10}{.28} = .36$$

$$p(C|E) = \frac{p(C \cap E)}{p(E)} = \frac{.03}{.28} = .11$$

$$p(S|E) = \frac{p(S \cap E)}{p(E)} = \frac{.15}{.28} = .53$$

The results of Example 7.6.1 show that from the data given a person who enjoys his work is most likely to be in sales, even though a substantially larger fraction of the people in management enjoy their work than do the people in sales (50 to 30%). The explanation, of course, is that the data show there are many more people engaged in sales work than are in management.

Another application of Bayes' theorem in the social sciences is in testing of alternate hypotheses. Social scientists frequently propose different explanations for a given phenomenon, each of which predicts different probabilities for the result of a particular experiment. A probability is assigned to each hypothesis; that is, the relative likelihood of the validity of each hypothesis (explanation of the phenomenon) is assigned a numerical measure. Then the experiment is performed and the result observed. The probabilities assigned to the individual hypotheses are thereby tested.

Example 7.6.2

Suppose that a psychologist is studying the learning process and the way a series of experiences affect the rate at which a task is mastered. He has three different ideas about how experience is used by a person; that is, three alternative hypotheses. He devises an experiment in which a person learns to master a simple task by practicing just part of the task. His different hypotheses make predictions about how many times a person must practice before he masters the task. Let us designate the three hypotheses as H_1, H_2, and H_3. If H_1 is correct, the probability that a person can master the task by practicing two or fewer times is 0.3; the probability that it takes three practice operations is 0.5; and the probability that it will take more than three is 0.2.

If H_2 is correct, the probability that a person can master the task with two or fewer practice operations is 0.5. The probability for three practice operations is 0.4, and the probability for more than three is 0.1.

If H_3 is correct, the probability that the person requires two or fewer practice operations is 0.2, the probability for three is 0.3, and the probability for more than three is 0.5.

Let us designate the event that it takes two or fewer practice operations as A; three practice operations, as B; and more than three, as C. Let us also assume that on the basis of previous work the probability that Hypothesis 1 is correct is 0.4, that Hypothesis 2 is correct is 0.3, and that Hypothesis 3 is correct is 0.3. That is,

$$p(H_1) = 0.4 \qquad p(H_2) = 0.3 \qquad p(H_3) = 0.3$$

We can diagram the situation as shown in Figure 7.6.4.

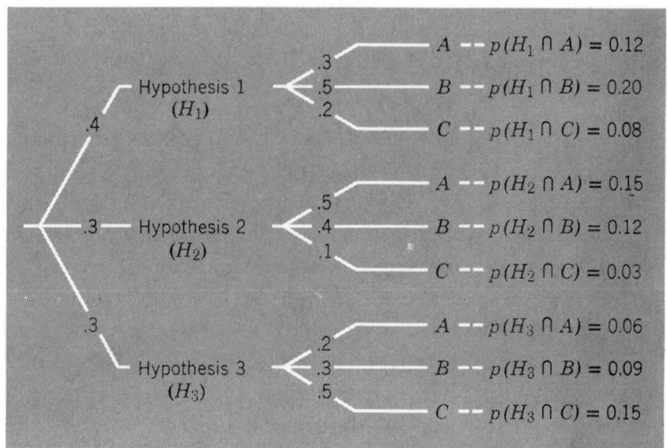

Figure 7.6.4

When the experiment is performed, let us assume that 10 people are tested, and they all take four or more practice operations to master the task. That is, event C occurs. (In an actual experiment we can expect more variations in outcome. However, this requires statistical methods along with the concept of probability for proper analysis. We, therefore, simplify the results to illustrate only the role that probability plays.)

Using the values listed in Figure 7.6.4 we see that the probability C will occur is given by

$$\begin{aligned} p(C) &= p(H_1 \cap C) + p(H_2 \cap C) + p(H_3 \cap C) \\ &= .08 + .03 + .15 = .26 \end{aligned}$$

From Bayes' theorem

$$p(H_1 | C) = \frac{.08}{.26} = .31$$

$$p(H_2 | C) = \frac{.03}{.26} = .11$$

$$p(H_3 | C) = \frac{.15}{.26} = .58$$

Thus, on the basis of the results of this hypothetical experiment, Hypothesis 3 is the most likely to be correct.

As we have mentioned earlier in general discussions, the process of developing an understanding of people involves experiments similar to the one in this example. The results are used to suggest further experiments, whose results would confirm or suggest modifications for the hypothesis. At each stage the mathematical methods that we have discussed would be used in the analysis of the result.

Let us return now to a consideration of the business problem that we proposed at the beginning of this section.

Example 7.6.3

A manufacturing company produces a small electronic computer used as a desk-top calculator by accountants and statisticians, among others. Every one of these machines is tested before delivery to a customer by using it to do a prescribed set of computations. These computations were carefully chosen to reveal any defects in the operation of the machines. As we shall see, they do not do this with complete accuracy.

The company records show that 60% of the machines operate properly as built. That is, the production process has been carried out properly using good components. Thirty percent of the machines do not perform properly as produced, but can be made to operate by adjusting some of their components. Ten percent of the machines contain defective components and must be returned to the production department for repair. The testing procedure, however, is not perfect and sometimes gives an incorrect diagnosis of a machine's performance. Of the machines that are built properly and have no defects, the test correctly shows 90% to be okay. Nine percent are said to require adjustment, and 1% are said to have defective components.

The test misses the deficiencies of 15% of the 30% of the machines that require adjustment and passes them for shipment. It properly diagnoses the problem with 75% of them and incorrectly rejects 10% as requiring rework by the production department.

Of the 10% of the machines having defective components, 5% actually pass the test; 25% are said to need adjustment only; and 70% are correctly found to need repair.

The situation can be diagrammed as shown in Figure 7.6.5.

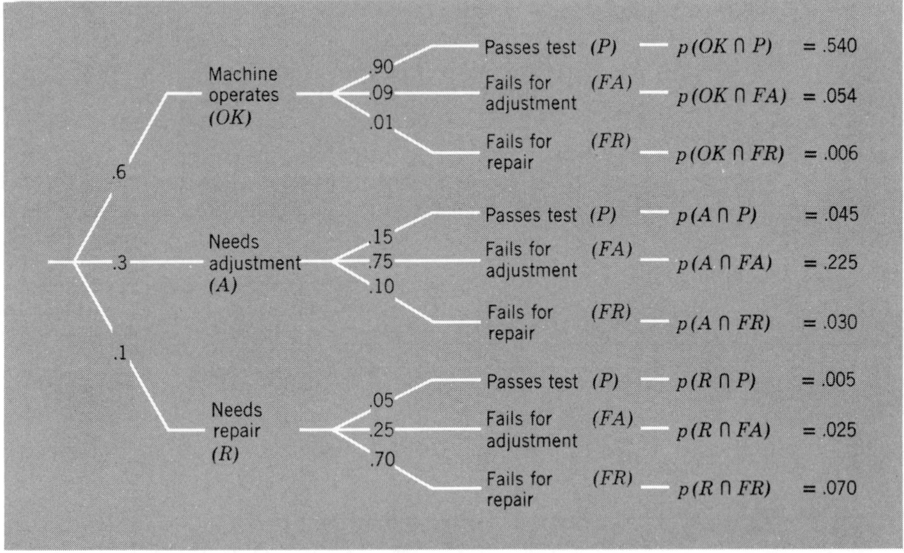

Figure 7.6.5

The company wants to know the probability that if the machine passes the test, it will not perform properly and require field service by a technician, and also the probability that it must be returned to the factory for repair. $p(A|P)$ is the probability the machine needs adjustment given that it has passed the test; and $p(R|P)$ is the probability that the machine needs repair given that it passed the test. (These are the events that make unhappy customers.)

$$\begin{aligned} p(P) &= p(OK \cap P) + p(A \cap P) + p(R \cap P) \\ &= .540 + .045 + .005 \\ &= .590 \end{aligned}$$

The probability the machine needs adjustment even though it passed the test is

$$p(A \mid P) = \frac{p(A \cap P)}{p(P)}$$

$$= \frac{.045}{.590} = .076$$

The probability the machine needs repair even though it passed the test is

$$p(R \mid P) = \frac{p(R \cap P)}{p(P)}$$

$$= \frac{.005}{.590} = .008$$

The probability that the machine is going to require either adjustment or repair after it is shipped to the customer is the sum of these two probabilities or 0.084. That is, approximately 8.4% of the machines are going to require some sort of repair service after being delivered to a customer.

The foregoing example is an illustration of a situation that will require a management decision on whether a new testing procedure is needed. In a later section we shall see one way to evaluate the alternatives on which the decision can be based.

EXERCISE 7.6

1. Two urns, A and B, contain balls of three different colors. Urn A contains six red, eight green, and six yellow balls. Urn B contains one red, three green, and six yellow balls. One of the urns is chosen at random and a ball drawn.

   a. If the ball is yellow, what is the probability it was drawn from Urn A?
   b. If the ball is green, what is the probability it was drawn from Urn B?

2. Three urns, A, B, and C, contain green and white balls. Urn A contains four green and six white; Urn B, six green and fourteen white; and Urn C, nine green and three white. An urn is selected at random and a ball drawn.

   a. If the ball is green, what is the probability it was drawn from Urn B?
   b. If the ball is white, what is the probability it came from Urn A?

3. Three urns, A, B, and C, contain red, white, yellow, and green balls. The number of balls of each color in each urn is shown in the table.

	Red	White	Yellow	Green
A	5	3	2	5
B	8	2	7	3
C	2	4	3	1

An urn is selected at random and a ball drawn.

a. If the ball is red, what is the probability that it came from Urn A? Urn C?

b. If the ball is white, what is the probability that it came from Urn A? Urn C?

4. The student body of a college is made up of 60% men and 40% women. Seventy percent of the men and 80% of the women are registered to vote. A letter is written to the student newspaper and signed, "Registered Voter." What is the probability that it was written by a woman?

5. In a community organization 20% of the members are 25 years old or younger, and the remainder are over 25. A poll is taken to learn whether the members want to undertake a boycott of two companies in the community that have been dumping waste material in a nearby river. Eighty percent of the members 25 and younger favor the boycott, and 40% of the members over 25 are for it. A member of the organization is selected at random and asked his views of the boycott. He favors it.

a. What is the probability that he is 25 or younger?

b. What is the probability that he is over 25?

6. An appliance dealer is interested in the effectiveness of his advertising in a local newspaper. He asks each person who looks at appliances in his store whether or not he has seen his ad. Twenty-five percent have; 75% have not. Of the 25% who have seen his ads, 20% buy an appliance; 80% do not. Of the ones who have not seen the ads, 10% buy; 90% do not. What is the probability that a person who buys an appliance has seen the dealer's ads?

7. A company employs many workers, whom it must train to do a highly technical job. Its experience indicates that about 30% of the people who apply for these jobs can actually do them. Each applicant is given an aptitude test. Experience with the test shows that 80% of the people that can do the work pass the test, and 40% of the people who cannot do the work pass.

a. If a person passes the test, what is the probability that he can do the work?

b. If a person does not pass the test, what is the probability that he can do the work?

8. Two automobiles are sitting in front of a house. One has the key in the ignition. The other has the ignition locked, but the key is one of three keys lying on a table in the house.

A man selects one of the keys at random from the table, leaves the house, and gets into one of the automobiles without knowing which one has the key in it. He is observed to start the car and drive away. What is the probability that he chose the car with the key left in the ignition?

9. There are two candidates, A and B, in an election. Of those voting 40% are Republicans, 45% are Democrats, and 15% are Independents. Sixty percent of the Republicans voted for A; and 40% for B. A and B each received the vote of 50% of the Democrats, but 30% of the Independents voted for A while 70% voted for B.
   a. Who won the election?
   b. If it is known that a particular voter voted for B, what is the probability he is an Independent? A Republican? A Democrat?

10. In a survey 64 women and 36 men are interviewed and asked their opinion on price controls. Thirty-six of the women favor them, 24 are opposed, and 4 have no opinion. Of the men, 18 favor them, 12 are opposed, and 6 have no opinion. One of the interview cards is selected at random. It shows that the person interviewed opposed price controls. What is the probability that this person is a man?

11. Statistics show that one out of every 100 people is afflicted with a certain disease. A test used to detect the disease correctly shows the presence of the disease 90% of the time. However, in 1% of the tests on people who do not have the disease, the presence of the disease is incorrectly indicated. A person is tested and the test indicates the presence of the disease. What is the probability that the person actually is afflicted?

12. In a large manufacturing company 5% of the employees work in the executive departments of the company, 20% are in sales, and 75% are in the production department. Of those in the executive department, 50% earn more than \$20,000 per year. In the sales department 30% earn more than \$20,000 per year; and in production, 10% earn more than \$20,000 per year. If it is known that a person earns more than \$20,000 per year,
   a. What is the probability that he is in one of the executive departments?
   b. What is the probability that he is in production?

13. A company buys one of the parts that it uses in its products from three suppliers, A, B, and C. Fifty percent are purchased from A, 30% from B, and 20% from C. Two percent of the parts from A are defective; 4% from B are defective; and 5% from C are defective. The parts from the three different suppliers are not distinguishable after they are removed from the package. If a part is found to be defective, what is the probability that it came from A? From B? From C?

14. A sociologist obtains the following results in a survey of the amount of education of people living in urban, suburban, and rural communities.

	Urban Community	Suburban Community	Rural Community
Percent of population	40%	50%	10%
Not complete high school	42%	20%	30%
High school graduate	50%	70%	65%
College graduate	8%	10%	5%

a. If a person selected at random has not finished high school, what is the probability that he lives in an urban area?
b. If a person selected at random is a college graduate, what is the probability that he lives in an urban area?
c. What is the probability that a person selected at random has not finished high school and lives in the suburbs?

15. In a survey of job attitudes, 10% of those interviewed were manual laborers (M); 35% were "blue collar" workers (BC); 50% were "white collar" workers (WC); and 5% were professionals (P). Of the manual laborers 50% disliked their work (D); 40% were neutral toward it (N); and 10% enjoyed it (E). Of the blue collar workers 40% disliked their work, 40% were neutral toward it, and 20% enjoyed it. Of the white collar workers 35% disliked their work, 45% were neutral, and 20% enjoyed it. Among the professionals 20% disliked their work, 30% were neutral, and 50% enjoyed it.
a. If one of those surveyed is selected at random, what is the probability that he enjoys his work?
b. If one of those surveyed is selected at random and found to dislike his work, what is the probability that he is a manual laborer? A blue collar worker? A white collar worker?

7.7 RANDOM VARIABLES AND EXPECTED VALUE

7.7.1 Random Variables

In a great many situations involving probabilities, knowledge of the probabilities of the different events possible does not give enough information for making decisions. The events themselves may well have widely different degrees of importance. As a simple example let us suppose that an acquaintance comes to us with a proposition. An airplane carrying a large sum of money has crashed in a remote mountain area, and the exact location of the crash scene is not known. The company that owns the airplane has offered a reward of $50,000 to anyone who locates the airplane and brings back the money. The acquaintance is an experienced hiker and wants to explore the area where the crash is believed to have occurred. He needs $100 for food and supplies. He is willing to split the reward with you if you will supply the money. He estimates that he has one chance in a thousand to find the airplane. The question is, purely on the basis of the information you have, is the proposition a good one? Since this is a one-time situation, the probabilities do not mean a great deal to you; but if you were in the business of financing such activities, does this particular proposition have sufficient expectation of success to justify your investment?

In making your decision you must assign appropriate values to success and failure. In this case, of course, the value of success is $25,000. The cost of failure is \$100. Such an assignment of values to the events that can result from an experiment is called a **random variable.** This is an unusual and not particularly descriptive term. The "random" comes from the association with stochastic (random) processes. The term "variable" implies that assigned values change from one event to another.

In the example described above, the "random variable" is the monetary gain (or loss) associated with the occurrence of each of the possible events. In other situations the random variable can assign any set of values that we choose to the events. A student, for example, could consider taking a course as an "experiment." He could estimate probabilities for getting the various grades. If the course is worth four units, then according to the usual accounting procedure, an A in the course has a value of 16 grade points; a B, 12 grade points; and so on. The assignment of values to the different grades (events) is a random variable. The idea is that each of the events possible for an experiment is assigned a numerical value. (Some, of course, may be zero.) It follows from this that a random variable is, in fact, a **function.** It consists of a set of pairings—each event with its "value." The same value can be assigned to more than one event, but no event can have more than one value. In formal terms:

Definition 7.7.1

Random Variable: A rule or formula by which a set of numbers is associated with the events resulting from an experiment is called a **random variable.** The only restriction to the assignment is that there can be but one number associated with each event.

Definition (Alternate) 7.7.1A

Random Variable: A random variable is a **function** in which the **domain** is the **set of all events** associated with an experiment and the **range** is a **set of numbers.**

7.7.2 Expected Value

The concept of a random variable is not particularly useful by itself. It is usually employed in the computation of the **expected value** of an "experiment." The expected value is a "weighted average" of the numerical values assigned by a random variable. It is obtained by finding the sum of the products of the probability of each possible event times its assigned value. We formalize this concept in the following definition:

Definition 7.7.2

Expected Value: In an experiment composed of the events, E_1, E_2, E_3, . . . , E_k, which have assigned values of V_1, V_2, V_3, . . . , V_k, respectively, the **expected value** of the experiment, (EV), is given by

$$EV = p(E_1) \cdot V_1 + p(E_2) \cdot V_2 + p(E_3) \cdot V_3 + \ldots + p(E_k) \cdot V_k$$

It is assumed in the above definition that $E_1 \cup E_2 \cup E_3 \cup \ldots \cup E_k = S$, where S is the sample space of the experiment.

Considering the "lost treasure" proposition that we described earlier in the section, we can answer the question of whether this is a sound business venture by computing the **expected value.** Recall that the probability of success, $p(S)$ was set at .001 and the value of success, (V_s) was \$25,000. The probability of failure, $p(F)$, is, therefore, .999; and the value of failure, (V_f) is $-\$100$ (\$100 loss). The expected value, EV, is given by

$$\begin{aligned} EV &= p(S) \cdot V_s + p(F) \cdot V_f \\ &= (.001)(25{,}000) + (.999)(-100) \\ &= 25 - 99.90 = -\$74.90 \end{aligned}$$

We see that the expected value is a loss of \$74.90. On a strictly business basis, this proposition is not a sound one.

Strict dollar return is not the only way that we can assign values to an event. In the foregoing example let us suppose that you have a project that you very much want to do and that requires an investment of approximately \$25,000. Because of the great difficulty of obtaining so large a sum, let us suppose you estimate that \$25,000 is 10,000 times as useful to you as \$100. Thus, you assign a value of 10,000 to success in the venture ($V_s = 10{,}000$) and a value of -1 to failure ($V_f = -1$). Now the expected value becomes

$$\begin{aligned} EV &= p(S) \cdot (V_s) + p(F) \cdot (V_f) \\ &= (.001)(10{,}000) + (.999)(-1) = 10 - .999 = 9.001 \end{aligned}$$

In this case the expected value (in usefulness) is 9.001 and is larger than the value that you have placed on the 100 dollars. With these assigned values, the proposition becomes a good investment.

Let us consider further the meaning of the statement earlier in the section that the expected value is a "weighted average" of the numerical values assigned by a random variable. Let us consider an experiment with n possible outcomes. To simplify the arithmetic, let us assume that these n outcomes are equally probable. We now **partition** the n outcomes into a set of k mutually exclusive events, $E_1, E_2, E_3, \ldots, E_k$. The union of these events is the entire sample space, S. Let e_1 represent the number of outcomes in Event, E_1; e_2, the number of outcomes in Event, E_2, and so on. Thus

$$p(E_1) = \frac{e_1}{n}; \quad p(E_2) = \frac{e_2}{n}; \quad p(E_3) = \frac{e_3}{n}; \ldots; \quad p(E_k) = \frac{e_k}{n}$$

We assign values, $V_1, V_2, V_3, \ldots, V_k$, to the respective events. From Definition 7.7.2,

$$EV = p(E_1) \cdot V_1 + p(E_2) \cdot V_2 + p(E_3) \cdot V_3 + \ldots + p(E_k) \cdot V_k$$

Since $p(E_1) = \frac{e_1}{n}$, $p(E_2) = \frac{e_2}{n}$, and so on.

$$\begin{aligned} EV &= \frac{e_1}{n} V_1 + \frac{e_2}{n} V_2 + \frac{e_3}{n} V_3 + \ldots + \frac{e_k}{n} V_k \\ &= \frac{e_1 V_1 + e_2 V_2 + e_3 V_3 + \ldots + e_k V_k}{n} \end{aligned} \tag{7.7.1}$$

e_1 represents the number of outcomes in Event E_1 and, therefore, the value V_1 is added e_1 times in finding the sum in the numerator of Equation 7.7.1. Similarly, V_2 is added e_2 times; V_3, e_3 times; and so on. The e's are the "weighting factors" for the different values. Since the events are a partition of the outcomes,

$$e_1 + e_2 + e_3 + \ldots + e_k = n$$

In summing the V's, there are exactly n terms. When this sum is divided by n, the average value of the V's is obtained.

The expected value can be viewed as the average of the values of the random variable that we would obtain if the experiment was repeated a large number of times. It is important to recognize that the expected value is not necessarily one that can be obtained. In the "treasure hunt" the investor cannot possibly lose \$74.90 – the expected value. He will either gain \$24,900 or lose \$100. For the student estimating his chances for different grades, let us suppose that he decides the probability for an A is 0.1; for a B, 0.3; for a C, 0.5; and for a D, 0.1. Then the expected value of his grade points is

$$EV(GP) = .1(16) + .3(12) + .5(8) + .1(4)$$
$$= 9.6$$

There is no way for him to obtain 9.6 grade points; he will actually obtain 16 or 12 or 8 or 4. The expected value is a **relative measure** of the value of the course in grade points.

We have seen an example of using expected value as a measure of the return on an investment. It can be used in other contexts as well.

Example 7.7.1

A group of five items includes one that is defective. The items must be tested in order to find the defective one. How many should we **expect** to test before we find the defective one? The events of interest are

$E_1 =$ the first item tested is the defective one.
$E_2 =$ the second item tested is the defective one.
$E_3 =$ the third item tested is the defective one.
$E_4 =$ the fourth item tested is the defective one.
$E_5 =$ the fifth item tested is the defective one.

Since we are concerned with the **number of tests** that we must perform, we assign **one** as the value of E_1, **two** as the value of E_2, **three** as the value of E_3, **four** as the value of E_4, and **five** as the value of E_5. We have, where EV represents the expected number of items that we must test,

$$EV = p(E_1)\cdot(1) + p(E_2)\cdot(2) + p(E_3)\cdot(3) + p(E_4)\cdot(4) + p(E_5)\cdot(5)$$

$p(E_1)$ is 1/5, the probability of selecting the defective item from the group of five.

E_2 requires that the first item selected is good, and the second, defective. Thus,

$$p(E_2) = \left(\frac{4}{5}\right) \cdot \left(\frac{1}{4}\right)$$
$$= \frac{1}{5}$$

By similar computations we find that $p(E_3)$, $p(E_4)$, and $p(E_5)$ are all also equal to 1/5. Therefore

$$EV = \left(\frac{1}{5}\right) \cdot (1) + \left(\frac{1}{5}\right) \cdot (2) + \left(\frac{1}{5}\right) \cdot (3) + \left(\frac{1}{5}\right) \cdot (4) + \left(\frac{1}{5}\right) \cdot (5)$$
$$= \frac{15}{5} = 3$$

This says that "on the average" we can expect to find the defective item on the third test.

The expected number of successes in a sequence of Bernoulli trials is of interest in a number of situations. For example, if it is known that 60% of the people living in a large city favor a proposal to build a new stadium, what is the number of people **expected** to favor it when a group of n persons is selected at random? We shall compute the answer for a group of five and use that as an indicator of the general result.

Since 60% of the people are said to favor the stadium, the probability that a person selected at random will do so is .6. We call this a "success" in a sequence of Bernoulli trials. If we choose a group of five people, we have as possibilities:

$E_0 =$ 0 people in favor
$E_1 =$ 1 person in favor
$E_2 =$ 2 people in favor
$E_3 =$ 3 people in favor
$E_4 =$ 4 people in favor
$E_5 =$ 5 people in favor

The **expected** number of people in favor is the expected value of a random variable that assigns as the value of an event, the number of people in favor of the stadium if that event occurs. That is,

$V_0 =$ value of $E_0 = 0$
$V_1 =$ value of $E_1 = 1$
$V_2 =$ value of $E_2 = 2$
$V_3 =$ value of $E_3 = 3$
$V_4 =$ value of $E_4 = 4$
$V_5 =$ value of $E_5 = 5$

Then, where EV represents the expected number of people in favor of the stadium,

$$EV = V_0 \cdot p(E_0) + V_1 \cdot p(E_1) + V_2 \cdot p(E_2) + V_3 \cdot p(E_3) + V_4 \cdot p(E_4) + V_5 \cdot p(E_5)$$

From the formulas developed in Section 7.5,

$$p(E_0) = C(5,0) \cdot p^0 q^5$$

where $p = 0.6$, the probability of "success," and $q = 0.4$, the probability of "failure." Similarly,

$$\begin{aligned}
p(E_1) &= C(5,1) \cdot pq^4 \\
p(E_2) &= C(5,2) \cdot p^2q^3 \\
p(E_3) &= C(5,3) \cdot p^3q^2 \\
p(E_4) &= C(5,4) \cdot p^4q \\
p(E_5) &= C(5,5) \cdot p^5q^0
\end{aligned}$$

Evaluating these probabilities, we have

$$\begin{aligned}
p(E_0) &= (1) \cdot (.6)^0 \cdot (.4)^5 = .010 \\
p(E_1) &= (5) \cdot (.6) \cdot (.4)^4 = .077 \\
p(E_2) &= 10 \cdot (.6)^2 \cdot (.4)^3 = .231 \\
p(E_3) &= 10 \cdot (.6)^3 \cdot (.4)^2 = .345 \\
p(E_4) &= (5) \cdot (.6)^4 \cdot (.4) = .259 \\
p(E_5) &= (1) \cdot (.6)^5 \cdot (.4)^0 = .078
\end{aligned}$$

For the expected value we have

$$\begin{aligned}
EV &= 0(.010) + 1(.077) + 2(.231) + 3(.345) + 4(.259) + 5(.078) \\
&= 0 + .077 + .462 + 1.035 + 1.036 + .390 \\
&= 3
\end{aligned}$$

That is, we expect to find three out of five people in favor of the stadium.

We note that $(0.6)(5) = 3$. **The probability of success times the number of trials is equal to the expected value.** It can be shown (by rather complex algebra) that, in general, in a sequence of n Bernoulli trials in which the probability of success is p, the **expected number** of successes is given by np. Note, however, that it is by no means certain, nor necessarily more likely than not, that the actual number of successes is the expected number. In the preceding example the probability of getting the expected number of successes (3 out of 5) was only 0.345 – less than an "even" chance. The expected number, however, is the "most likely" number—the probability for the expected number is greater than for any other number.

7.7.3 Use of Expected Value in Decision Making

To illustrate how the concept of expected value can be used in making a business decision let us consider again the situation in Example 7.6.3. In that example a manufacturing company used a test procedure that sometimes gave incorrect results concerning the operation of the machines that it produced. The diagram of the test results from that example is shown again in Figure 7.7.1.

The three events marked with an asterisk are those in which the test procedure gives the correct analysis of the operation of the machine. In the remaining events the incorrect test results cause unnecessary expense to the com-

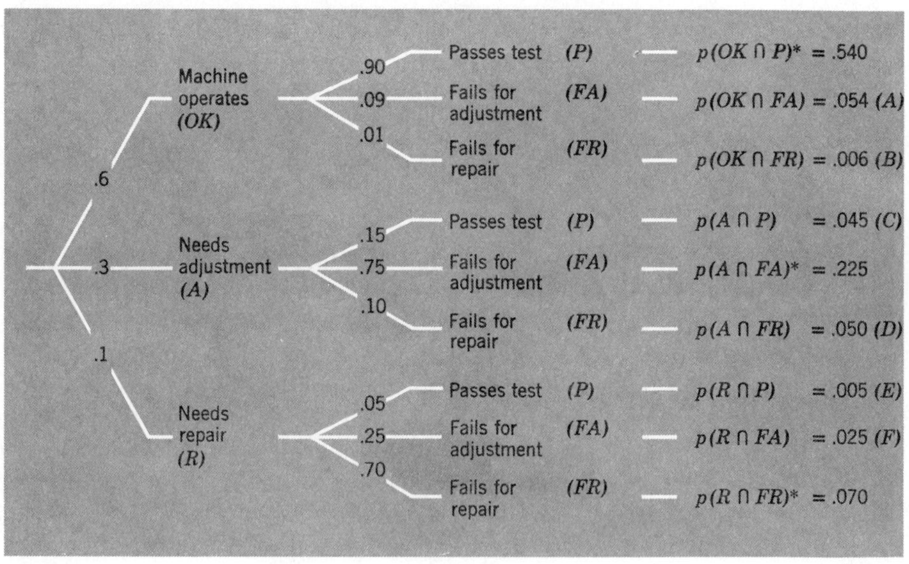

Figure 7.7.1

pany. The **average** amount of that unnecessary expense **for each machine** can be computed by assigning each of these events a value equal to the cost of the unnecessary operations resulting from the incorrect testing. Each of these events is marked with a letter in Figure 7.7.1.

The event marked (A) is the diagnosis of a good machine as requiring adjustment. Let us suppose that the cost of the work done during the adjustment is \$20. Thus, $V_A = 20$.

The event marked (B) is the diagnosis of a good machine as needing repair. Let the cost of taking the machine apart looking for defective components and reassembling it be \$50. Thus, $V_B = 50$.

The event marked (C) is the passing as OK a machine that needs adjustment. When the customer receives the machine, it will not operate properly; and the company must send a technician out to adjust it. Let the average cost of such a service call be \$60. That is, $V_C = 60$.

The event marked (D) is incorrectly returning for repair a machine that needs only adjustment. Let the additional cost of this work over the adjustment cost be \$30. Thus, $V_D = 30$.

The event marked (E) is incorrectly passing a machine that contains faulty components. Such a machine will not operate properly when received by a customer, and it must be returned to the factory for repair. The service technician cannot do this work on a field service call. Let the cost of returning the machine and repairing it average \$120. Thus, $V_E = 120$.

The event marked (F) incorrectly diagnoses as needing adjustment a machine that actually must have defective components replaced. Let us suppose that the additional work that this causes costs \$50. Thus, $V_F = 50$.

We are now ready to compute the additional costs caused by errors in the testing process. Using the probabilities for each of the events, A to F, shown in Figure 7.7.1 and the values (costs) of these events listed above, we have, where EV represents the unnecessary costs,

$$EV = p(A)\cdot V_A + p(B)\cdot V_B + p(C)\cdot V_C + p(D)\cdot V_D + p(E)\cdot V_E + p(F)\cdot V_F$$
$$= (.054)(20) + (.006)(50) + (.045)(60) + (.030)(30) + (.005)(120) + (.025)(50)$$
$$= 1.08 + .30 + 2.70 + .90 + .60 + 1.25 = 6.83$$

This means that the cost of each machine is increased by an average of \$6.83 because of the inadequate test procedure.

Let us suppose that the company produces 50,000 machines each year. The expected total additional cost is \$341,500 (50,000 · \$6.83).

These facts are brought to the attention of the engineering department. After study the chief engineer reports that for a cost of \$25,000 a new test procedure can be developed and put into operation. The new test procedure will take longer to perform and will add \$5.00 to the cost of testing each machine. Taking into account the 50,000 machines produced in one year, the additional cost per machine for development of the test procedure and performing it will average \$5.50 per machine.

The new test procedure will not be perfect either. The errors are greatly reduced, however, and the following errors can be expected:

On machines that are satisfactory, 1% will be rejected as requiring adjustment, but no machines will be incorrectly rejected as requiring repair.

Of the machines that need adjustment, 2% will be incorrectly passed, but none will be rejected as needing repair.

All of the machines that have defective components will have that fact revealed by the test.

The question is, will the new procedure save enough money to make its development worthwhile?

To answer this question we compute the expected value of the additional costs occurring as a result of the new test procedure. We have a new set of probabilities for the "stochastic process" of testing the machines. These are shown in Figure 7.7.2. We have assumed that the relative number of good machines, machines that need adjustment, and machines that need repair is the same as before.

As before, the events marked * are those in which the test correctly determines the condition of the machine.

The event (G) incorrectly diagnoses a properly operating machine as needing adjustment. Using the same costs as before this unnecessary adjustment is assumed to cost \$20. Thus, $V_G = 20$.

The event (H) incorrectly passes a machine that needs adjustment. This error means that a technician must make a field service call to adjust the machine after the customer receives it. The cost of this call is assumed to be \$60, as before. That is, $V_H = 60$.

We have, therefore, as additional costs incurred by testing errors,

$$EV = p(G)\cdot V_G + p(H)\cdot V_H$$

These are the only two additional costs involved. Evaluating them, we have

$$EV = (.006)(20) + (.006)(60) = .12 + .36 = .48$$

Thus, the average additional cost of this new test is \$5.50 + .48 = \$5.98.

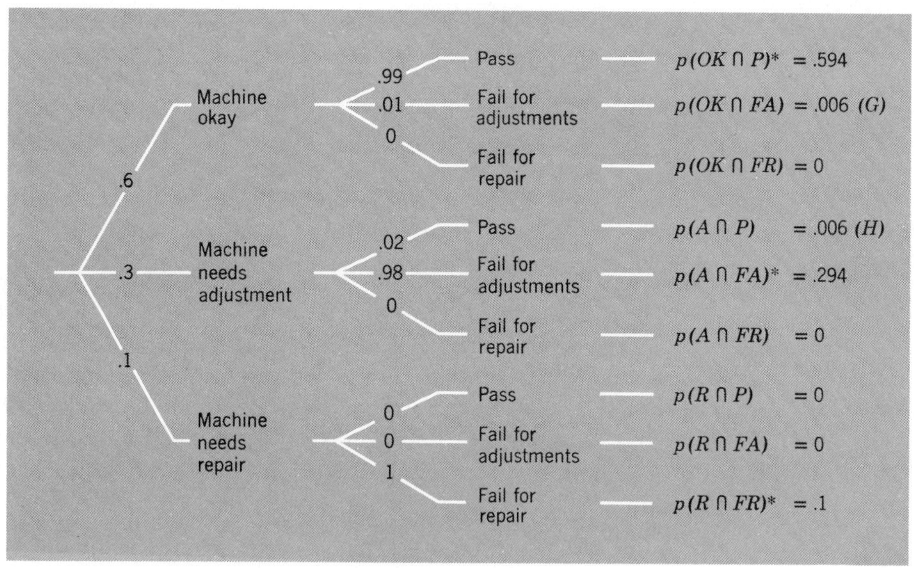

Figure 7.7.2

Since the added cost incurred by the first testing procedure is \$6.83, developing the new procedure will save \$6.83 − 5.98 = \$.85 per machine. For the 50,000 machines produced, this is a total saving of \$42,500 for the first year.

In an actual situation several alternative proposals would probably be made, each with a different cost per instrument and a different test effectiveness. Each would be analyzed in a similar manner, and the one effecting the greatest savings chosen for implementation.

In summary, an experiment can be "evaluated" by the use of a "random variable." The **random variable is a function** that assigns to each event in the experiment a value, which is obtained from the nature of the situation. The "expected value" of the experiment is the sum of the values for each event, found by multiplying the value of the event by the probability of its occurrence. Thus, the expected value can be taken as a **relative measure** of the "value" of any stochastic process.

EXERCISE 7.7

1. A person flips two "fair" (equal probability for heads or tails) coins. He receives \$1 for each "heads" flipped. What is the expected value of the "experiment"?
2. A person rolls a single die. Each spot on the face showing is worth one point. What is the expected value in points of each roll?
3. A person rolls a pair of dice. Each spot showing is worth one point. What is the expected value of each roll?

4. An organization holds a "raffle" in which the prizes are 10 turkeys worth $10 each. One thousand tickets are sold and each ticket can win only one prize. What is the expected value of each ticket?

5. A carnival booth features a "Pot of Gold," which is an urn containing colored balls, one blue, two red, three yellow, four green, and 40 white. A customer gets to reach into the urn (which he cannot see inside) and select a ball. If it is blue, he gets $10; if red, $5; if yellow, $2; if green, $1; and if white, 50¢. What is the expected value of each draw?

6. A second carnival booth is set up and features two "Pots of Gold." Two urns are used, and the customer can select a ball from either one. One of the urns contains one blue ball, three red balls, three yellow balls, four green balls, and nine white ones. The other urn has one blue, two yellow, six green, and 71 white. As before, the blue balls pay $10; the red, $5; the yellow, $2 and the green, $1; but the white balls pay only 25¢. What is the expected value of each draw at this booth?

7. A game is set up in which a player tosses a "fair" coin three times. He receives $1 for each head that appears, but if no heads appear (i.e., three tails), he must pay $10.
   a. What is the expected value of the game?
   b. How much should he pay when three tails appear to make it a "fair" game? (A fair game is one whose expected value is 0.)

8. An automatic lathe in a factory produces small parts for a machine manufactured there. The lathe does not work very well, and 20% of the parts that it produces are too large. If five of these parts are selected at random,
   a. What is the expected number that will be too large?
   b. What is the probability that there will be exactly that many parts too large in the sample?

9. A machinist at the factory of Problem 8 wishes to show one of the defective parts from the automatic lathe to his foreman. He selects parts one by one from a box and checks the size until he finds a defective one. What is the probability that he will find a defective part after checking no more than three parts?

10. Statistics for a recent period show that for men age 30, the probability of living for the next five years is .9954. Considering only the amount paid out on claims, what is the expected cost to an insurance company of a five-year term life insurance policy for $1000 on a man 30 years of age?

11. A charitable organization is conducting a fund-raising campaign. They prepare a letter that they plan to send to a select mailing list asking for donations. The experience of fund raisers has shown that they have a probability of .001 of getting a contribution of $1000 from the recipient of a letter. The probability for $500 is .003; for $200, .006; for $100, .01; for $50, .03; for $25, .07; for $10, .10; for $5, .15; and for $1, .20. The remainder do not respond. It costs 50¢ each to prepare and mail the letters.

a. What is the expected value of each letter sent?
b. If they need $25,000, how many letters should they send out?

12. A family unit is a group of people living in one household. In a survey of family size a field worker collected the following information.

Number of Family Units	Number of members of family unit
40	1
300	2
500	3
800	4
400	5
260	6
150	7
50	8

What is the average size of a family unit?

7.8
PROBABILITIES AND CERTAINTIES; ODDS

7.8.1
Effect of Knowledge on Probability

In the previous sections we studied some of the properties of the concept of probability. We have seen how the probabilities for events can be combined to obtain the probabilities for combinations of events, and how probabilities can be computed from the nature of some types of situations. In other situations we assumed values for probabilities, which are normally obtained by statistical methods. In these latter cases it is important to keep in mind that **in applying statistical probabilities we assume we know no more about the situation than was known in obtaining the statistics**. For example, in Problem 10 of Exercise 7.5 the question was asked, "If three people are to be hired and the probability that an applicant is hired is .6, what is the probability that the fifth person in line to apply for the jobs will be hired?" The .6 probability for hiring an applicant is quite clearly a statistical probability. It measures the relative frequency that a job applicant has the qualifications deemed necessary by the employer. Thus, in solving the problem we assume that we **know nothing about the applicants other than that they are in line** waiting to be interviewed.

The fifth person in line, in assessing his chances for being employed, probably would try to obtain additional information by "sizing up" those ahead of him in line. Any additional information that he can obtain affects the probability of his being hired. If the jobs require a good deal of physical strength and there is a frail, elderly person in line, the probability that person is hired is considerably less than .6. Conversely a strong virile-looking person whom we know to have a good employment record has a probability greater than .6 of being hired. A person can also use a knowledge of his own capabilities to reevaluate his own probability of being hired if he makes it to the head of the line.

Statistical probabilities are often misapplied by using them in situations about which more is known than was known in obtaining the statistics. Let us use an example of computed probabilities to see how knowledge of a situation affects the probabilities of various events.

Example 7.8.1

Let us select from a deck of playing cards the four aces and the four kings. We now have a "deck" of eight cards. We select two cards from the deck at random. We wish to know the probability that both cards selected are aces. We shall consider the question in four different situations, in each of which we have a little bit more information.

First, what is the probability that both cards are aces, based on the number of cards in the deck?

We can compute the probability in two ways: Selecting two cards from eight can be done in $C(8,2)$ ways. There are, thus, $C(8,2)$ outcomes to our "experiment." Selecting two aces from four can be done in $C(4,2)$ ways. Thus,

$$p(2A) = \frac{C(4,2)}{C(8,2)} = \frac{6}{28} = \frac{3}{14}$$

Alternately, the probability of choosing an ace on the first draw is 4/8. The probability of drawing an ace on the second draw, given an ace was drawn on the first draw, is 3/7. Thus

$$p(2A) = \frac{4}{8} \cdot \frac{3}{7} = \frac{3}{14}$$

Second, what is the probability of drawing two aces given that one of the cards is an ace? This question would arise in a situation in which the two cards were drawn, and one of them was exposed just long enough for it to be recognized as an ace.

Note that we do not know whether it is the first or the second card drawn that is the ace. To evaluate the probability that both cards are aces, we must consider the elements of the sample space.

There are three events possible for drawing two cards—drawing two aces, drawing one ace and one king, and drawing two kings.

There are $C(4,2) = 6$ different combinations of two aces we can draw.

There are $4 \cdot 4 = 16$ different combinations of one ace and one king we can draw.

There are $C(4,2) = 6$ different combinations of two kings we can draw.

These make up the $C(8,2) = 28$ different outcomes for the experiment. In this case we know that one of the cards is an ace. Thus, the six outcomes of drawing two kings are removed from consideration. The remaining $6 + 16 = 22$ outcomes are equally probable, and six of them are the event we are concerned with. Thus

$$p(2A|A) = \frac{6}{22} = \frac{3}{11}$$

Our knowledge has raised the probability for an event **consistent with that knowledge**. If the exposed card is a king, the probability of getting two aces is, of course, zero.

Third, what is the probability that we draw two aces given that one of the cards is a black ace? In the momentary exposure of the card, it was noted, not only that it is an ace, but also that the markings are black.

We must now look more closely at the elements of the sample space. Of the six outcomes in which two aces are drawn, five are still possible in the light of the new information. It is not possible that the two cards are the two red aces.

Of the outcomes in which there is one ace and one king, we have eliminated those in which the one ace is red. Thus, there are $2 \cdot 4 = 8$ possible outcomes remaining.

Our "possible" sample space is now restricted to $5 + 8 = 13$ outcomes. Since five of these are "favorable," we have

$$p(2A|BA) = \frac{5}{13}$$

Finally, what is the probability of drawing two aces given that one is the ace of clubs? The one card was exposed long enough for us to see clearly what card it is.

This new information imposes more restrictions on the sample space. There are now just three possible outcomes with two aces—the ace of clubs plus any one of the other three aces.

There are four possible outcomes consisting of an ace and a king—the ace of clubs plus any of the four kings.

The total remaining sample space contains $3 + 4 = 7$ elements. Three are "favorable," so we have

$$p(2A|CA) = \frac{3}{7}$$

In the previous example we have seen how knowledge about a situation affects the probability of a given event. This is consistent with the nature of probability, which is a measure of the relative likelihoods of things we do not know for sure. Note that in each case in the example the additional knowledge gained was consistent with the event of interest. If at any time we obtained the information that one of the cards is a king, the probability that both are aces immediately becomes zero.

Note also that as we gained more and more **specific** information, we reduced the uncertainty about part of the experiment and, thus, the number of different outcomes possible.

In applying these ideas to reevaluating possibilities we must be certain that the knowledge we gain does indeed reduce uncertainty and not just shift it from one aspect of the experiment to another.

Example 7.8.2

A problem frequently found in texts on probability is the following, which has several variations in setting:

Three people, A, B, and C, are seeking a choice assignment, which is to be made by lot. Each of the three, therefore, has an equal probability (1/3) of receiving it. The drawing is made, but the announcement of the result is withheld so that it can be made at a special ceremony. A goes to the person who knows the result and asks to be told the name of one of the other two (B or C) who is not going to be selected. A suggests that if neither of them is to be selected, one of their names can be chosen at random to tell him.

Since A knows already that one of them is not going to be selected, it will not be giving away the secret. The person who knows the result refuses on the basis that this would effectively narrow the selection to two people, A and one of the others. Thus, A's chance would increase from 1/3 to 1/2. The question is, is this reasoning sound—does the probability for A change as a result of his obtaining the information?

This is a good example of a situation in which it is easy to be led away from a correct analysis by an erroneous assumption. The problem is one of conditional probability. The experiment as proposed by A consists of two stages: selection of the person for the assignment, and telling A that either B or C was not selected. The tree diagram for the experiment is shown in Figure 7.8.1.

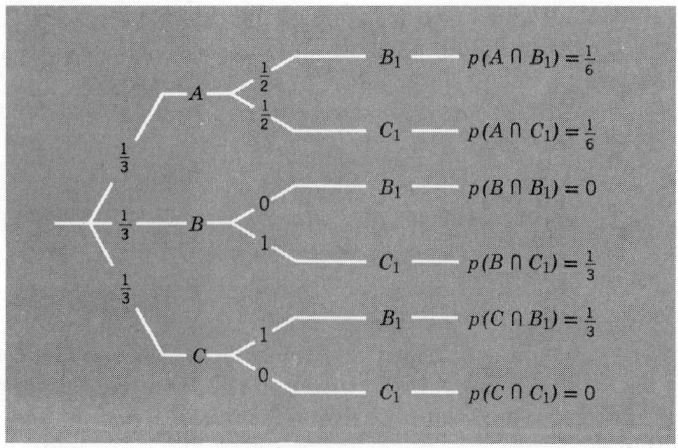

Figure 7.8.1

We are told that the selection was made by lot, and the probability that each is selected is 1/3, as shown in the diagram. We are concerned with whether the information requested by A will shed any light on the identity of the selectee. That is, we must reevaluate $p(A)$ given that he is told either that B or that C was not selected. Let B_1 represent the event that A is told that B was not selected; and C_1, the event that A is told that C was not selected. If A is the one selected, there is equal probability for B_1 and C_1. That is,

$$p(B_1|A) = \frac{1}{2} \qquad p(C_1|A) = \frac{1}{2}$$

If B was selected, C_1 is certain to happen; if C was selected, B_1 is certain. From this information we compute the probabilities for each of the possible sequences, $A \cap B_1$, $A \cap C_1$, and so on. These probabilities are shown in Figure 7.8.1. Since

$$B_1 = (A \cap B_1) \cup (B \cap B_1) \cup (C \cap B_1)$$

$$C_1 = (A \cap C_1) \cup (B \cap C_1) \cup (C \cap C_1)$$

and these are mutually exclusive events,

$$p(B_1) = \frac{1}{6} + 0 + \frac{1}{3} = \frac{1}{2}$$

$$p(C_1) = \frac{1}{6} + \frac{1}{3} + 0 = \frac{1}{2}$$

Using Bayes' theorem, we can compute

$$p(A|B_1) = \frac{p(A \cap B_1)}{p(B_1)} = \frac{1/6}{1/2} = \frac{1}{3}$$

$$p(A|C_1) = \frac{p(A \cap C_1)}{p(C_1)} = \frac{1/6}{1/2} = \frac{1}{3}$$

Thus

$$p(A) = p(B_1) \cdot p(A|B_1) + p(C_1) \cdot p(A|C_1) = \frac{1}{2} \cdot \frac{1}{3} + \frac{1}{2} \cdot \frac{1}{3} = \frac{1}{3}$$

This is the same probability for A's being selected as before.

In the example above we obtained the perhaps surprising result that A's probability of being selected is not affected by the disclosure that he requested. This appears to contradict what was said earlier, that additional information consistent with an outcome increases the probability of that outcome. The apparent paradox here is explained by the fact that the disclosure is not really additional information. It trades one uncertainty for another. The key lies in fact that A does not know whether the choice of the person named as the non selectee was made at random or by reason of his being the only nonselectee possible. The error in analysis was made in assessing the probabilities of the two remaining possibilities (A and one other) after A's question is answered. It was arbitrarily assumed that both these possibilities are equally probable, each with a probability of 1/2. This is not true. The person not named is now twice as likely as before to be the one chosen; that is, his probability is now 2/3.

Example 7.8.2 is an illustration of comments made in Chapter 2 concerning the use of abstract symbols to guard against being confused by the meanings of statements. In this problem the solution follows directly as soon as we translate the situation into symbols and apply the laws of probability.

7.8.2 Odds

Before concluding our discussion of probability, let us take up the concept of **odds**. The "odds" for an experiment are a way of stating the **relative** probability between **two events**. They are used in two ways: as a statement of the relative probability between the two outcomes of an experiment that can have only two outcomes, or as a statement of the relative probability that a given event will or will not occur.

Let us first consider an experiment in which only two outcomes, A and B, are possible. We say that the "odds" on A happening instead of B are "two-to-one" (or event A is "favored" two-to-one over event B) if the probability of A is two times as great as the probability of B. That is, if $p(A) = 2p(B)$, the "odds" favor A over B by two-to-one.

Where A and B are the only two events possible,

$$p(A) + p(B) = 1$$

Thus

$$2p(B) + p(B) = 1$$

$$p(B) = \frac{1}{3} \qquad p(A) = \frac{2}{3}$$

We say the odds are "even," or the outcome is "even money," when the probability of the two events is the same. That is, when $p(A) = p(B)$, they each have a value of 1/2; and the odds on either A or B are "even."

In general, if the odds on event A over event B are "a to b,"

$$p(A) = \frac{a}{a+b} \qquad p(B) = \frac{b}{a+b}$$

If event A is favored over event B by odds of five to three,

$$p(A) = \frac{5}{5+3} = \frac{5}{8} \qquad p(B) = \frac{3}{5+3} = \frac{3}{8}$$

When we consider odds on a single event, we compare the probability the event will happen to the probability that it will not. If the odds that event A will happen are three to one, this means the probability of event A is

$$p(A) = \frac{3}{3+1} = \frac{3}{4}$$

This same situation is sometimes expressed as, "The odds in **favor** of event

A are three to one." The expression, "The odds are five to one **against** event A," is also used. This means the probability that A **will not** happen is

$$p(A') = \frac{5}{5+1} = \frac{5}{6}$$

Thus, the probability that A **will** happen is

$$p(A) = \frac{1}{1+5} = \frac{1}{6}$$

The probabilities in an experiment with several outcomes are sometimes expressed in terms of **odds**. When this is done, it is important that the odds are expressed properly so that the probability of the entire sample space is exactly equal to one.

For example, let us suppose that in an experiment the events A, B, C, and D are mutually exclusive and that $A \cup B \cup C \cup D = S$. We could state the probabilities for the result of this experiment as either "odds for" each event or "odds against" each event.

Event	Odds For	Odds Against
A	1 to 2	2 to 1
B	1 to 4	4 to 1
C	1 to 5	5 to 1
D	3 to 7	7 to 3

From these odds we can compute:

$$p(A) = \frac{1}{1+2} = \frac{1}{3}$$

$$p(B) = \frac{1}{1+4} = \frac{1}{5}$$

$$p(C) = \frac{1}{1+5} = \frac{1}{6}$$

$$p(D) = \frac{3}{3+7} = \frac{3}{10}$$

Note that

$$p(A) + p(B) + p(C) + p(D) = \frac{1}{3} + \frac{1}{5} + \frac{1}{6} + \frac{3}{10} = 1$$

This means that the quoted odds correctly take into account all of the possible outcomes of the experiment.

Example 7.8.3

Odds are frequently used in gambling. Betting on horse races is legal in many states, and newspapers often have lists of the "probable odds" on the horses entered in the races. Note that rarely, if ever, will the odds listed satisfy the criterion discussed above: that the sum of the probabilities equivalent to the odds should be one.

In a newspaper selected at random, the following "probable odds" were listed for the horses in one of the races:

Horse Number	Probable Odds (Against)	Equivalent Probability of Winning
1	$4-1$	$\frac{1}{5}$ (.200)
2	$7-2$	$\frac{2}{9}$ (.222)
3	$8-1$	$\frac{1}{9}$ (.111)
4	$5-2$	$\frac{2}{7}$ (.286)
5	$8-1$	$\frac{1}{9}$ (.111)
6	$4-1$	$\frac{1}{5}$ (.200)

Figure 7.8.2

If we add the probabilities for winning for each horse, we have

$$p(S) = .200 + .222 + .111 + .286 + .111 + .200 = 1.130$$

The total probability that the race will be won by one of the horses is quoted to be greater than one. Clearly, this is an impossible result. The reason is that the "probable odds" are a statement of the **betting** odds—the amount of money won by betting on the winning horse. Odds of four-to-one mean a payment of \$4 for every \$1 bet in the event that horse wins. Thus, the expected value of this experiment can be computed. If we use the odds to compute the value of a \$1 bet on each horse, we have

Horse Number	Value
1	4
2	3.5
3	8
4	2.5
5	8
6	4

Figure 7.8.3

Since the probabilities assigned to each horse's winning total more than one, these probabilities must be too large. Let us assume that each probability is too large in the same proportion. That is, we reduce each probability in the same proportion, so that the total probability that a horse will win is one. This means we must multiply each probability by 1/1.130. We obtain the following revised probabilities:

Horse Number	Probability of Winning
1	(.200)(1/1.130) = .177
2	(.222)(1/1.130) = .197
3	(.111)(1/1.130) = .098
4	(.286)(1/1.130) = .253
5	(.111)(1/1.130) = .098
6	(.200)(1/1.130) = .177
	1.000

Figure 7.8.4

We now compute the expected value of this "experiment" using the values from Figure 7.8.3 and the probabilities from Figure 7.8.4.

$$EV = (.177)(4) + (.197)(3.5) + (.098)(8) + (.253)(2.5) + (.098)(8) + (.177)(4)$$

$$= .708 + .690 + .784 + .632 + .784 + .708$$

$$= 4.306$$

The expected value of betting \$1 on **each horse** in the race is \$4.306. The cost is \$5 (you get back the \$1 you bet on the winner). Thus, the expected return is \$4.306 − \$5 = −\$.694. This loss, of course, is the profit of the gambling operation operator.

EXERCISE 7.8

1. Convert the following odds to probabilities:

a. *A* is favored over *B* by odds of 3 to 2.
b. *A* is favored over *B* by odds of 10 to 1.
c. The odds for event *A* are 5 to 1.
d. The odds against event *A* are 9 to 5.

2. Convert the following probabilities into odds:

a. $p(A) = \frac{3}{4}$ $\quad p(B) = \frac{1}{4}$

b. $p(A) = \frac{1}{5}$ $\quad p(B) = \frac{4}{5}$

c. $p(A) = \frac{1}{2}$

d. $p(A) = \frac{1}{10}$

3. A group of 10 employees of a company meet to discuss their working conditions. They decide to select a spokesman by lot to present their case. Ten pieces of paper, nine blank, and one marked with an "x," are placed in a box; and each person draws a paper. The one with the x is the spokesman.

a. If Joe is to be seventh to draw, what is the probability that he will be the spokesman?

b. During the drawing each person exposes his paper as soon as he draws. If no one has obtained the x by the time Joe's turn comes, what is the probability that he will be spokesman?

4. A person flips a "fair" coin three times.

a. What are the odds that he will flip three heads?

b. What are the odds that he will flip exactly two tails?

5. A person is rolling dice. He rolls five straight "sevens." What are the odds that the next roll comes up "seven"?

6. An urn contains 20 balls: six red, four white, five yellow, and five green. A person draws two balls from the urn and does not replace them. Both balls are red. What are the odds that he will get a red ball on the next draw?

7. In an election with four candidates, A, B, C, and D, a local political "expert" quoted the following odds:

A: 5 to 3 against winning
B: 3 to 1 against winning
C: 11 to 5 against winning
D: 15 to 1 against winning

a. What are the probabilities that each will win according to this "expert"?

b. If he offered to accept bets at his quoted odds, would these be "fair" bets; that is, the expected value of a bet on each candidate is zero?

8. The odds that event A will occur are 3 to 1. The odds that event B will occur are 2 to 1. What are the odds that both A and B will occur if A and B are independent?

9. A person is rolling a pair of dice.

a. What is the probability that he will roll an "eight"?

b. He rolls the dice one at a time. The first die comes up "three." What is the probability that he will roll an "eight"?

10. Statistics show that 50% of new businesses fail during their first year of operation. Four new businesses start in a city at about the same time:

a. What are the odds in favor of their all being in operation after one year?

b. What are the odds in favor of one-half of them being in operation after a year?

8

markov chains

8.1 SIMPLE MARKOV PROCESSES

In the preceding chapter we studied briefly the concept of **stochastic processes**—processes in which the exact course of action cannot be predicted with certainty. We used a sequence of Bernoulli trials as an example of a stochastic process. Recall that in a Bernoulli sequence each trial has only two possible outcomes, and each trial is independent of preceding trials. We direct our attention now to another type of stochastic process, one in which there can be any finite number of outcomes and in which the probabilities for the outcomes depend on the result of the preceding stage of the process, but not on earlier stages. Such a process is called a Markov chain process.

In essence a Markov chain process occurs when a system can exist in one of several different conditions, or **states**. There is a finite number of these states. A "trial" in the process consists of subjecting the system to a set of circumstances capable of causing a change in the state occupied by the system. Whether the system does or does not change state and, if it does, the state it changes to, is governed by a set of probabilities, called **transition probabilities**. The same set of probabilities applies to each succeeding trial in the process.

For example, let us suppose there are two companies supplying a product that a large group of people purchases each week. The buying habits of the people follow a pattern. For any week, 75% of the people purchasing A's product purchase it again the next week, and 25% switch to B's product. Of those who purchase B's product, 60% purchase B's product again next time and 40% switch to A's. This situation can be considered as a Markov chain process with two states. These states are the conditions of being A's customer or B's customer. The percentages of "stayers" and "switchers," when expressed as fractions, are the transition probabilities. Each week's purchases constitute a trial. If we know the percentage of people purchasing each company's product, we can use the transition probabilities to find the market share for each company for the next week. Using these market shares we can find the market shares for the following week, and so on.

Let us consider a slightly more complicated example. Suppose a company provides armored car service in a large city. As part of their security program they have divided the city into three zones, A, B, and C. Each week the personnel are reassigned to work in the different zones. The assignments are made on a "lottery" basis in which the following conditions apply:

An employee in Area A has a probability of .3 of staying at A, a probability of .7 of being transferred to B, and no workers from A are transferred to C.

An employee in Area B has a .5 probability of being transferred to A, and .5 probability of being transferred to C. No employees remain at B for a second consecutive week.

For employees in Area C the probability is .2 for being transferred to A, .3 of being transferred to B, and .5 of staying at C.

Let us consider each assignment of personnel as a "trial" in which the outcome is governed by probabilities. First we look at the nature of the conditions governing the transfers. Note that the probability of being transferred to each of the areas depends only on the area in which the employee is assigned at the time of the transfer. No consideration is given to the amount of time the employee has spent in each area in the past. The only effect of past assignment is the effect that the previous week's assignment has on the employee's present location.

Note also that the policy describes a sequence of transfers, all to be made with the same set of probabilities. The pattern of transfers stays the same from week to week.

The transfer process described is an example of a Markov chain process, which is defined formally as follows:

Definition 8.1.1

Markov Chain: An experiment consisting of a sequence of trials is a **Markov chain** if, and only if, it has the following properties:

i. The outcome of each trial is one of a finite set of possible outcomes.
ii. The probability of each outcome depends upon the outcome of the immediately preceding trial, but none earlier.

There are a wide variety of situations that can be classified as Markov chains—much wider than suggested by the simple examples above. We shall examine several examples in different fields; but first let us use our example to investigate some of the properties of Markov chains.

The outcomes of the sequence of trials are called "states," and the trials themselves are often referred to as "transitions." The term "state" cannot be defined precisely. We can regard it as descriptive of a situation that exists at a given time or place and that is subject to change. The states of an individual system, however, can be described and defined; and we shall be concerned only with systems in which there are a finite number of definably different "states." In the preceding example, assignment to one of the areas is a "state." In any sequence the same set of states remain as possible outcomes for each transition. The probability of transfer from any state to another can be determined, and each of these probabilities stays constant throughout the sequence of transitions. These probabilities are called **transition probabilities**. (Remember that one of the possible "new" states is a recurrence of the same state. The particular system studied determines each of the probabilities.) In some systems a state may be reached in which there is no possibility for further changes; that is, the system remains in this state permanently. Such a Markov chain is called an **absorbing** Markov chain. We shall study examples of this later.

The possible outcomes of a single trial in a Markov chain can be represented in a diagram (called a **transition diagram**).

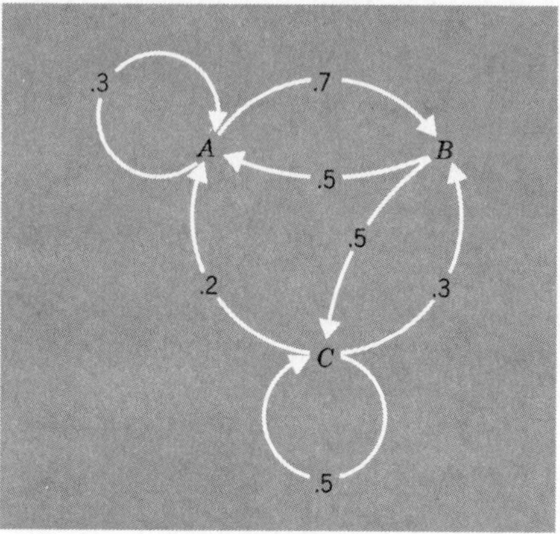

Figure 8.1.1

Figure 8.1.1 represents the transfer policy diagram. The arrows show the possible transitions from one state to another, and the numbers represent the probability for that particular transition to occur.

If we wish to study what happens in several successive experiments, we can use a tree diagram, similar to those which we used earlier in studying conditional probabilities.

Figure 8.1.2 shows the possible transfers for a person starting in Area A in the example. As an illustration of the use of the diagram, we find the probability that a person starting at A will be assigned to C after three weeks. With the diagram we can trace the possible sequences that will lead to C after three weeks and compute the probability of each. In each case the probability of a sequence is the product of the probabilities of each step in the sequence. The total probability of arriving at C is the sum of the probabilities of the different sequences leading there, since they constitute mutually exclusive events. As can be seen in the diagram, there are two paths leading to C; they are labeled (1) and (2) in the diagram. In path 1 the probability is $(.3)(.7)(.5) = .105$. In path 2 the probability is $(.7)(.5)(.5) = .175$. Therefore, the probability of starting at A and being in C after three weeks is $.105 + .175 = .280$.

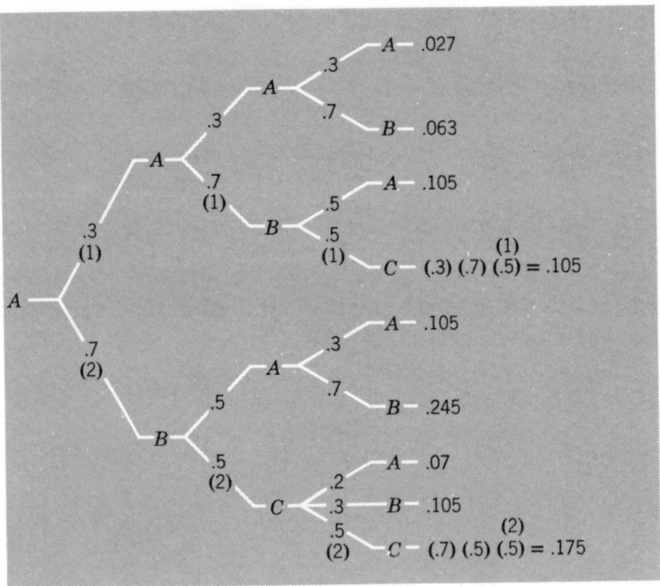

Figure 8.1.2

The tree diagram display of the sequence of transitions is useful in some cases, but it can get complex and confusing, particularly if we try to include the transfers of the people who start in Areas B and C. The entire process can be displayed most conveniently in a **matrix**, in which the rows represent the initial states, the columns represent the outcomes, and the elements of the matrix are the transition probabilities. The matrix for our example becomes

$$\begin{array}{r r} & \text{To:} \\ & \begin{array}{ccc} A & B & C \end{array} \\ \text{From: } \begin{array}{c} A \\ B \\ C \end{array} & \begin{bmatrix} .3 & .7 & 0 \\ .5 & 0 & .5 \\ .2 & .3 & .5 \end{bmatrix} \end{array}$$

This matrix shows that the probability of staying in Area A (row 1, column 1) is .3; the probability of transferring from A to B (row 1, column 2) is .7; and from A to C (row 1, column 3) is 0; and so on, matching the verbal description of the transfer policy.

A matrix of this type is called a **transition matrix**. A transition matrix can be formed for each Markov chain process. It is always a square matrix since the number of possible states (outcomes) resulting from a transition is the same as the number of initial states. As we shall see, there are a number of operations that we can perform with a transition matrix. These provide a great deal of information about the system that the matrix represents.

If we represent the transfer policy matrix of our example by T, we can obtain the probabilities for two weeks' transfers by finding T^2—by multiplying T by itself. In general, the result of n steps in the chain will be given by T^n.

Let us find T^3 and compare this with the information given by the tree diagram above.

$$T^2 = \begin{bmatrix} .3 & .7 & 0 \\ .5 & 0 & .5 \\ .2 & .3 & .5 \end{bmatrix} \begin{bmatrix} .3 & .7 & 0 \\ .5 & 0 & .5 \\ .2 & .3 & .5 \end{bmatrix} = \begin{bmatrix} .44 & .21 & .35 \\ .25 & .50 & .25 \\ .31 & .29 & .40 \end{bmatrix}$$

$$T^3 = T^2T = \begin{bmatrix} .44 & .21 & .35 \\ .25 & .50 & .25 \\ .31 & .29 & .40 \end{bmatrix} \begin{bmatrix} .3 & .7 & 0 \\ .5 & 0 & .5 \\ .2 & .3 & .5 \end{bmatrix} = \begin{bmatrix} .307 & .413 & .280 \\ .375 & .250 & .375 \\ .318 & .337 & .345 \end{bmatrix}$$

The probability of starting at A and transferring to C after three weeks is shown in row 1, column 3 of T^3. As shown above, this probability is .280. This is the same probability that we computed in the tree diagram. However, the matrix shows also the probability of starting in any of the areas and ending in any of the areas. All of the probabilities are available at a glance. In succeeding sections we shall investigate additional information that can be obtained by using the transition matrix.

(Remember that computers can be programmed to do the matrix multiplication which, admittedly, gets a little tedious. There are also shortcuts to getting higher powers of matrices, but they are beyond the scope of this book.)

EXERCISE 8.1

1. Draw a transition diagram for the Markov chains with transition probabilities given by the following matrices:

   a. $\begin{bmatrix} 0 & 1 \\ 1 & 0 \end{bmatrix}$ b. $\begin{bmatrix} \frac{1}{2} & \frac{1}{2} & 0 \\ 0 & 1 & 0 \\ \frac{1}{2} & 0 & \frac{1}{2} \end{bmatrix}$ c. $\begin{bmatrix} \frac{1}{2} & \frac{1}{4} & \frac{1}{4} \\ \frac{1}{2} & \frac{1}{4} & \frac{1}{4} \\ \frac{1}{4} & \frac{1}{2} & \frac{1}{4} \end{bmatrix}$

2. Draw transition diagrams for the following Markov chains:

   a. $\begin{bmatrix} .6 & .2 & .2 \\ .3 & .4 & .3 \\ .1 & .5 & .4 \end{bmatrix}$ b. $\begin{bmatrix} .2 & .5 & .3 \\ 0 & .4 & .6 \\ 0 & 1 & 0 \end{bmatrix}$

3. Give the matrix of transition probabilities corresponding to each of the following transition diagrams:

a.

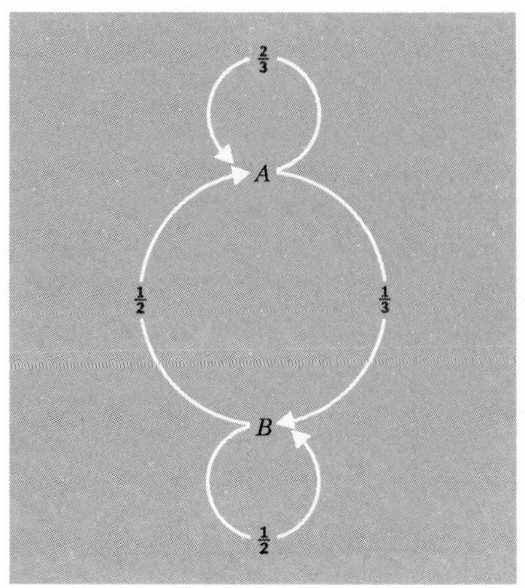

b.

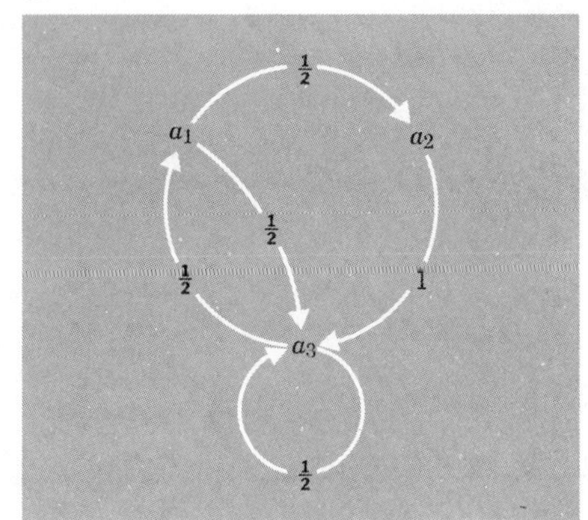

4. Form transition matrices for the following Markov chains:

a.

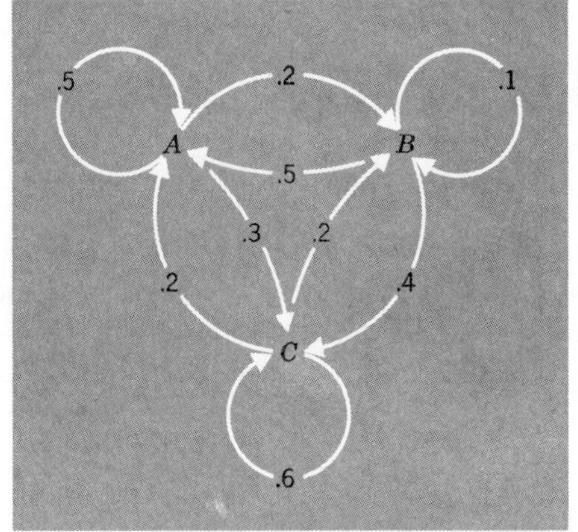

b.

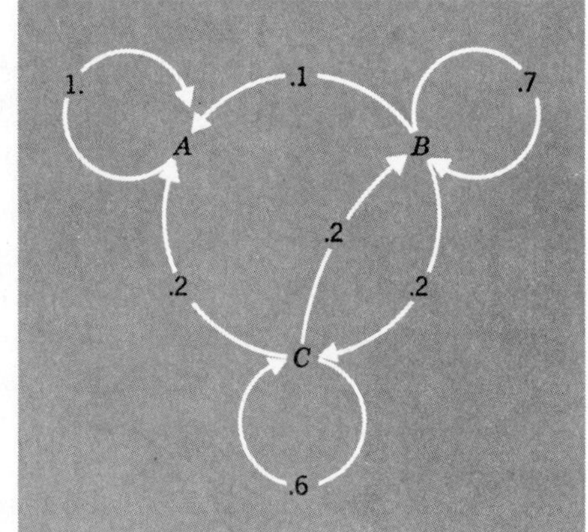

c.

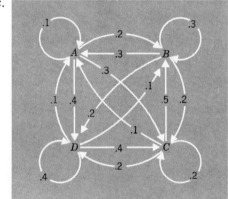

For each of the following (a) draw a transition diagram and (b) form a transition matrix.

5. Perform these operations for the first example in this section, where the customer switches between companies A and B.

6. A district that votes Democratic in a presidential election has a 70% chance of voting Democratic again, and a 30% chance of voting Republican. On the other hand, a district that votes Republican has an 80% chance of voting the same way in the next election and only a 20% chance of switching parties.

7. Each year 10% of the people living in the city move to the suburbs, and 2% move to rural areas. Of the people living in the suburbs, 8% move to the city and 3% to the country. Twelve percent of the people in rural areas move to the city, and 5% move to the suburbs. In each case the remainder stay where they are.

8. A study of the transmission of political views from father to son showed the following:

If the father is politically conservative, 40% of the sons are conservative, 30% moderate, and 30% liberal.

If the father is politically moderate, 20% of the sons are conservative, 50% moderate, and 30% liberal.

If the father is politically liberal, 20% of the sons are conservative, 40% moderate, and 40% liberal.

9. Three companies, A, B, and C, dominate the market in supplying income tax service. A survey of their customers showed the following:

Sixty percent of A's customers intend to return to A next year; 25% intend to switch to B; and 15% switch to C.

Fifty percent of B's customers intend to return to B next year; 20% intend to switch to A; and 30% to C.

Seventy percent of C's customers intend to return to C next year; 10% intend to switch to A; and 20% to B.

10. Let us assume that a man's occupation can be classified as professional, paraprofessional, or nonprofessional. Assume that 70% of the sons of professional men are also professionals; 20% are paraprofessionals; and 10% are nonprofessional. Of the sons of paraprofessionals, 60% are also paraprofessionals, 20% are professional, and 20% are nonprofessional. Fifty percent of the sons of nonprofessionals are also nonprofessional, 30% are paraprofessional and 20% professional.

11. A portion of the income tax returns are selected at random for checking by the Internal Revenue Service. Let us assume that the following schedule applies: Of the returns of people whose return was not checked the previous year, 35% are selected for a "quick check" to verify that the tax paid is within "expected limits" for the income reported, and 10% are selected for a thorough check. The remaining 55% get by for another year. Of those whose returns were given a quick check the previous year, 20% had minor errors or irregularities. These are given a quick check again. Ten percent are selected for a thorough check, and the remaining 70% are not checked. Of those given a thorough check the previous year, 30% are chosen for a quick check, 10% are given another thorough check, and the remainder receive no check.

12. Three companies, A, B, and C, share the market for a product. Statistics have shown that each year 60% of A's customers continue to buy A's product, while 30% switch to B and 10% to C. Of B's customers 10% switch to A, 70% stay with B, and 20% switch to C. Of C's customers 20% switch to A, 20% to B, and 60% stay with C.

8.2 PROBABILITY VECTORS

In the previous section we saw how changes in a system capable of existing in several different states can be studied using matrices to display the probabilities for change from one state to another. Let us now examine these matrices more closely to see what they mean and how they can be used.

First, some symbology. We shall use the symbols a_1, a_2, a_3, and so on, to represent the various possible states of a system. The probabilities for change from one state to another will be represented by the symbols p_{11} (read "p sub one-one"), p_{12}, p_{13}, and so on, which represent respectively, the probability of staying in state a_1, (p_{11}); the probability of changing from state a_1 to state a_2, (p_{12}); the probability of changing from state a_3 to state a_2, (p_{32}); and so on. The probability for **being** in a given state will be represented by p's with single subscripts – p_1 (the probability of being in state a_1 after 0 changes; that is, at the start), p_2 (the probability of starting in state a_2); and so on. To represent the probability of being in a given state after a given number of transitions we shall use the single subscript p's with the number of transitions in parentheses above the p. $p_1^{(1)}$ will represent the probability of being in state a_1 after one transition; $p_2^{(3)}$ will represent the probability of being in state a_2 after three transitions; $p_j^{(n)}$ will represent the probability of being in state a_j after n transitions; and so on.

At any time the status of a system can be represented by a **probability vector** in which each component is the probability for the system to be in each respective state—for example, $[p_1^{(\ )}, p_2^{(\ )}, p_3^{(\ )}]$ is the probability vector for a system with three possible states. The number in the parentheses above the p's tells how many transitions have taken place in the Markov sequence.

Important note: Since the components of this probability vector represent probabilities, they must all be numbers between 0 and 1 inclusive. Further, since they represent all the possibilities for the system, their sum must be one.

As we have seen in the first section, the transition matrix displays the transition probabilities among the various states:

$$\begin{array}{rc} & \text{To state:} \\ & \begin{array}{ccc} a_1 & a_2 & a_3 \end{array} \\ \text{From state:} \begin{array}{c} a_1 \\ a_2 \\ a_3 \end{array} & \begin{bmatrix} p_{11} & p_{12} & p_{13} \\ p_{21} & p_{22} & p_{23} \\ p_{31} & p_{32} & p_{33} \end{bmatrix} \end{array}$$

(We shall use three-state systems for most of our examples. The same ideas can be expanded to any number of states. The transition matrix in a Markov chain will always be a **square** matrix, however, because there are always the same number of states possible after a transition as before.)

The sum of the entries in each **row** of a transition matrix is equal to one. In fact, **each row of a transition matrix is the probability vector for the system after a single transition, assuming that the system started in the state represented by that row.**

The sum of the entries in each column do not necessarily equal one. The relative value of these sums does give an indication about the probability vectors for the system after many transitions, but there is no way to use these sums in computations.

Let us return again to the matrix for the transfer policy example of the previous section. Note that the sum of each row is one. Further, note that the first row is made up of the transfer probabilities for a person assigned to Area

A; row 2 consists of the probabilities for transfer from Area B; and row 3, the probabilities for transfer from Area C.

$$\begin{bmatrix} .3 & .7 & 0 \\ .5 & 0 & .5 \\ .2 & .3 & .5 \end{bmatrix}$$

Using our previous study of probabilities for mutually exclusive events, we can compute the probabilities for being in each of the states after a transition:

$$\text{For state } a_1 \quad p_1^{(1)} = p_1 p_{11} + p_2 p_{21} + p_3 p_{31}$$

$$\text{For state } a_2 \quad p_2^{(1)} = p_1 p_{12} + p_2 p_{22} + p_3 p_{32}$$

$$\text{For state } a_3 \quad p_3^{(1)} = p_1 p_{13} + p_2 p_{23} + p_3 p_{33}$$

In words, for state a_1, the probability of being in state a_1 after one transition ($p_1^{(1)}$) is the probability of being in state a_1 and staying there ($p_1 p_{11}$) plus the probability of being in state a_2 and transferring to state a_1 ($p_2 p_{21}$) plus the probability of being in state a_3 and transferring to a_1 ($p_3 p_{31}$).

Note that the new probability vector $[p_1^{(1)}, p_2^{(1)}, p_3^{(1)}]$ is given by the product of the initial probability vector $[p_1, p_2, p_3]$ and the transition matrix. In general, the probability vector after n transitions will be given by

$$[p_1^{(n)}, p_2^{(n)}, p_3^{(n)}] = [p_1, p_2, p_3]\ T^n$$

where T^n is the nth power of the transition matrix, symbolized T.

Returning to our example again, let us assume that a new employee is assigned to Area A. The initial probability vector for this employee is $[1, 0, 0]$. After the first transfer, as we have seen, his assignment probability vector is $[.3, .7, 0]$, and after three transfers it is $[.307, .413, .280]$. To find the assignment probabilities after the next three transfers, we can take this latter vector and multiply it by T^3:

$$[.307, .413, 280] \begin{bmatrix} .307 & .413 & .280 \\ .375 & .250 & .375 \\ .318 & .337 & .345 \end{bmatrix}$$

In this new probability vector the probabilities for being assigned to each of the three offices is just about the same. In fact, if we were to continue to find the probability vector after more and more transitions, we find that it would get closer and closer to $[1/3, 1/3, 1/3]$. This suggests that there is a sort of equilibrium condition that is reached in which an employee is equally likely to be assigned to each of the three areas. If we start with the probability vector $[1/3, 1/3, 1/3]$ as the employee's initial probability vector, we find that his probability vector stays the same after any number of transitions. This can be shown by multiplying this vector by the single transition matrix:

$$\left[\frac{1}{3}, \frac{1}{3}, \frac{1}{3}\right] \begin{bmatrix} .3 & .7 & 0 \\ .5 & 0 & .5 \\ .2 & .3 & .5 \end{bmatrix} = \left[\frac{.3}{3} + \frac{.5}{3} + \frac{.2}{3}, \ \frac{.7}{3} + \frac{.3}{3}, \ \frac{.5}{3} + \frac{.5}{3}\right]$$

$$= [1/3, 1/3, 1/3]$$

A probability vector that remains unchanged in a transition is called a **fixed point**, or **fixed vector**, in the system. A process, or system, that has a fixed point is called a **regular Markov process**. One characteristic of the transition matrix of a regular Markov process is that **some power** of the matrix, (not necessarily the matrix itself), will have all positive elements—none of them zero. Such a process has an equilibrium condition, which is characterized by the fixed vector. This equilibrium condition is approached by successive state changes—that is, successive state changes, or transitions, bring the system nearer and nearer to the equilibrium condition. This approach to equilibrium takes place no matter what the initial probability vector is.

Markov processes are regular except for two cases. In one, some or all of the state changes take place in a fixed pattern (e.g., state a_1 always changes to state a_3 and a_3 to a_1) to form a "loop" in the system. The matrices for such processes have a section consisting of 1's and 0's. In the other exception, called an absorbing processs there is at least one state for which there are no transitions out of that state. The row in the matrix for such states has a "one" in the diagonal and zeros in the other columns. We shall study such processes in greater detail later.

It is often of interest to find the fixed vector for a regular Markov process. One way, of course, is to raise the transition matrix to a high power (multiply it by itself many times) and then see what values a probability vector approaches for a large number of transitions. This is a cumbersome process, however, and we shall develop a somewhat simpler method.

If the system has only two states, it is a fairly simple process to calculate the fixed vector. If there are only two states, we know the transition matrix must be of the form,

$$\begin{bmatrix} 1-a & a \\ b & 1-b \end{bmatrix}$$

(Remember that the sum of the entries in each row must be one.)

The matrix must also satisfy the condition

$$[f_1, f_2] \begin{bmatrix} 1-a & a \\ b & 1-b \end{bmatrix} = [f_1, f_2]$$

where $[f_1, f_2]$ is the fixed vector. This condition can also be expressed as

$$f_1(1-a) + f_2 b = f_1 \qquad f_1 a + f_2(1-b) = f_2$$

Both of these equations simplify to

$$f_1 a = f_2 b$$

Since $f_1 + f_2 = 1$, we have

$$b(1-f_1) = f_1 a$$

from which we can obtain

$$f_1 = \frac{b}{a+b} \quad \text{and} \quad f_2 = \frac{a}{a+b} \qquad (8.2.1)$$

We can check this result as follows:

$$\left[\frac{b}{a+b}, \frac{a}{a+b}\right] \begin{bmatrix} 1-a & a \\ b & 1-b \end{bmatrix} = \left[\frac{b}{a+b}(1-a) + \frac{a}{a+b} \cdot b,\ \frac{b}{a+b} \cdot a + \frac{a}{a+b}(1-b)\right]$$

$$= \left[\frac{b}{a+b} - \frac{ba}{a+b} + \frac{ab}{a+b}, \quad \frac{ba}{a+b} + \frac{a}{a+b} - \frac{ab}{a+b}\right]$$

$$= \left[\frac{b}{a+b}, \frac{a}{a+b}\right]$$

Application of this result to a situation involving transmission of information by "word of mouth" produces an interesting result.

Example 8.2.1

Suppose that the President of the United States tells person A whether he intends to run for reelection. A relays the news to B, who relays the message to C, and so on, always to some new person. Let us assume that there is a probability p, greater than 0, that a person somewhere along the line, when he gets the message, will change it before telling the next person. (Statistically this means that at least one of the people that hears this news and relays it will change it either from "No" to "Yes" or vice versa.) What is the probability that the nth person to hear the message will be told that the President will run?

We can consider this to be a two-state Markov chain, with the states being "Yes" and "No." The process is in state "Yes" at time n if the nth person is told the President will run, and "No" if he is told the President will not run. The matrix of transition probabilities is

$$\begin{array}{c} \\ \text{Yes} \\ \text{No} \end{array} \begin{array}{c} \begin{array}{cc} \text{Yes} & \text{No} \end{array} \\ \begin{bmatrix} 1-p & p \\ p & 1-p \end{bmatrix} \end{array} = T$$

The matrix T^n gives the probabilities that the nth man is given a certain answer, assuming the President said "Yes" (first row) or assuming he said "No" (second row). We are assuming the probability of a person changing the message from "Yes" to "No" is the same as for changing from "No" to "Yes." From the nature of the matrix (no zero entries) we know it is a regular Markov process and there will be a fixed point. With Equation 8.2.1 we can calculate this fixed point as [1/2, 1/2]. Thus, the probabilities for the nth man being told "Yes" or "No" approach 1/2, regardless of what the President actually said as long as there is a probability greater than zero that the message will be switched along the way. If many people are given the message by the method described, we can expect half of them will be told "Yes" and the other half "No."

If the probability, a, that a person will change the news from "Yes" to "No" is different from the probability, b, he will change it from "No" to "Yes," then the matrix of probabilities becomes

$$\begin{array}{c} \\ \text{Yes} \\ \text{No} \end{array} \begin{array}{c} \begin{array}{cc} \text{Yes} & \text{No} \end{array} \\ \begin{bmatrix} 1-a & a \\ b & 1-b \end{bmatrix} \end{array}$$

As we have seen above, the fixed vector in this case is $[b/(a+b), a/(a+b)]$. Again, the probability of the nth person being told "Yes" is independent of what the President said and depends only on the people who are transmitting the message.

Let us now consider the situation in which there are three states. From the definition of a fixed vector we have

$$[f_1, f_2, f_3]\begin{bmatrix} p_{11} & p_{12} & p_{13} \\ p_{21} & p_{22} & p_{23} \\ p_{31} & p_{32} & p_{33} \end{bmatrix} = [f_1, f_2, f_3]$$

where $[f_1, f_2, f_3]$ represents the fixed probability vector of the system. If we do the multiplication indicated, we find that this matrix equation can be expressed as three linear equations:

$$\begin{aligned} p_{11}f_1 + p_{21}f_2 + p_{31}f_3 &= f_1 \\ p_{12}f_1 + p_{22}f_2 + p_{32}f_3 &= f_2 \\ p_{13}f_1 + p_{23}f_2 + p_{33}f_3 &= f_3 \end{aligned} \qquad (8.2.2)$$

We recognize this as a set of linear equations that can be solved by the methods of Chapter 4. If we do so, however, we find that there is an infinite number of solutions. That is, we obtain the situation discussed in Section 4.5 in which the last row of the reduced matrix contains all zeros. This result can be handled by including the consideration

$$f_1 + f_2 + f_3 = 1$$

imposed by the fact that $[f_1, f_2, f_3]$ is a probability vector, and its components must total one. Using this relationship and any two of the three equations of Equation 8.2.2 (recall from Chapter 4 that one equation of the set is redundant) we have

$$\begin{aligned} f_1 + \quad f_2 + \quad f_3 &= 1 \\ p_{11}f_1 + p_{21}f_2 + p_{31}f_3 &= f_1 \\ p_{12}f_1 + p_{22}f_2 + p_{32}f_3 &= f_2 \end{aligned} \qquad (8.2.3)$$

This set of equations is still not ready for solution because the right-hand side includes some of the unknowns. We must rearrange the equations to obtain

$$\begin{aligned} f_1 + \qquad\qquad f_2 + \quad f_3 &= 1 \\ (p_{11} - 1)f_1 + \qquad p_{21}f_2 + p_{31}f_3 &= 0 \\ p_{12}f_1 + (p_{22} - 1)f_2 + p_{32}f_3 &= 0 \end{aligned} \qquad (8.2.4)$$

Solving this system of equations for the values of f_1, f_2, and f_3 will give the fixed probability vector for the system.

Example 8.2.2

Three companies, A, B, and C, introduce a new item on the market at the same time. Each company initially gets approximately one-third of the market (each sells about the same amount as the others.)

During the first year the following occurs:

A keeps 75% of its customers, loses 15% to B and 10% to C.
B keeps 60% of its customers, loses 30% to A and 10% to C.
C keeps 80% of its customers, loses 5% to A and 15% to B.

Assuming this trend continues, what share of the market is each likely to have at the end of the next year? What is the long-range prediction if the same trend continues?

The transition matrix is

$$\begin{array}{c} \\ A \\ B \\ C \end{array} \begin{array}{c} \begin{array}{ccc} A & B & C \end{array} \\ \begin{bmatrix} .75 & .15 & .10 \\ .30 & .60 & .10 \\ .05 & .15 & .80 \end{bmatrix} \end{array}$$

Since the elements of this matrix are all positive, it is a regular transition matrix; and it will have a fixed point: the long-range prediction. For the results at the end of the first year we have

$$[.33,.33,.33] \begin{bmatrix} .75 & .15 & .10 \\ .30 & .60 & .10 \\ .05 & .15 & .80 \end{bmatrix} = [.37,.30,.33]$$

The vector [.37, 30, .33] gives each company's share at the end of the first year. To get the market shares for the end of the second year, we multiply this vector by the transition matrix again. This product is [.382, .285, .333]. To find the long-term prospects—the fixed vector—we set up a set of equations of the form of Equation 8.2.4. We have

$$\begin{aligned} f_A + \qquad f_B + \quad f_C &= 1 \\ (.75-1)f_A + \qquad .30f_B + .05f_C &= 0 \\ .15f_A + (.60-1)f_B + .15f_C &= 0 \end{aligned}$$

which can be simplified to

$$\begin{aligned} f_A + \quad f_B + \quad f_C &= 1 \\ -.25f_A + .30f_B + .05f_C &= 0 \\ .15f_A - .40f_B + .15f_C &= 0 \end{aligned}$$

(Note that in setting up these equations the coefficients of the fixed vector components are obtained from the **columns** of the transition matrix, not its rows.)

The methods of Chapter 4 yield the solution (the decimals are approximate values, rounded off in the third place).

$$[f_A, f_B, f_C] = [.394, .273, .333]$$

The student should verify this result either by solving the system of equations or by multiplying this vector times the transition matrix, or both.

This result means that the long-range prospects are for Company A to have 39.4% of the market; Company B, 27.3%; and Company C, 33.3%.

Long-term and short-term predictions such as those obtained in the example above are used to determine company policy on new products and sales campaigns. Note that we cannot reasonably expect the same transition probabilities to continue. However, the assumption that they will continue can be used to obtain a projection of where "present trends" are heading.

EXERCISE 8.2

1. A Markov process has the following transition matrix:

$$\begin{array}{cc} & \begin{array}{cc} a_1 & a_2 \end{array} \\ \begin{array}{c} a_1 \\ a_2 \end{array} & \begin{bmatrix} .7 & .3 \\ .4 & .6 \end{bmatrix} \end{array}$$

The system has equal likelihood of being in states a_1 and a_2 at the start.

a. Find the probability vector after one transition.
b. Find the probability vector after two transitions.
c. Find the fixed vector for this process.

2. A Markov process has the following transition matrix:

$$\begin{array}{cc} & \begin{array}{ccc} a_1 & a_2 & a_3 \end{array} \\ \begin{array}{c} a_1 \\ a_2 \\ a_3 \end{array} & \begin{bmatrix} \frac{1}{2} & \frac{1}{2} & 0 \\ 0 & \frac{1}{2} & \frac{1}{2} \\ \frac{1}{2} & 0 & \frac{1}{2} \end{bmatrix} \end{array}$$

a. If the process starts in state a_1, what is the probability vector after two transitions?
b. What is the fixed vector for this system?

3. Use the information given in the example at the start of Section 8.1 relating to buying habits for a product.

a. If a person buys A's product one week, what is the probability that he buys B's product the next week? The week following that?
b. What is the long-term projection for the percentage of people buying A's product and those buying B's?

4. The voting district described in Problem 6 of Exercise 8.1 is carried by the Democrats in a presidential election.

a. What are the probabilities for the outcomes of the next two elections?
b. What are the long-term prospects for this district: Will it swing over to the Republicans if the voting patterns remain the same?

5. If the demographic patterns of Problem 7 of Exercise 8.1 continue for a long time, what percent of the people will live in the city? The suburbs? the country?

6. Using the probabilities for political leanings of Problem 8 of Exercise 8.1, if a man is a political conservative, what is the probability that his grandson will be a liberal?

7. With the customer intentions given in Problem 9 of Exercise 8.1, what is the expected share of the market for each of these companies for the next year assuming each has an equal share of the market this year? What is the long-term projection for the market share of each?

8. Use the occupational pattern described in Problem 10 of Exercise 8.1 and find the probability that the grandson of a nonprofessional will be a professional. If this pattern continues, what will the eventual distribution of occupations be?

9. Assume the schedule for checking income tax returns given in Problem 11 of Exercise 8.1 continues for an indefinite period.
   a. What percentage of the tax returns does the IRS expect to give quick checks and what percent thorough checks each year?
   b. If a person's tax return is given a thorough check, what is the probability his return will receive a thorough check at least once in the next three years?

10. In the market situation described in Problem 12 of Exercise 8.1, Company *A* put on an advertising blitz the first year that the product was sold and captured 70% of the market. Companies *B* and *C* shared the remainder equally.
   a. The performance of *A*'s product was poor and their market share decreased. Assuming the buying pattern described continues, how many years does it take for *A*'s share of the market to fall below 1/3?
   b. Company *B* put on an aggressive campaign of service and advertising with the goal of outselling its two competitors—selling more than either of them. With the buying pattern described, how many years will it take to accomplish this?
   c. Company *B* wants eventually to outsell its two competitors combined—have more than one-half the market. With the specified trends will they achieve this goal? What is their eventual share of the market?

8.3 ABSORBING MARKOV CHAINS

Some of the systems described by Markov chains have one or more states that are "permanent"—once the system reaches such a state it cannot leave it. A Markov chain with this property is called an **absorbing Markov chain.**

A simple example of an absorbing Markov chain is the game of matching pennies. In one version of this game the two players each flip a coin; if the exposed sides of the coins are both the same, one player takes both coins; if the exposed sides are different, the other player takes the coins. The game usually continues until one player or the other has all the pennies. If we consider the game from the viewpoint of one of the players and designate the number of coins that he has as a "state," the process can be regarded as a Markov chain. Since the game ends when the player has either none or all of the coins—no further state changes are possible—it is an absorbing Markov chain.

Let us consider a game in which each player starts with two pennies (to keep the matrix of reasonable size, not for financial reasons). It is not necessary for each player to start with the same number of coins. However, as we shall see, the relative number of coins that each possesses at the start affects the probabilities for who wins all the money. In this example there are five states.

	States (Number of coins	Player A (after flip) 0	1	2	3	4
Player A (before flip)	0	1	0	0	0	0
	1	$\frac{1}{2}$	0	$\frac{1}{2}$	0	0
	2	0	$\frac{1}{2}$	0	$\frac{1}{2}$	0
	3	0	0	$\frac{1}{2}$	0	$\frac{1}{2}$
	4	0	0	0	0	1

The entries show that when A has any number of coins other than zero or four, there is a probability of one-half that he will have one less; and a probability of one-half that he will have one more after the flip. No other transitions are possible. (A process of this type in which all of the transitions are into "adjacent" states is often called a **random walk**. It has many applications in various fields.)

Notice that the transition matrix of an absorbing Markov chain is characterized by having at least one row in which the element of the principal diagonal of the matrix is one, and all the other entries are zero. **Such rows represent absorbing states**. Note also that **the one must be on the diagonal**—the state after the transition is the same as before. There can be rows with 1's in other columns. These represent mandatory state changes, but not absorbing states.

Comment: In an absorbing chain with a finite number of states, the probability of eventual absorption is one. That is, the system is certain to reach an absorbing state sooner or later.

The questions of interest in an absorbing chain are

a. On the **average**, how many times will the system be in each nonabsorbing state?

b. On the **average**, how long will it take (how many transitions) for the system to reach an absorbing state?

c. What is the probability the system will end up in a given absorbing state, assuming there is more than one?

To illustrate how these questions can be answered, let us consider our example above. First, we rearrange the matrix so that the absorbing states come first. For generality, let us call the state that A has zero pennies, a_1; A has one penny, a_2; and so on. The transition matrix becomes

$$T = \begin{array}{c} \\ a_1 \\ a_5 \\ a_2 \\ a_3 \\ a_4 \end{array} \begin{array}{c} \begin{array}{ccccc} a_1 & a_5 & a_2 & a_3 & a_4 \end{array} \\ \begin{bmatrix} 1 & 0 & 0 & 0 & 0 \\ 0 & 1 & 0 & 0 & 0 \\ \frac{1}{2} & 0 & 0 & \frac{1}{2} & 0 \\ 0 & 0 & \frac{1}{2} & 0 & \frac{1}{2} \\ 0 & \frac{1}{2} & 0 & \frac{1}{2} & 0 \end{bmatrix} \end{array}$$

The order of the rows can be changed whenever convenient in a transition matrix. It must be remembered, however, that there must also be a corresponding change in the order of the columns. **The states represented by the rows and columns must be in the same order.** This means, of course, that the entries in each row must be changed when the order of the rows is changed.

We shall now perform various operations with the transition matrix without offering proof that the results we obtain are valid. The proofs require concepts in matrix theory that are of much less interest here than are the applications that we can make of the results.

First, we partition the transition matrix into four parts according to the following pattern:

$$T = \begin{array}{c} \\ \text{Absorbing states} \\ \text{Nonabsorbing states} \end{array} \begin{array}{c|c} \text{Absorbing states} & \text{Nonabsorbing states} \\ I & O \\ \hline R & Q \end{array}$$

The matrix, I, in the upper-left corner is an identity matrix. In this example it has two rows and two columns. The matrix, O, in the upper right corner always contains all zeros. In this example it has two rows and three columns. The matrix, R, in the lower left has its rows and columns apportioned by the particular system—the number of rows equals the number of nonabsorbing states; and the number of columns, the number of absorbing states. In our example there are three rows and two columns. The matrix, Q, in the lower-right corner always is a square matrix; and the number of rows and columns equals the number of nonabsorbing states. In our example there are three rows and three columns. After partitioning, the example matrix becomes:

$$T = \begin{array}{c} I \\ \\ \\ \\ \\ R \end{array} \left[\begin{array}{cc|ccc} 1 & 0 & 0 & 0 & 0 \\ 0 & 1 & 0 & 0 & 0 \\ \hline \frac{1}{2} & 0 & 0 & \frac{1}{2} & 0 \\ 0 & 0 & \frac{1}{2} & 0 & \frac{1}{2} \\ 0 & \frac{1}{2} & 0 & \frac{1}{2} & 0 \end{array}\right] \begin{array}{c} O \\ \\ \\ \\ \\ Q \end{array}$$

It can be shown that after a sequence of n transitions, the matrix will be of the form

$$T^n = \begin{array}{c|c} I & O \\ \hline * & Q^n \end{array}$$

Note that the identity matrix remains unchanged, and that the matrix Q is multiplied only by itself—the rest of the matrix has no effect on this section.

We will not need the matrix in the section indicated by the $*$. **The elements of Q^n give the probability for being in each of the nonabsorbing states after n steps, each row giving the probabilities for the case in which the sequence started in the state that it represents.** In our example

$$Q^2 = \begin{bmatrix} \frac{1}{4} & 0 & \frac{1}{4} \\ 0 & \frac{1}{2} & 0 \\ \frac{1}{4} & 0 & \frac{1}{4} \end{bmatrix} \qquad Q^3 = \begin{bmatrix} 0 & \frac{1}{4} & 0 \\ \frac{1}{4} & 0 & \frac{1}{4} \\ 0 & \frac{1}{4} & 0 \end{bmatrix}$$

$$Q^4 = \begin{bmatrix} \frac{1}{8} & 0 & \frac{1}{8} \\ 0 & \frac{1}{4} & 0 \\ \frac{1}{8} & 0 & \frac{1}{8} \end{bmatrix}$$

Thus, the probability of being in state a_4 after four steps is 1/8 if the system started in state a_2; this probability is 0, if it started in state a_3; and 1/8 if it started in state a_4, and so on. Note that the probabilities are getting smaller as the number of steps increases. This reflects the increasing probability of reaching one of the absorbing states as the number of transitions increases.

Let us return now to section Q of the original matrix T. We form a new matrix, $I - Q$, where I is the identity matrix with the appropriate number of rows and columns. In our example, this matrix is

$$I - Q = \begin{bmatrix} 1 & 0 & 0 \\ 0 & 1 & 0 \\ 0 & 0 & 1 \end{bmatrix} - \begin{bmatrix} 0 & \frac{1}{2} & 0 \\ \frac{1}{2} & 0 & \frac{1}{2} \\ 0 & \frac{1}{2} & 0 \end{bmatrix} = \begin{bmatrix} 1 & -\frac{1}{2} & 0 \\ -\frac{1}{2} & 1 & -\frac{1}{2} \\ 0 & -\frac{1}{2} & 1 \end{bmatrix}$$

It can be shown from matrix theory that this matrix will always have an inverse. This inverse matrix, $(I - Q)^{-1}$, is called the **fundamental matrix** of the absorbing chain. In our example

$$(I - Q)^{-1} = \begin{bmatrix} \frac{3}{2} & 1 & \frac{1}{2} \\ 1 & 2 & 1 \\ \frac{1}{2} & 1 & \frac{3}{2} \end{bmatrix}$$

The elements of this matrix, $(I - Q)^{-1}$, give the answer to question (a) above. That is, recalling that the rows refer to the starting state, the columns to the new states, and that we rearranged rows and columns with the result that the first row and columns of $(I - Q)^{-1}$ refer to state a_2; the second, to state a_3; and the third, to state a_4, the matrix shows that if we start in state a_3 (the middle row), the system can be **expected** to be in state a_2 once, state a_3 twice, and a_4 once before absorption. These are average numbers, and any given sequence may not follow this expectation exactly.

Question (b) above is answered by taking the **sum of the entries in a given row**. That is, if we start in state a_2, the average number of steps before absorption will be $3/2 + 1 + 1/2 = 3$. If we start in state a_3, we can expect to take four steps before absorption, and so on.

Question (c) is answered by finding the product $(I - Q)^{-1}R$. In our example we have

$$(I - Q)^{-1}R = \begin{bmatrix} \frac{3}{2} & 1 & \frac{1}{2} \\ 1 & 2 & 1 \\ \frac{1}{2} & 1 & \frac{3}{2} \end{bmatrix} \begin{bmatrix} \frac{1}{2} & 0 \\ 0 & 0 \\ 0 & \frac{1}{2} \end{bmatrix} = \begin{matrix} \\ a_2 \\ a_3 \\ a_4 \end{matrix} \begin{matrix} \begin{matrix} a_1 & a_5 \end{matrix} \\ \begin{bmatrix} \frac{3}{4} & \frac{1}{4} \\ \frac{1}{2} & \frac{1}{2} \\ \frac{1}{4} & \frac{3}{4} \end{bmatrix} \end{matrix}$$

That is, if we start in state a_2, the probability of being absorbed into state a_1 is 3/4; and into state a_5, 1/4, and so on. If we recall that state a_2 means that Player A has one penny, and Player B three pennies; and that state a_1 means that Player A has zero pennies, and state a_5 means that A has four pennies, we see that A's chance of winning all the pennies is proportional to the fraction of the total pennies in the game that he starts with. (Moral: Do not match pennies with a person with a handful of pennies, unless you have a larger handful.)

The process that we have illustrated here with a simple example can be generalized into any number of absorbing and nonabsorbing states using the procedure described above for partitioning the matrix and finding the fundamental matrix.

To summarize, the answers to the questions (a), (b), and (c) above about absorbing Markov chains can all be obtained using the fundamental matrix for the process, $(I - Q)^{-1}$. The entries in this matrix are the expected numbers of times the system is in each state before absorption, depending on the starting state. The starting state determines the row; the state passed through, the column.

The **expected**, or **average**, number of steps before absorption is given by the sum of the entries in the row representing the starting state.

The matrix $(I - Q)^{-1}R$ gives the probability of absorption into each absorbing state, depending on the starting state.

Let us consider a somewhat more complex example.

Example 8.3.1

Large corporations study their executive personnel policies to make certain that they will provide a balanced supply of junior, middle-level, and senior executives without excessive hiring and training of new personnel. Among the types of studies made is a study of transfers and promotions. These can be considered as an absorbing Markov process.

In a hypothetical situation let us consider the executives as divided into three principal groups: junior (J), middle-level (M), and senior (S). These groupings can be considered as states for the executive personnel. There are two absorbing states: leaving the company through resignation or dismissal (L), and retirement (R).

Let us suppose that company records show that each year 80% of the junior executives remain junior executives for the next year; 10% leave the company; 2% retire (usually for disability); 8% are promoted to middle-level positions; and none become senior executives. For the middle-level executives, 80% remain at the same level; 5% leave the company; 8% retire; 2% are reassigned to junior executive positions; and 5% are promoted to senior positions. In the senior executive positions, 5% leave the company; 20% retire; 3% are reassigned to middle-level positions; 72% remain at the same level; and none are reassigned to junior positions.

The percentages expressed in their equivalent fraction form can be considered as the probabilities for an executive in a position at any of the levels. We have, therefore, the following transition matrix:

$$\begin{array}{c} \\ L \\ R \\ \\ J \\ M \\ S \end{array} \begin{array}{c} \begin{array}{ccccc} L & R & J & M & S \end{array} \\ \left[\begin{array}{cc|ccc} 1 & 0 & 0 & 0 & 0 \\ 0 & 1 & 0 & 0 & 0 \\ \hline .1 & .02 & .8 & .08 & 0 \\ .05 & .08 & .02 & .8 & .05 \\ .05 & .2 & 0 & .03 & .72 \end{array}\right] \end{array}$$

The matrix, Q, which gives the transition probabilities among the nonabsorbing states, is

$$Q = \begin{array}{c} \\ J \\ M \\ S \end{array} \begin{array}{c} \begin{array}{ccc} J & M & S \end{array} \\ \begin{bmatrix} .8 & .08 & 0 \\ .02 & .8 & .05 \\ 0 & .03 & .72 \end{bmatrix} \end{array}$$

We can raise this matrix to the fifth power to see the prospects for executives at each level after five years, assuming they have not been "absorbed" either by leaving the company, or by retirement. Matrix multiplication, as learned in Chapter 4, yields

$$Q^5 = \begin{array}{c} \\ J \\ M \\ S \end{array} \begin{array}{c} \begin{array}{ccc} J & M & S \end{array} \\ \begin{bmatrix} .33 & .17 & .02 \\ .04 & .36 & .09 \\ 0 & .05 & .22 \end{bmatrix} \end{array}$$

An interpretation of this matrix shows that a junior executive (row 1) has a probability of approximately 1/3 of remaining as a junior executive after five years. His probability of being promoted to a middle-level position in that period is slightly larger than 1/6. He has a probability of 1/50 of making it to a senior position in that period of time. Similar evaluations can be made for executives at the other two levels.

We find the fundamental matrix, $(I - Q)^{-1}$, by subtracting Q from the three-by-three identity matrix and computing its inverse, again using the methods of Chapter 4. We find

$$(I - Q)^{-1} = \begin{bmatrix} 5.32 & 2.14 & .37 \\ .54 & 5.36 & .96 \\ .06 & .58 & 3.68 \end{bmatrix}$$

Adding the entries in the first row of this matrix shows that **on the average** a junior executive works for the company for 7.83 years before either leaving or being retired. The average tenure of a middle-level executive is 6.86 years; and for a senior executive, 4.32 years.

If we multiply $(I - Q)^{-1}$ times the matrix, R, we find the probabilities for "absorption" for the different levels. We have

$$(I-Q)^{-1}R = \begin{matrix} J \\ M \\ S \end{matrix} \begin{bmatrix} 5.32 & 2.14 & .37 \\ .54 & 5.36 & .96 \\ .06 & .58 & 3.68 \end{bmatrix} \overset{\begin{matrix} L & \quad R \end{matrix}}{\begin{bmatrix} .1 & .02 \\ .05 & .08 \\ .05 & .2 \end{bmatrix}}$$

$$= \begin{matrix} J \\ M \\ S \end{matrix} \overset{\begin{matrix} L & \quad R \end{matrix}}{\begin{bmatrix} .65 & .35 \\ .37 & .63 \\ .22 & .78 \end{bmatrix}}$$

This matrix shows that 65% of the junior executives leave the company, and 35% remain until retirement. As might be expected, the probabilities for the middle-level and senior executives show that substantially more are retired, and fewer leave for other reasons. Sixty-three percent of the middle-level and 78% of the senior executives remain with the company until retirement.

These matrices provide information about the personnel situation in a hypothetical company. If they represented an actual case, it could be a cause for concern, for example, that a junior executive remains with the company an average of less than eight years.

The foregoing example is only one of many ways in which the concept of an absorbing Markov chain can be applied. When the matrices are properly interpreted, they yield a great deal of information with relatively little effort.

EXERCISE 8.3

1. State whether the following transition matrices are for absorbing or nonabsorbing Markov chains:

a. $\begin{bmatrix} 1 & 0 \\ .75 & .25 \end{bmatrix}$ b. $\begin{bmatrix} .75 & .25 \\ 1 & 0 \end{bmatrix}$

c. $\begin{bmatrix} 0 & 1 \\ .75 & .25 \end{bmatrix}$ d. $\begin{bmatrix} .75 & .25 \\ 0 & 1 \end{bmatrix}$

e. $\begin{bmatrix} .5 & .5 & 0 \\ 0 & 1 & 0 \\ 0 & .5 & .5 \end{bmatrix}$ f. $\begin{bmatrix} 0 & 1 & 0 \\ .5 & .5 & 0 \\ 0 & .5 & .5 \end{bmatrix}$

g. $\begin{bmatrix} \frac{1}{3} & \frac{1}{3} & \frac{1}{3} & 0 \\ 0 & 1 & 0 & 0 \\ 0 & 0 & 1 & 0 \\ 0 & \frac{1}{4} & \frac{1}{4} & \frac{1}{2} \end{bmatrix}$ h. $\begin{bmatrix} \frac{1}{3} & \frac{1}{3} & \frac{1}{3} & 0 \\ 0 & 0 & 1 & 0 \\ 0 & 1 & 0 & 0 \\ 0 & \frac{1}{4} & \frac{1}{4} & \frac{1}{2} \end{bmatrix}$

2. Find R, Q, $(I-Q)^{-1}$, and $(I-Q)^{-1}R$ for the absorbing matrices of Problem 1.

3. An object moves back and forth along a line that is numbered 1, 2, 3, 4, 5. It has a probability of 1/3 of moving one unit to the right and a probability of 2/3 of moving one unit to the left. Let the states of this Markov chain be 1, 2, 3, 4, 5, with 1 and 5 being absorbing states. Find the transition matrix. Find, R, Q, $(I - Q)^{-1}$, and $(I - Q)^{-1}R$. Use these results to find the mean number of times the process is in state three before absorption. What is the probability that if the process starts in state three, it will be absorbed in state five?

4. Consider the penny-matching example in which each participant starts with two pennies: stage a_3. The matrix Q^3 shows that after three flips there is a probability of 1/4 of being in state a_2 and a probability 1/4 of being in state a_4. The probability of all the possibilities must add to one. What are the other possibilities? Compute their probabilities.

5. Suppose the company of Example 8.3.1 instituted changes in its salary structure and promotion practices so that only 6% of the junior executives leave the company; 2% retire; 80% remain junior executives; but 12% are promoted to middle-level positions. In the middle-level, 5% leave; 8% retire; 2% are reassigned to junior positions; but 75% remain at the middle-level, and 10% are promoted to senior positions. At the senior level, the number leaving the company is reduced to 2%; 20% retire; 3% are reassigned to middle-level positions; but 75% continue at the same level.
   a. Form the transition matrix for the new situation.
   b. Find the five-year prospects for a junior executive by finding Q^5. **Hint:** $Q^5 = (Q^2)^2\,Q$.
   c. Find the fundamental matrix, $(I - Q)^{-1}$.
   d. Find the percentage leaving the company and retiring for each level of position.

6. Three tanks fight a three-way duel. Tank A has probability 1/2 of destroying the tank it fires at. Tank B has probability 1/3 of destroying its target tank, and Tank C has probability 1/6 of destroying its target tank. The tanks fire at the same time, and each tank fires at the strongest opponent not yet destroyed.
   a. Consider all the possible results of the shots and set up the eight states of the system in terms of the tanks that survive any round of firing.
   b. Compute each of the transitional probabilities from one state to another.
   c. Assume that the firing continues until no targets remain with each tank firing simultaneously with the other(s). Consider the battle as a Markov chain process and form the transition matrix.
   d. Find $(I - Q)^{-1}$, $(I - Q)^{-1}R$, and the expected number of steps to absorption. Interpret the results: that is, what are the possible outcomes of the battle and their probabilities?

8.4
RANDOM WALK

In the preceding section we discussed briefly the game of matching pennies and mentioned that it was an example of a process called a **random walk.** A random walk is a Markov chain process in which transitions can occur only between "adjacent" states. In the process we shall consider—a one-dimension random walk—we can visualize the states of the system as arranged in a line. The transitions occur as the result of a "step" in either direction from a given state—thus, the name "random walk." The probability of taking a step in one "direction" is the same no matter which state the system is in. A diagram for a six-state process is shown below:

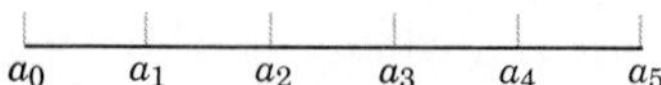

We assume the probability of moving one step to the right is p. That is, the probability is p for a transition from a_2 to a_3, a_3 to a_4, and so on. The probability for moving one step to the left is q, where $q = 1 - p$.

The problem arises of what happens when the system moves to one of the states at the end of the line. When the system is in state a_0, it cannot move to the "left"; and when in state a_5, it cannot move to the "right." There are two types of random walk: reflecting and absorbing. In a "reflecting" walk the transition from state a_0 must be to a_1; and from a_5, it must be to a_4. In an "absorbing" walk the end states are absorbing states, and no further transitions are possible. We shall consider only the absorbing type.

The transition matrix for a six-state absorbing random walk is shown below:

$$\begin{array}{c} \\ a_0 \\ a_1 \\ a_2 \\ a_3 \\ a_4 \\ a_5 \end{array} \begin{array}{c} \begin{array}{cccccc} a_0 & a_1 & a_2 & a_3 & a_4 & a_5 \end{array} \\ \begin{bmatrix} 1 & 0 & 0 & 0 & 0 & 0 \\ q & 0 & p & 0 & 0 & 0 \\ 0 & q & 0 & p & 0 & 0 \\ 0 & 0 & q & 0 & p & 0 \\ 0 & 0 & 0 & q & 0 & p \\ 0 & 0 & 0 & 0 & 0 & 1 \end{bmatrix} \end{array}$$

In most applications of an absorbing random walk the interest focuses on which of the two absorbing states the system will reach first for various starting states and values for p and q (the transition probabilities). In a system with many "steps" the transition matrix becomes quite large, but most of the entries are zeros. It is actually simpler to work with the absorption probabilities directly than to work with the fundamental matrix for the system. If we can find the probability of reaching one of the absorbing states, we can obtain the probability of reaching the other by subtracting from 1. Since the system must reach one or the other absorbing state, the sum of the probabilities for each must be one. We start with the probability of reaching the state, a_0, and call this the "zero-state" probability. We expect this to be different for different starting states, so we designate the zero-state probability if we start in state a_i as x_i. Thus, x_2 represents the zero-state probability if the system starts in state a_2, and so on.

In general, there are N states other than the zero state, including a_N, the other absorbing state. Let us first consider a relatively simple situation in which $N = 5$. We shall develop the general relationships by extension of this case.

We have a total of six zero-state probabilities when $N = 5$: x_0, x_1, x_2, x_3, x_4, and x_5. x_0 is the zero-state probability if the system is in state a_0. Since absorption into state a_0 is certain in this case, $x_0 = 1$. Similarly, x_5 is the probability of reaching state a_0 if the system is in state a_5. Since a_5 is an absorbing state, x_5 must be zero. Notice that once a system reaches one of the non-absorbing states, its future prospects are exactly the same as though the system had started in that state. That is, the zero-state probability for any state is going to be the same regardless of whether the system starts in that state, or reaches it after a series of transitions. This key consideration says that the zero-state probability for state a_2 is the sum of the zero-state probability for state a_1 times the probability of a transition from state a_2 to a_1 plus the zero-state probability of state a_3 times the probability of a transition from a_2 to a_3. (Recall that the only transitions possible from state a_2 are to a_1 or a_3.) Using p as the probability of a "step to the right" (a_2 to a_3) and q as the probability of a "step to the left" (a_2 to a_1), we have in symbols,

$$x_2 = qx_1 + px_3$$

In writing this equation we have not evaluated x_2, but merely have expressed it in terms of x_1 and x_3. We now continue our algebraic manipulations in the hope of obtaining expressions in terms of x_0 and x_5, whose values we know.

We note that $p + q = 1$. We can, therefore, multiply x_2 in the above equation by the expression, $(p + q)$, without changing its value. Thus,

$$(p + q)x_2 = qx_1 + px_3$$

We can rearrange this equation in a sequence of algebraic operations, as follows:

$$px_2 + qx_2 = qx_1 + px_3 \qquad \text{Expand left side of equation}$$

$$px_2 - px_3 = qx_1 - qx_2$$

$$p(x_2 - x_3) = q(x_1 - x_2) \qquad \text{Factor each side}$$

$$\frac{p}{q}(x_2 - x_3) = x_1 - x_2 \qquad \text{Divide each side by } q$$

To simplify the equation, we replace the quantity p/q with r, which is the ratio of the probability of moving a step right to the probability of moving a step left. The following derivation is valid for any value of r, except $r = 1$: that is, except for a system in which the probability of moving right and left is equal. The equation we are considering becomes, after reversing the order of the equation,

$$x_1 - x_2 = r(x_2 - x_3) \qquad (8.4.1)$$

By the same process used to obtain Equation 8.4.1 we obtain the following:

$$x_0 - x_1 = r(x_1 - x_2) \quad (8.4.2A)$$

$$x_1 - x_2 = r(x_2 - x_3) \quad (8.4.2B)$$

$$x_2 - x_3 = r(x_3 - x_4) \quad (8.4.2C)$$

$$x_3 - x_4 = r(x_4 - x_5) \quad (8.4.2D)$$

Recall that $x_5 = 0$. Thus, Equation 8.4.2D can be simplified to

$$x_3 - x_4 = rx_4$$

We use this value for $(x_3 - x_4)$ in equation 8.4.2C:

$$x_2 - x_3 = r \cdot rx_4 = r^2x_4$$

Similar substitutions in each of the other of Equations 8.4.2 yields the following set of equations:

$$\begin{aligned} x_4 &= x_4 \\ x_3 - x_4 &= rx_4 \\ x_2 - x_3 &= r^2x_4 \\ x_1 - x_2 &= r^3x_4 \\ x_0 - x_1 &= r^4x_4 \end{aligned} \quad (8.4.3)$$

We can add all of the "left-sides" of these equations together, and the sum is equal to the sum of all the "right-sides." Thus,

$$\begin{aligned} &x_4 + (x_3 - x_4) + (x_2 - x_3) + (x_1 - x_2) + (x_0 - x_1) \\ &\quad = x_4 + rx_4 + r^2x_4 + r^3x_4 + r^4x_4 \end{aligned}$$

Note that on the left side of this equation all of the terms cancel except x_0. Since $x_0 = 1$, we have

$$1 = x_4 + rx_4 + r^2x_4 + r^3x_4 + r^4x_4$$

or

$$1 = (1 + r + r^2 + r^3 + r^4)x_4 \quad (8.4.4)$$

It is known from algebra that

$$1 - r^5 = (1 - r)(1 + r + r^2 + r^3 + r^4)$$

(The student should perform the multiplication indicated on the right side of this equation to verify the relationship.) This equation can be rearranged into the form,

$$1 + r + r^2 + r^3 + r^4 = \frac{1 - r^5}{1 - r}$$

If this expression is substituted into Equation 8.4.4, we have

$$1 = \frac{1 - r^5}{1 - r} x^4$$

which can be "solved" for x_4:

$$x_4 = \frac{1 - r}{1 - r^5} \tag{8.4.5}$$

We return now to the set of Equations 8.4.3. Note that adding the first two of these yields

$$x_4 + (x_3 - x_4) = x_4 + rx_4$$

or

$$x_3 = x_4(1 + r)$$

Replacing x_4 by its equal from Equation 8.4.5, we obtain

$$\begin{aligned} x_3 &= \frac{1 - r}{1 - r^5}(1 + r) \\ &= \frac{1 - r^2}{1 - r^5} \end{aligned} \tag{8.4.6}$$

If we add the first three of Equations 8.4.3, we have

$$x_4 + (x_3 - x_4) + (x_2 - x_3) = x_4 + rx_4 + r^2x_4$$

from which we can obtain

$$x_2 = x_4(1 + r + r^2)$$

Substituting for x_4 as before

$$x_2 = \frac{1 - r}{1 - r^5}(1 + r + r^2)$$

$$= \frac{1 - r^3}{1 - r^5} \tag{8.4.7}$$

By a similar process,

$$x_1 = \frac{1 - r^4}{1 - r^5} \tag{8.4.8}$$

We now have a value for each of the zero-state probabilities in terms of the ratio, r, of the probability of moving right (**away** from the zero state) to the probability of moving left (**toward** the zero state).

These results were obtained using $N = 5$. From the form of the equations we can infer the following: We note the exponent of r in the denominator of each equation is 5; that is, N. If i is the number of the starting state (not counting zero),

$$x_i = \frac{1 - r^{(N-i)}}{1 - r^N} \tag{8.4.9}$$

Equation 8.4.9 can be considered as a general result in which x_i is the probability of being absorbed into one of the absorbing states (i.e., the system reaches that state before it reaches the other absorbing state). i is the number of states the system is away from the absorbing state (start counting with the absorbing state as zero). r is the ratio of the probability of moving away from the absorbing state to the probability of moving toward the absorbing state.

Note that Equation 8.4.9 is meaningless when $r = 1$ – equal probability of moving toward or away from the absorbing state. When $r = 1$ Equation 8.4.2D can be simplified to $x_4 = x_3/2$. By successive substitutions in Equations 8.4.2C, 8.4.2B, and 8.4.2A, it is found that $x_1 = 4/5$. Further substitution yields the results that $x_2 = 3/5$, $x_3 = 2/5$, and $x_4 = 1/5$. From these results, recalling that $N = 5$, it can be inferred that when $r = 1$,

$$x_i = \frac{N - i}{N} \tag{8.4.10}$$

8.4.1 Gambler's Ruin

An interesting application of a random walk is an analysis of the probabilities relating to gambling in a casino. We consider a situation in which a person plays one game repeatedly until either he loses all his money (is "ruined") or he wins a predetermined amount. (The resources of the usual casino are so great that "breaking the bank" is, as we shall see, virtually impossible.)

To simplify the computations we shall consider only those games with "even-money" payoffs—the amount that can be won is the same as the amount bet. We shall further simplify by considering only bets of $1 each time. Thus, each "play" is a transition, or step, in a random walk. The amount of money the player has at any time in the process is the state. This increases or decreases by one for each "play." The "i" in the random walk formula is the amount the player starts with. $N - i$, which we shall call B, is the predetermined amount to be won. Thus, $N = i + B$, is the total amount of money involved. As soon as the player has either N dollars total, or no money left, the game ends. The zero-state probability, x_i, is the probability the player loses all his money—his "ruin probability." Using these symbols, we see that the ruin probability is given by the formula

$$x_i = \frac{1 - r^B}{1 - r^N} \tag{8.4.11}$$

In all the games in a casino the expected value of each play is a small negative number—the "average" result of each play is that the player loses a small portion of his bet. In the even-money games this means that the probability the player wins, p, is slightly less than one-half. However, playing a game repeatedly makes the probability the player will lose his "stake" very nearly one. Table 8.4.1 shows a few typical values for the ruin probability, x_i. Here i is the amount the player starts with. B $(= N - i)$ is the amount he wishes to win. p is the probability of winning a single play of the game. The values shown for p are slightly smaller than .5 – the games represented are very nearly, but not quite, "fair." Some of the values shown may startle you. For example, in "American" roulette the probability of winning if you bet "red," "black," "odd," or "even" is approximately .475 (actually 18/38). If we look at the table at the value for p equal to .475 – the second section of the table—we get a look at your chances. Suppose you start with $20 and are satisfied to win $5 – you quit as soon as your total holdings reach $25. Under

Table 8.4.1
Ruin-Probabilities for $p = .45, .475, .49, .495$

$p = .45$ $r = .818$

		B (N − i)				
		1	5	10	20	50
i	1	.550	.905	.973	.997	1
	5	.260	.732	.910	.988	1
	10	.204	.666	.881	.984	1
	20	.185	.638	.868	.982	1
	50	.182	.633	.866	.982	1

$p = .475$ $r = .905$

		B (N − i)				
		1	5	10	20	50
i	1	.525	.871	.948	.986	.999
	5	.211	.622	.813	.942	.998
	10	.142	.506	.731	.911	.996
	20	.108	.428	.665	.881	.994
	50	.096	.395	.634	.866	.994

$p = .49$ $r = .961$

		B (N − i)				
		1	5	10	20	50
i	1	.510	.850	.926	.969	.994
	5	.184	.550	.731	.871	.972
	10	.110	.402	.599	.788	.951
	20	.069	.287	.472	.690	.921
	50	.045	.204	.363	.586	.881

$p = .495$ $r = .980$

		B (N − i)				
		1	5	10	20	50
i	1	.505	.842	.918	.961	.989
	5	.175	.525	.699	.838	.948
	10	.100	.367	.550	.731	.905
	20	.058	.242	.402	.599	.839
	50	.031	.143	.259	.438	.731

i at the left the fourth line shows the values for starting with \$20. The second column (under 5, for B) gives the probability of going broke before you win the \$5: .428. This means that your probability of winning the \$5 is 1 − .428, or .572. Note, however, how quickly the ruin probability increases as you increase the amount you wish to win. If you want to "double your money"—win \$20 — the probability of **losing** your \$20 becomes .881. That is, the probability that you will succeed is only .119, less than one chance in eight.

Let us examine the ruin probability formula more closely (Equation 8.4.11 above). We consider the case when r is less than one: when your probability of winning an individual game is less than the probability of losing. The denominator in the formula, $1 - r^N$, is always less than one, and so the probability of ruin is always greater than the numerator, $1 - r^B$. This estimate does not depend on how much you start with, only on p and B. Since r is less than one, r^B can be made to be very nearly zero just by making B large enough. Hence, you are virtually certain to be ruined when playing against someone with lots of resources if you agree to play until one or the other goes broke. In this case B is the amount that your opponent has. For example, suppose that your opponent wants the probability of your ruin to be .999 — that is, there is only one chance in a thousand of his losing. To do this he must make sure that r^B is less than .001. If $p = .495$, a very nearly "fair" game, your opponent needs only \$346 to have probability .999 of ruining you (assuming a \$1 bet per game) no matter how much money you have. If $p = .48$, he needs only \$87. Even if $p = .499$, \$1727 is all he needs. (The numbers, 346, 87, and 1727 are the exponents for r that make r^B less than .001).

In our discussion above, we excluded the situation in which $p = .5$, a "fair" game; that is, one in which you have equal probability of winning or losing. The results in this case are given by Equation 8.4.10. Computing your prospects in this situation is left as an exercise.

8.4.2 Expected Duration of Walk

Another consideration in a random walk is the number of steps taken before one of the absorbing states is reached. The development of these formulas is beyond the scope of this text, and we shall present them without proof. For a random walk in which $p \neq q$ and which starts at state i, the **expected** number of steps before absorption is

$$S_i = \frac{i}{q-p} - \frac{N}{q-p} \cdot \frac{r^{(N-i)} - r^N}{1 - r^N} \qquad (8.4.12)$$

In this equation p, q, r, and N have the same meanings as before.

When p and q both equal one-half, the expression giving the expected number of steps required to reach one of the absorbing states is

$$S_i = i(N - i) \qquad (8.4.13)$$

EXERCISE 8.4

1. A system has five states, the lowest and highest of which are absorbing. At any of the three nonabsorbing states the probability of a transition to a higher state is .4; and, to a lower state, .6.
   a. If the system starts at the middle (third) state, what is the probability that it will be absorbed into the lowest state?
   b. If it starts at the fourth state (next to highest), what is the probability that it will reach the highest state?

2. A system has 11 states, a zero state and 10 others, with the zero state and the highest state both absorbing. The probability of a transition to a higher state is .45. If the system starts in the middle state, what is the probability that it will reach the highest state?

3. An urn has 49 red balls and 51 white balls. A game is played in which a ball is drawn, its color noted, and then replaced in the urn. If the ball is red, you win; if it is white, you lose. If you have \$1 and your opponent has 50¢ and you wager 10¢ on each game, what is the probability that you will lose your dollar?

4. One of the red balls is removed from the urn of Problem 3 and replaced with a white ball. With the same distribution of money and the same wagers, what is the probability that you will win your opponent's 50¢?

5. Consider a situation in which exactly 60 customers are served by two competing companies, A and B. Suppose that periodically a customer will change supplier. Fifty-two and one-half percent of the time this switch is from A to B and 47.5% of the time from B to A. A starts with 50 of the customers. What is the probability that A loses all its customers to B?

6. In a game of matching pennies ($p = 1/2$) you start with 25¢ and your opponent has 50¢. What is your probability of winning all the pennies?

7. Four people, A, B, C, and D, each have five pennies. A proposes that he will match pennies with B until one or the other wins all the pennies. Then the winner will match with C until one or the other wins all the pennies; and the winner of that game will match with D until one person has all the pennies. What is the probability of winning for A? B? C? D?

8. Consider the relative merits of "conservatism" and "plunging" when the odds are against you. You are playing a game in which your chance of winning is 1/4. You are down to your last dollar. Your opponent is willing to risk \$3. What is your probability of losing your dollar (the play continues until one or the other has lost all his money) if:
   a. You bet \$1 on each game?
   b. You bet all your money each time?

9. In an isolated region ecological considerations limit the number of a certain species to 2000. A mutant appears among the species. The situation is such that in each generation another mutant displaces one of the "normal" species or vice versa. The probability that a mutant displaces

a "normal" specimen is 2/3, and the probability one of the mutants is displaced is 1/3. Consider the situation from the point of view of the "normal" species trying to eradicate the mutants, starting with 1999 "normal" specimens and one mutant. Compute the probability that the entire species is eventually displaced by the mutants. (Assume $1/2^{2000}=0$.)

10. Consider the situation of Problem 7. What is the total number of "flips" expected before one person wins all the pennies?

8.5
MARKOV PROCESSES IN GENETICS

Geneticists are concerned, among other things, with the transmission of various traits and characteristics by inheritance. One of the simple mechanisms for such transmission is by pairs of genes, each of which may be of two types, say G and g. An individual may have any one of the following combinations: GG, Gg, gG, or gg. Gg and gG are genetically the same. Usually, the individuals with GG and with Gg combinations are virtually the same with respect to the particular characteristics. In such cases the G gene is said to dominate the g gene. An individual is called **dominant** with respect to this characteristic if he has GG genes, **recessive** if he has gg genes, and **hybrid** if he has Gg or gG mixture.

In the mating of two animals, the offspring inherits one of its pair of genes from each of the parents. The basic assumption in genetics is that these genes are selected at random from the parents; that is, there is equal likelihood of one or the other of a parent's pair of genes being transmitted to the offspring. The offspring of two dominant (GG) parents must, therefore, be dominant—there are only G genes available for transmission. The offspring of two recessive parents must, similarly, be recessive. The offspring of one dominant and one recessive parent must be hybrid (Gg). If a dominant and a hybrid mate, the offspring must get a G gene from the dominant parent and has a probability of 1/2 of getting a G or a g gene from the hybrid parent. Thus, there is a probability of 1/2 the offspring will be dominant and 1/2 it will be hybrid. Similarly, in the mating of a recessive and a hybrid, there is a probability of 1/2 of getting a recessive offspring and of 1/2 of getting a hybrid. In the mating of two hybrids, the offspring has a probability of 1/2 for getting a G or a g from each parent. The probabilities are, therefore, 1/4 for GG, 1/2 for Gg, and 1/4 for gg.

Example 8.5.1

Let us consider a process of continued crossings in which we start with an individual of **unknown** genetic make-up and cross it with a **hybrid**. The offspring of this mating is again crossed with a hybrid; and so on. The resulting process is a Markov chain. The states in this process are "dominant," "hybrid," and "recessive." The transition probabilities are

$$P = \begin{array}{c} \\ D \\ H \\ R \end{array} \begin{array}{c} \begin{array}{ccc} D & H & R \end{array} \\ \begin{bmatrix} \frac{1}{2} & \frac{1}{2} & 0 \\ \frac{1}{4} & \frac{1}{2} & \frac{1}{4} \\ 0 & \frac{1}{2} & \frac{1}{2} \end{bmatrix} \end{array}$$

using the probabilities for different pairs of genes from the previous section. The initial states in this process are the possibilities for the genetic make-up of the "unknown" parent. The other parent is assumed to be a hybrid. (Continued mating with a dominant parent will be considered later.)

The matrix, P, represents a **regular** Markov chain: P^2 has only nonzero components. This means, as we have seen earlier, that there will be a fixed probability vector, which represents an equilibrium condition. That is, there will be a fixed probability after many generations for offspring of each genetic type. By solving the equation

$$[x, y, z](P) = [x, y, z]$$

for x, y, and z, we find this fixed probability vector to be $p = [1/4, 1/2, 1/4]$, where the components are, respectively, the probability for the offspring of a mating to be dominant, hybrid, and recessive. The method for solving the equation was discussed in Section 8.2.

The results here mean that no matter what genetic type the original "unknown" parent was, after repeated crossings with a hybrid parent there is a probability of 1/4 that the offspring will be dominant, 1/2 that it will be hybrid, and 1/4 that it will be recessive.

Example 8.5.2

If we keep crossing the offspring with a dominant, the result is quite different. The transition matrix is

$$S = \begin{array}{c} \\ D \\ H \\ R \end{array} \begin{array}{c} \begin{array}{ccc} D & H & R \end{array} \\ \begin{bmatrix} 1 & 0 & 0 \\ \frac{1}{2} & \frac{1}{2} & 0 \\ 0 & 1 & 0 \end{bmatrix} \end{array}$$

This is an absorbing Markov chain with one absorbing state, D.

Using the material on absorbing Markov chains developed in Section 8.3.1, we have

$$Q = \begin{bmatrix} \frac{1}{2} & 0 \\ 1 & 0 \end{bmatrix} \qquad R = \begin{bmatrix} \frac{1}{2} \\ 0 \end{bmatrix}$$

so that

$$I - Q = \begin{bmatrix} \frac{1}{2} & 0 \\ -1 & 1 \end{bmatrix}$$

and

$$(I-Q)^{-1} = \begin{bmatrix} 2 & 0 \\ 2 & 1 \end{bmatrix}$$

The absorption probabilities are

$$(I-Q)^{-1}(R) = \begin{bmatrix} 2 & 0 \\ 2 & 1 \end{bmatrix} \begin{bmatrix} \frac{1}{2} \\ 0 \end{bmatrix} = \begin{bmatrix} 1 \\ 1 \end{bmatrix}$$

This result is not surprising since there is only one absorbing state. The meaning of these results is that if we continue crossing the population with dominants, then after sufficiently many crossings we can expect only dominants as offspring. The **average** number of steps (generations) to absorption are given by the sums of the rows of $(I - Q)^{-1}$: two generations if we start with a hybrid crossed with a dominant, and three generations if we start with a recessive as the first unknown parent.

Example 8.5.3

Let us consider a more complicated example of an absorbing Markov chain. We start with two animals of opposite sex, mate them, select two of their offspring of opposite sex, mate them, and so on. To simplify the example we shall consider that the trait under consideration is independent of sex.

In this process the states are determined by pairs of animals. Hence, the states will be: $a_1 = (D,D)$, $a_2 = (D,H)$, $a_3 = (D,R)$, $a_4 = (H,H)$, $a_5 = (H,R)$, and $a_6 = (R,R)$. Since we have assumed that the trait being considered is independent of sex, the pairs above represent the possible genetic make-ups of the parents with no regard to their individual genetic make-ups. Thus, the pair (D,H) represents the situation in which one parent is dominant and the other hybrid with no regard for which parent is which. Clearly, states a_1 and a_6 are absorbing—if we cross two dominants or two recessives, the offspring must be of the same type as the parents. The remaining transition probabilities are obtained in the following manner: If we start in state a_2, one parent is dominant and the other is hybrid. Thus, one parent has GG genes, and the other has Gg. The probability that an offspring is dominant is 1/2; and hybrid, 1/2. Two offspring are to be selected for the next mating. The probability that they are both dominant (state a_1) is $(1/2)(1/2) = 1/4$. Thus, the probability for a transition from state a_2 to state a_1 is 1/4. The probability for remaining in state a_2—selecting one dominant and one hybrid offspring for mating—is $2(1/2)(1/2) = 1/2$. The probability for selecting two hybrids (a transition to state a_4) is $(1/2)(1/2) = 1/4$. There is no way to get recessive parents out of this mating. The other transition probabilities can be computed in a similar manner. The complete matrix becomes, (listing the absorbing states first)

$$
T = \begin{array}{c c} & \begin{array}{cccccc} a_1 & a_6 & a_2 & a_3 & a_4 & a_5 \end{array} \\ \begin{array}{c} a_1\ (D,D) \\ a_6\ (R,R) \\ a_2\ (D,H) \\ a_3\ (D,R) \\ a_4\ (H,H) \\ a_5\ (H,R) \end{array} & \begin{bmatrix} 1 & 0 & 0 & 0 & 0 & 0 \\ 0 & 1 & 0 & 0 & 0 & 0 \\ \frac{1}{4} & 0 & \frac{1}{2} & 0 & \frac{1}{4} & 0 \\ 0 & 0 & 0 & 0 & 1 & 0 \\ \frac{1}{16} & \frac{1}{16} & \frac{1}{4} & \frac{1}{8} & \frac{1}{4} & \frac{1}{4} \\ 0 & \frac{1}{4} & 0 & 0 & \frac{1}{4} & \frac{1}{2} \end{bmatrix} \end{array}
$$

Calculating the fundamental quantities for an absorbing chain, we have

$$
Q = \begin{array}{c c} & \begin{array}{cccc} a_2 & a_3 & a_4 & a_5 \end{array} \\ \begin{array}{c} a_2 \\ a_3 \\ a_4 \\ a_5 \end{array} & \begin{bmatrix} \frac{1}{2} & 0 & \frac{1}{4} & 0 \\ 0 & 0 & 1 & 0 \\ \frac{1}{4} & \frac{1}{8} & \frac{1}{4} & \frac{1}{4} \\ 0 & 0 & \frac{1}{4} & \frac{1}{2} \end{bmatrix} \end{array} \qquad R = \begin{bmatrix} \frac{1}{4} & 0 \\ 0 & 0 \\ \frac{1}{16} & \frac{1}{16} \\ 0 & \frac{1}{4} \end{bmatrix}
$$

$$
I - Q = \begin{bmatrix} \frac{1}{2} & 0 & -\frac{1}{4} & 0 \\ 0 & 1 & -1 & 0 \\ -\frac{1}{4} & -\frac{1}{8} & \frac{3}{4} & -\frac{1}{4} \\ 0 & 0 & -\frac{1}{4} & \frac{1}{2} \end{bmatrix}
$$

$$
(I - Q)^{-1} = \begin{bmatrix} \frac{8}{3} & \frac{1}{6} & \frac{4}{3} & \frac{2}{3} \\ \frac{4}{3} & \frac{4}{3} & \frac{8}{3} & \frac{4}{3} \\ \frac{4}{3} & \frac{1}{3} & \frac{8}{3} & \frac{4}{3} \\ \frac{2}{3} & \frac{1}{6} & \frac{4}{3} & \frac{8}{3} \end{bmatrix}
$$

The absorption probabilities are found to be

$$
A = (I - Q)^{-1}(R) = \begin{array}{c c} & \begin{array}{cc} a_1 & a_6 \end{array} \\ \begin{array}{c} a_2 \\ a_3 \\ a_4 \\ a_5 \end{array} & \begin{bmatrix} \frac{3}{4} & \frac{1}{4} \\ \frac{1}{2} & \frac{1}{2} \\ \frac{1}{2} & \frac{1}{2} \\ \frac{1}{4} & \frac{3}{4} \end{bmatrix} \end{array}
$$

The genetic interpretation of absorption is that after many inbreedings either the G or the g genes must disappear. It is also interesting to note that the probability of ending with only G genes, if we start from a given state is equal to the proportion of G genes in that state.

The average number of steps to absorption are, if we start in state a_2, $4\frac{5}{6}$; if we start in a_3, $6\frac{2}{3}$; for a_4, $5\frac{2}{3}$; and for a_5, $4\frac{5}{6}$. Hence, we see that if we start in a state other than (D,D) or (R,R), we can expect to reach one of these states in five or six steps. The matrix, $(I-Q)^{-1}$ gives more detailed information: how many times we can expect to have matings of the types (D,H), (D,R), (H,H), and (H,R) starting from a given nonabsorbing state. The matrix, A, gives the probabilities of ending up in state a_1 or a_6.

Suppose this theory of genetics were only a conjecture (as it once was, and a controversial one as well). The above calculations give a prediction of the outcomes of experiments that could be conducted. Any deviation from the predicted results outside the statistically expected fluctuations suggest a need to modify the hypothesis, or conjecture. As it happens, this particular hypothesis has been well substantiated by results of experiments.

EXERCISE 8.5

The first five problems in this exercise relate to the inheritance of colorblindness, which is a sex-linked characteristic. There are two genes, C and S; C tends to produce colorblindness and S, normal vision. The S gene is dominant. A man has only one gene; and if this is C, he is color-blind. A man inherits one of his mother's two genes, while a woman inherits one gene from each parent. Thus, a man may be of type C or S; while a woman may be of type CC, CS, or SS. Let us study a process of inbreeding similar to Example of 8.5.3.

1. List the states of the chain. (**Hint:** There are six.)

2. Compute the transition probabilities.

3. Show that the chain is absorbing, and interpret the absorbing states.

4. Find the fundamental matrix, $(I - Q)^{-1}$; and interpret it.

5. Find the average number of steps before absorption for each of the starting states.

6. Consider a model for occupational status:

Assume three classifications of occupation for a man: professional, white collar, and blue collar. Suppose that the sons of professional men have a probability of .65 to be professionals; .25 to be white collar; and .10 to be blue collar. For the sons of white collar workers, the probabilities are .3 for professional; .5 for white collar; and .2 for blue collar. The probabilities for the sons of blue collar workers are .25 for professional; .30 for white collar; and .45 for blue collar.

Assume every man has a son. Set up the transition matrix and compute the probabilities for each occupational classification for the son, grandson, and great-grandson of a white-collar worker.

7. Assume for Problem 6 above that instead of every man having a son, the probability a man has a son is .75. Form a new transition matrix, adding the state (absorbing) that a man has no son.

Find the probabilities that a white-collar worker has a great-grandson who is a blue-collar worker, a white-collar worker, and a professional.

answers to selected exercises

EXERCISE 1.1

1. a, d, f, g, i. c also, if there is agreement on which companies are "major."
3. $1 \in A$, $2 \in A$, etc.
$6 \in A$, and $6 \notin D$; etc.
$5 \in A$, and $5 \in B$; etc.
$8 \in B$, and $8 \notin C$; etc.
7. a. The set of expressions used in set algebra.
b. The set of first names of Presidents of the United States.
c. The set of odd integers.
d. The set of teams in the NFC.
9. a. $\{x|x$ is one of the first six letters in the alphabet$\}$.
c. $\{x|x$ is divisible by three and x is a positive number less than $16\}$.
10. b. $\subset$ d. $\in$
12. a. $x \in A$ c. $a \notin B$ or $a \in B'$ e. $B \subseteq A$
13. b. a is not an element of A.
e. The set A is an element of the set B.

EXERCISE 1.2

1. a. $\{a,b,c,d,e,f,g\}$ c. $\{a,b,h,i,j,k,l\}$
e. $\{f,g\}$ g. $\{a,b,c,d,e,f,g,h,i,j\}$
i. $\{a,b,c,d,e,k,l\}$ k. $\{f,g,h,i,j\}$
2. a. AT d. NT g. AT j. AT
3. a. $\subset$ c. $\subset$ e. $=$
5. $\emptyset \subseteq (A \cap B) \subseteq B \subseteq (A \cup B) \subseteq U$

7. a. The people earning less than \$10,000 per year who are college graduates.
   c. Those earning up to \$20,000 per year.
   e. The people earning more than \$20,000 per year who have no more than a high school education.
   g. $B \cap F$ i. $A \cup B$
8. a. $B \cap H$ d. $D \subseteq H$
   g. Those scoring above 50 on the exam.
   j. All those scoring between 25 and 75 are classified as file clerks.
9. a. $\{a,b\}, \{a\}, \{b\}, \emptyset$
   b. $\{a,b,c\}, \{a,b\}, \{a,c\}, \{b,c\}, \{a\}, \{b\}, \{c\}, \emptyset$
   c. (There are 16 total).

EXERCISE 1.3

1. a. {f,g} c. $\{c,d,e\}$ e. $\{a,b,c,d,e\}$
   g. $\{f,g\}$ i. $\{c,d,e,j\}$ k. $\{a,b,c,d,e\}$
3. a. $(A' \cup B') = (A \cap B)'$. $\therefore$ $(A' \cup B') \cap (A \cap B)$
   $= (A \cap B)' \cap (A \cap B) = \emptyset$
   c. $(A' \cup B')' = ((A \cap B)')' = A \cap B$
   e. $(A \cup B) \cap C = C \cap (A \cup B)$. $\therefore$ $[(A \cup B) \cap C] \cup [C \cap (A \cup B)]$
   $= (A \cup B) \cap C$
5. a. (I) ∪ (II) ∪ (III) ∪ (IV) ∪ (V) c. VI
   e. (III) ∪ (IV) ∪ (V) ∪ (VI) ∪ (VII)
6. b. $(A \cup C) - B$ d. $(A \cup B) - (A \cap B)$
7. a. Those earning more than \$10,000 who either bought a new car or own a color TV.
   c. Those earning \$10,000 or less who bought a new car and own color TV.
   e. Those earning \$10,000 or less who either purchased a new car or own a color TV, but not both.
8. b. $C \cap B'$ d. $A \cup B \cup C$
9. a. $M \cap (B \cup P)$ c. $(B \cup P) - (B \cap P)$ e. $M' \cap B \cap P'$
   g. The men who regularly read periodicals but not books.
   i. The women who read neither books nor periodicals regularly.

EXERCISE 1.4

1. a. 95 b. 75 c. 20
3. a. $n[(A \cap B) - C] = 15$ $n(A \cup B) = 69$ $n(A \cup B \cup C) = 84$
   b. 96
5. a. 29 c. 43
7. a. P_1: Yes answers: {1, 2, 3, 4}; No: {5, 6, 7, 8}
   P_2: Yes answers: {1, 2, 5, 6}; No: {3, 4, 7, 8}
   P_3: Yes answers: {1, 3, 5, 7}; No: {2, 4, 6, 8}
   b. 16
9. a. 2 c. 2

EXERCISE 2.1

3. a, b, c, e, g, i. j: If intended as a prediction—No.
If used as a premise in an argument—Yes.

3. a.

p	$-p$	$-(-p)$
T	F	T
F	T	F

c.

p	$-p$	$p \vee -p$
T	F	T
F	T	T

5. a. s: T v: T c. q: T p: T or F

7. **a.** $p \vee q$ c. $-p \wedge -q$ e. $-(p \wedge q)$

9. a. Either the economy is expanding or real income is rising.
c. Either the economy is not expanding or real income is not rising.
e. It is not true that either the economy is expanding or real income is rising.
g. The economy is not expanding and real income is not rising.

11. c and f. d and h. e and g.

EXERCISE 2.2

1. a. Hypothesis: You are 18 or over.
Conclusion: You are eligible to vote.
c. Hypothesis: A student's GPA is 3.2 or higher.
Conclusion: He qualifies for the Dean's list.
e. Hypothesis: You have a steady job.
Conclusion: You can buy a car on the installment plan provided you make the payments on time; **or**
Hypothesis: You have a steady job and make the payments on time.
Conclusion: You can buy a car on the installment plan.
g. Hypothesis: A person is a conservationist.
Conclusion: He is greatly concerned about our natural environment.

3. a. i. $(p \wedge q) \rightarrow r$ iii. $r \leftrightarrow (p \wedge q)$ v. $r \rightarrow (q \rightarrow p)$
b. i. If prices rise and there is inflation, wages rise.
iii. If either prices or wages rise, there is inflation.
v. If wages rise when prices rise, there is inflation.

5. a. If I drink coffee in the evening, I stay awake.
I drank coffee during the evening only if I stay awake that night.
Staying awake is a necessary condition for my having drunk coffee.
c. If a dog is normal, it has four legs.
A dog is normal only if it has four legs.
Having four legs is a necessary condition for a dog to be normal.

6. b. Converse: If he has a file number, he is a registered student.
Inverse: If he is not a registered student, he does not have a file number.
Contrapositive: If he does not have a file number, he is not a registered student.

7. a. Rain while the sun is shining.
c. A true implication whose converse is false.

8. a. $p \rightarrow q$ d. $p \rightarrow q$ g. $p \leftrightarrow q$

9. a. If you pay property taxes, you own your own home.
c. If you do not own your home, you do not pay property taxes.

11. a. If two angles of a triangle are unequal, the sides opposite are unequal. If two sides of a triangle are equal, the opposite angles are equal.
c. If two lines are not parallel, the alternate interior angles are not equal. If alternate angles are equal, the lines are parallel.

EXERCISE **2.3**

1. a.

p	q	$p \wedge q$	$-(p \wedge q)$
T	T	T	F
T	F	F	T
F	T	F	T
F	F	F	T

c.

p	q	$p \vee q$	$p \wedge q$	$-(p \vee q)$	$-(p \wedge q)$	$-(p \vee q) \vee -(p \wedge q)$
T	T	T	T	F	F	F
T	F	T	F	F	T	T
F	T	T	F	F	T	T
F	F	F	F	T	T	T

e.

p	q	$p \to q$	$-q$	$-p$	$-q \to -p$	$(p \to q) \wedge (-q \to -p)$
T	T	T	F	F	T	T
T	F	F	T	F	F	F
F	T	T	F	T	T	T
F	F	T	T	T	T	T

3. a.

p	q	r	$p \wedge q$	$p \wedge r$	$(p \wedge q) \vee (p \wedge r)$
T	T	T	T	T	T
T	T	F	T	F	T
T	F	T	F	T	T
T	F	F	F	F	F
F	T	T	F	F	F
F	T	F	F	F	F
F	F	T	F	F	F
F	F	F	F	F	F

c.

p	q	r	$p \to q$	$q \to r$	$(p \to q) \wedge (q \to r)$
T	T	T	T	T	T
T	T	F	T	F	F
T	F	T	F	T	F
T	F	F	F	T	F
F	T	T	T	T	T
F	T	F	T	F	F
F	F	T	T	T	T
F	F	F	T	T	T

5. a. T c. T e. T
7. a. Yes c. Yes e. No

EXERCISE 2.4

1. a.

p	q	$p \wedge q$	$(p \wedge q) \wedge p$	$[(p \wedge q) \wedge p] \rightarrow q$	
T	(T)	T	T	T	Valid
T	F	F	F	T	
F	T	F	F	T	
F	F	F	F	T	

c.

p	q	$p \rightarrow q$	$-q$	$-p$	
T	T	T	F	F	
T	F	F	T	F	
F	T	T	F	T	
F	F	T	T	(T)	Valid

d.

p	q	r	$p \rightarrow q$	$q \rightarrow r$	$p \rightarrow r$	
T	T	T	T	T	(T)	
T	T	F	T	F	F	
T	F	T	F	T	T	
T	F	F	F	T	F	
F	T	T	T	T	(T)	
F	T	F	T	F	T	
F	F	T	T	T	(T)	Valid
F	F	F	T	T	(T)	

3. a. $q \rightarrow p$ e. $q \leftrightarrow r$

5. p: prices raised.
q: sales volume increases.
r: layoffs in shipping department.
Argument:
Premises: $p \rightarrow -q$; $-q \rightarrow r$; $-r$
Conclusion: $-p$ Valid.

7. c: farm workers move to city.
s: farm workers move to suburbs.
j: workers commute to jobs.
Argument:
Premises: $c \vee s$; $s \rightarrow j$; $-j$
Conclusion: c Valid.

9. t: city lowers tax rate.
m: company moves its factory.
j: residents lose their jobs.
Argument:
Premises: $-t \rightarrow m$; $m \rightarrow j$; $-j$
Conclusion: t (tax rate lowered).

EXERCISE 2.5

3. Valid **5.** Not Valid **7.** Valid **9.** Valid
11. $-p$ **13.** $-s$ **15.** Not valid

EXERCISE 2.6

1. a. Every one who earns more than \$600 per year must file an income tax return.
c. All college graduates are considered literate for purposes of voter eligibility.
e. Any one who cannot use mathematics cannot be an engineer.

3. a. If a person is a musician, he is temperamental.
c. If a dog barks, he will not bite.
e. If it is an automobile, it produces smog.

4. b. Some liberals could be elected in that district.
d. There are students who have failed that course.

7. Valid argument. (Do you agree with the premises?)

9. a. The set of tone-deaf men is a subset of the set of left-handed men.
c. No; the theory does not say left-handed men are tone deaf.

EXERCISE 3.1

1. Possibilities; Relation: y is x's brother.
Function: y is x's mother.

3. a. (1,1), (2,2), (3,3), (4,4).
c. (2,1), (3,2), (4,3).

5. Function: y is the letter following x in the alphabet.
Relation: y precedes x in the alphabet.

7. a. Domain: {students in math class}
Range: {their sisters and brothers}
c. Domain: {students in math class}
Range: {numbers 17 to 65}
e. Domain: {Real numbers}
Range: {Real numbers}

9. $7a, 7d, 7e, 8b$

11. a. $(5, 7), (3, 3), (1, -1), (9, 15)$.
c. $(7, 11), (1, -1), (-2, -7), (1\frac{1}{2}, 0)$

13. $(b, a), (c, a), (d, c), (x, t), (v, u)$.

EXERCISE 3.2

1.

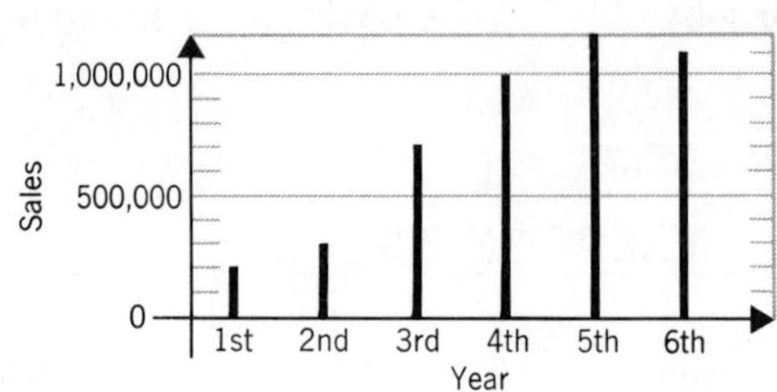

3.

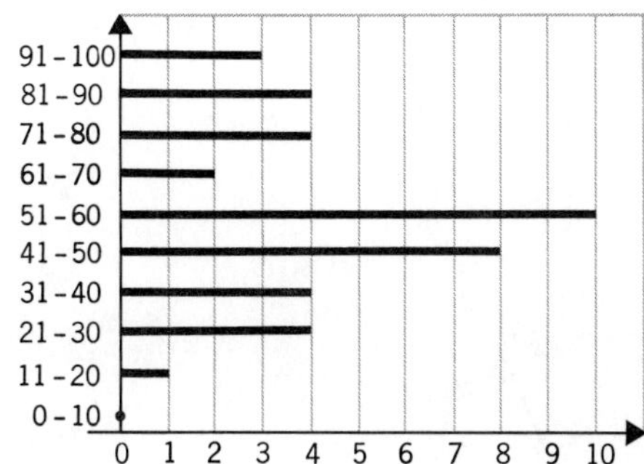

5. A: $(3, 2)$ C: $(-4, 0)$ E: $(3, -2)$

7. a. $f(0) = -2$ $f(1) = 1$ $f(-1) = -5$ $f(2) = 4$

c. $f(0) = 10/3$ $f(1) = 8/3$ $f(-1) = 4$ $f(2) = 2$

e. $f(0) = 2/3$ $f(1) = -1$ $f(-1) = 3$ $f(2) = -2$

9.

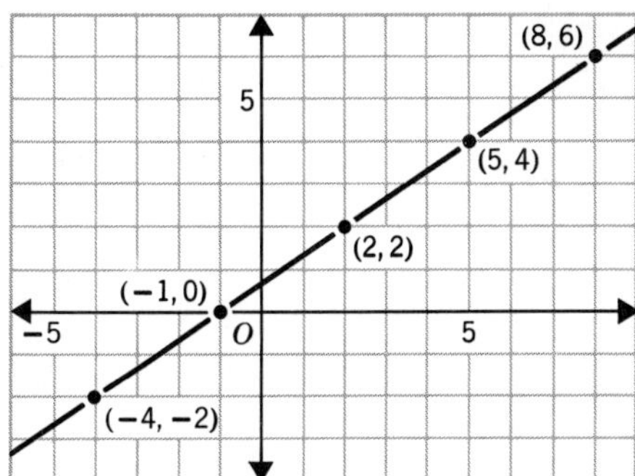

11.

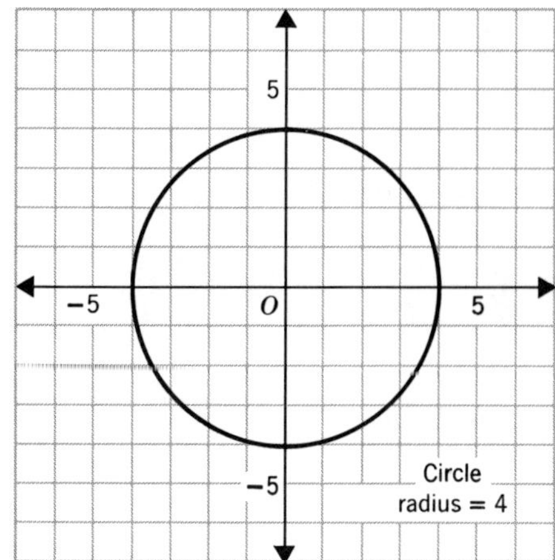

1.

a.

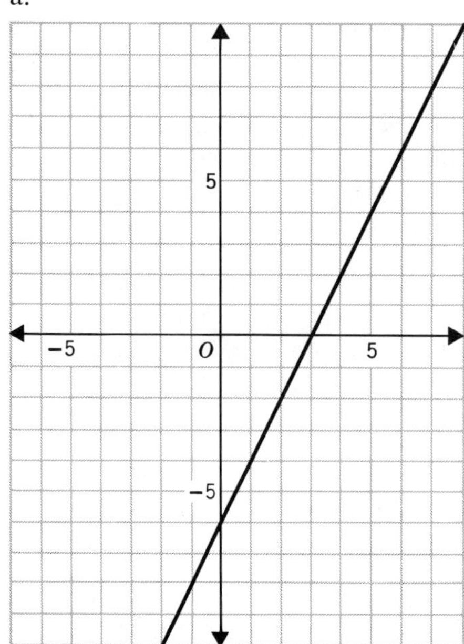

c.

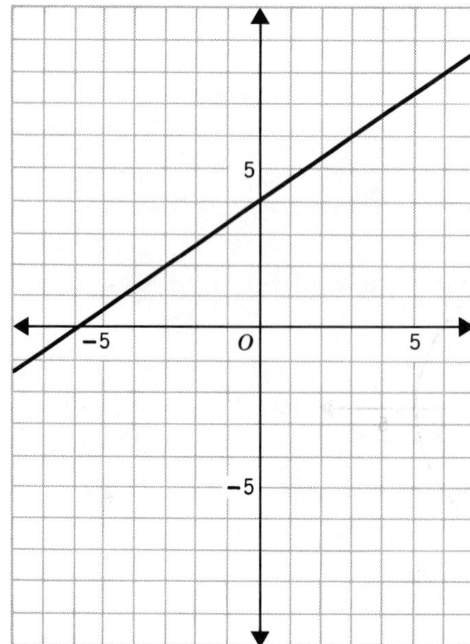

e.

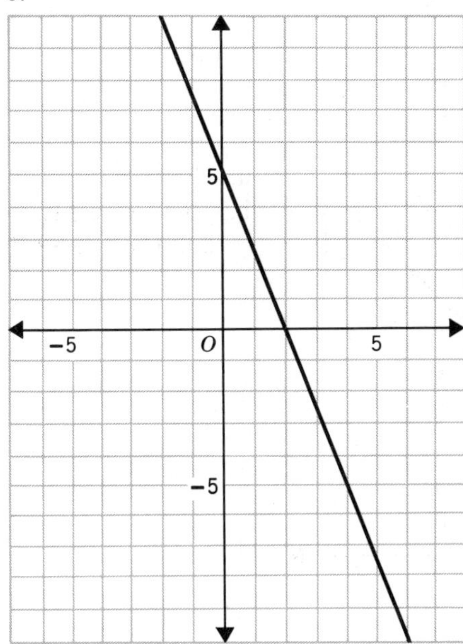

3. a. $m = 2, \quad b = -9/2$ c. $m = 3; \quad b = -6$ e. $m = 4/3 \quad b = -16/3$

5. a. 4 b. 9/8 **7.** a. \$8 b. \$16.60

9. a. $°C = \frac{5}{9}(°F) - (160/9)$ b. The change in Centrigrade temperature for each degree change in Fahrenheit.

11. a. $0 < x \leq 1000$: $y = .14x$

$1000 < x \leq 2000$:: $y = .15(x - 1000) + 140$; or $y = .15x - 10$

$2000 < x \leq 3000$: $y = .16(x - 2000) + 290$; or $y = .16x - 30$

etc.

b.

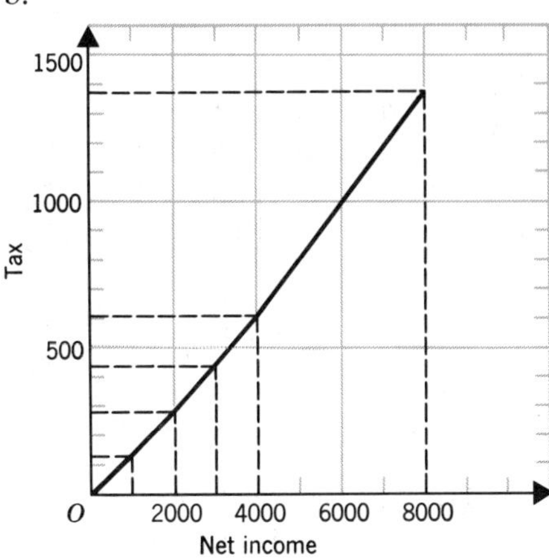

EXERCISE 3.4

1. a. $y = 250{,}000 - 10{,}000x$ b. \$150,000
3. \$348
5. \$455
7.

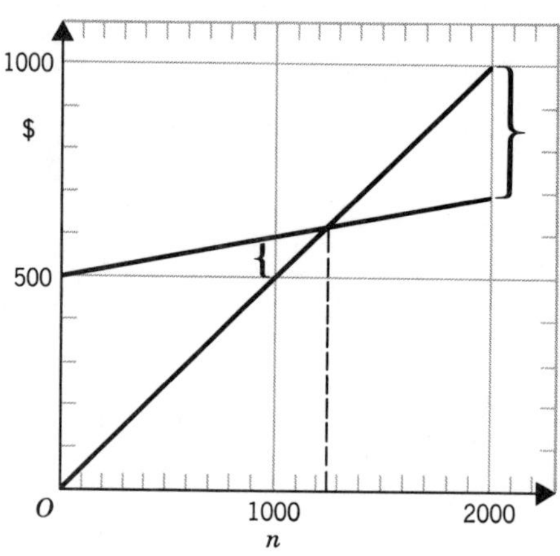

a. 1250
b. 1000: lose \$100
2000: make \$300

9.

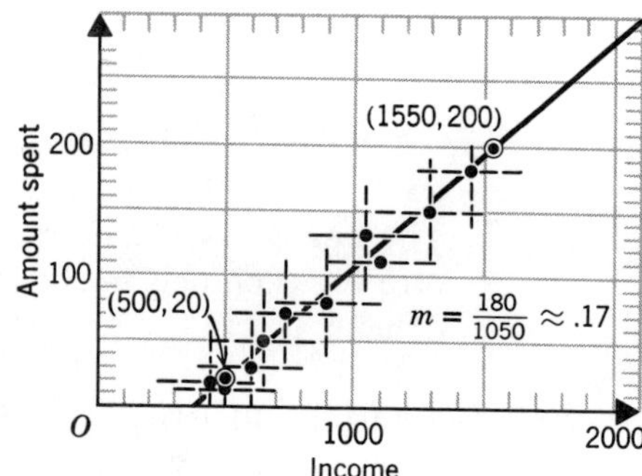

a. $y = .17x - 65$
b. Approximately \$275

EXERCISE 4.1

1. a. $a = 3$; $b = -2$; $c = 2$; $d = -2$
3. a. $A_{2\times 3}$ $C_{3\times 2}$ $D_{3\times 3}$ b. $a_{13} = 4$; $c_{22} = -1$; $f_{33} = -3$
5. a.

$$P + R = \begin{bmatrix} 2 & 12 & -8 \\ 1 & 4 & -4 \end{bmatrix} \qquad Q + S = \begin{bmatrix} 5 & 7 & -3 \\ 6 & -5 & 12 \\ 11 & 11 & -13 \end{bmatrix}$$

b.

$$Q - 2S = \begin{bmatrix} 2 & -2 & 3 \\ -21 & 16 & -6 \\ -1 & 5 & 2 \end{bmatrix} \qquad 3P - 2R = \begin{bmatrix} 1 & -9 & 6 \\ 18 & -13 & 8 \end{bmatrix}$$

7. $a = 2$; $c = -3$; $e = -1$; $w = 4$; $y = 1$.
9.

$$C = \begin{bmatrix} 1 & 2 & -6 \\ -1 & -2 & -11 \\ 9 & 11 & -4 \end{bmatrix}$$

11.

$$\begin{bmatrix} 2 & 1 & 2 & 2 \\ 3 & 5 & 5 & 3 \\ 3 & 1 & 0 & 1 \end{bmatrix}$$

13. She would look good in a bathing suit.

15. a.

	Receive			
	21″ B/W	25″ B/W	21″ C	25″ C
Standard	4	2	3	0
Deluxe	3	0	0	4
Regal	0	3	1	1

EXERCISE 4.2

1. a. -1

c. $\begin{bmatrix} 4 & -9 \\ 9 & -12 \end{bmatrix}$

e. $\begin{bmatrix} 2 & 6 & 6 \\ -1 & -3 & -3 \\ -2 & -6 & -6 \end{bmatrix}$

3. $AC = \begin{bmatrix} 3 & -4 \\ 2 & 9 \\ 2 & 14 \end{bmatrix}$ $AD = \begin{bmatrix} 0 & 3 & 1 \\ 14 & 9 & -4 \\ 20 & 12 & -6 \end{bmatrix}$ $EB = \begin{bmatrix} 23 & -3 \\ 7 & 11 \\ 4 & 10 \end{bmatrix}$

$DA = \begin{bmatrix} 9 & -7 \\ -2 & -6 \end{bmatrix}$ etc. Defined: AC, AD, BC, BD, CD, DA, DB, DE, EA, EB.

5. a. $\begin{bmatrix} -20 & 2 \\ -12 & -20 \\ 1 & -22 \end{bmatrix}$

8. a. $BAC = 75$

9. d. Matrix times T transposes columns 2 and 3.
T times matrix transposes rows 2 and 3.

11. $x_1 + 3x_2 = 0$
$2y_1 + 4y_2 = 0$

EXERCISE 4.3

1. $x =$ number of parts made per shift by first machine.
$y =$ number of parts made per shift by second machine.
$x = y + 4$ or $x - y = 4$
$x + y = 100$

3. $x =$ number of first type cartons needed.
$y =$ number of second type cartons needed.
A: $2x + 2y = 10$
B: $3x + y = 12$

5. $x =$ number of gallons 32¢ grade.
$y =$ number of gallons 40¢ grade.
Number of gallons purchased: $x + y = 10$.
Cost of purchase: $32x + 40y = 370$.

7. $x =$ number of red carnations.
$y =$ number of white carnations
$z =$ number of blue asters
Cost: $20x + 20y + 25z = 500$.

White/Blue: $y = 2z$ or $y - 2z = 0$.
Red: $x = y + z$ or $x - y - z = 0$.

9. $x =$ number of size 1 sheets of plywood
$y =$ number of size 2 sheets of plywood
$z =$ number of size 3 sheets of plywood
Type A: $x + 2y + 2z = 16$
Type B: $3x + 3y + z = 20$
Type C: $2x + 3y + 5z = 20$

11. $x =$ number of sheets of metal used with first die.
$y =$ number of sheets of metal used with second die.
$z =$ number of sheets of metal used with third die.
Metal available: $x + y + z = 100$
Number of pint lids: $8x + 10y + 6z = 800$
Number of quart lids: $6x + 8y + 6z = 600$
Number of gallon lids: $2x + 2y + 4z = 300$

EXERCISE 4.4

1. $x = 5$; $y = 3/2$
3. $x = 2$; $y = 1$
5. $x = 32/11$; $y = 14/11$
7. $x = 2$; $y = -3$; $z = -1$
9. $x = -3$; $y = 5$; $z = 2$
11. [−3, −4, 2]
13. [1, −2, −3)
15. [0, 1, 2]

EXERCISE 4.5

1. [−1, −4]
3. [1, 5]
5. [−5/2, 1/2, 3]
7. $\emptyset$
9. $x_1 = 2x_3$; $x_2 = 3 - 3x_3$; $x_3 =$ any number. or
[0, 3, 0] + k[2, −3, 1]; $k =$ any number.
11. [.33 .20, .47]

EXERCISE 4.6

2. b. Does not exist.
e. $\begin{bmatrix} \frac{9}{4} & \frac{3}{2} & \frac{5}{4} \\ \frac{3}{2} & 1 & \frac{1}{2} \\ \frac{1}{4} & \frac{1}{2} & \frac{1}{4} \end{bmatrix}$

3. a. [−11, 6]
c. [−25/4, −17/2, −31/4]
d. [−61/4, −19/2, −5/4]

4. b. $X = \begin{bmatrix} 0 & 1 \\ \frac{9}{5} & -\frac{7}{5} \end{bmatrix}$

EXERCISE 5.1

1. $a \leq b$ $a < c$ $c \geq b$

3. a.

−3 0 3

c.

0 2 4 6

5. a. $x \geq -1$ c. $x \geq -4$ e. $x \leq 2$

7. $x =$ number of cubic yards from Area A.
$y =$ number of cubic yards from Area B.
Sand requirement: $1\frac{1}{2}x + y \geq 60$
Rock requirement: $\frac{1}{2}x + y \geq 40$

9. $x =$ number of standard models produced.
$y =$ number of deluxe models produced.
Cutting time: $\frac{3}{4}x + y \leq 30$
Assembly time: $x + y \leq 36$
Finishing time: $x + 2y \leq 50$

EXERCISE **5.2**

1.

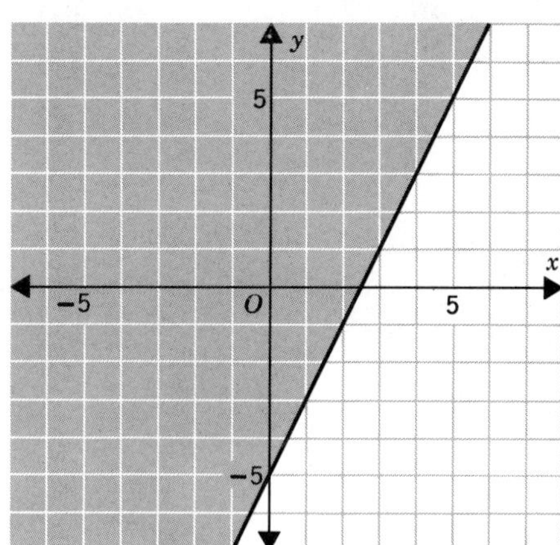

3.

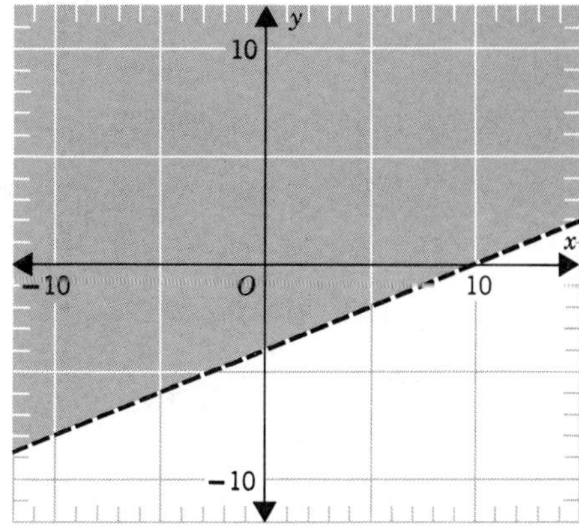

5.

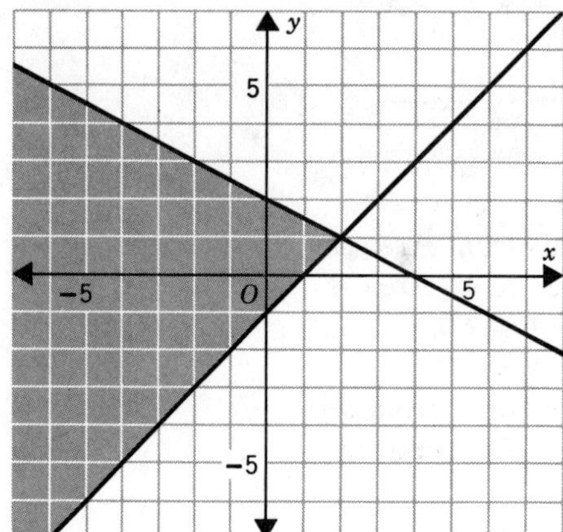

7.

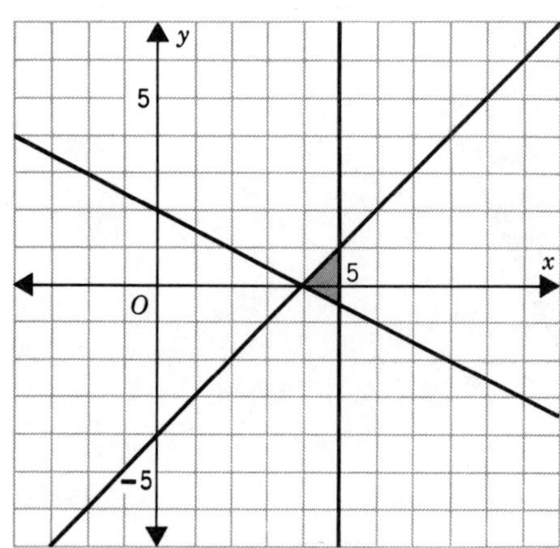

9.

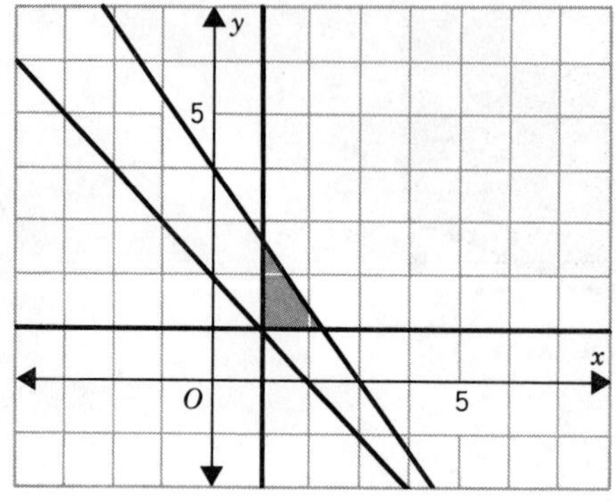

11.

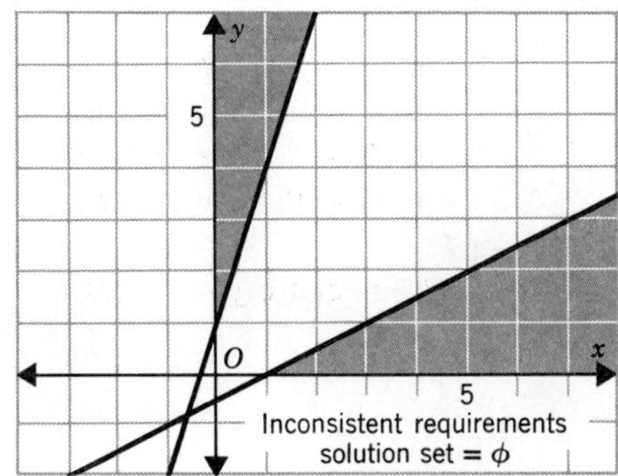
y
5
O
x
5
Inconsistent requirements
solution set = ϕ

13.

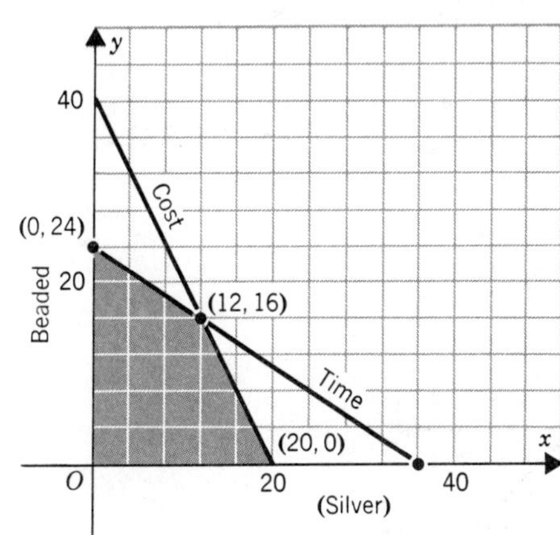
y
40
Cost
(0, 24)
Beaded
20
(12, 16)
Time
(20, 0)
x
O
20
40
(Silver)

15.

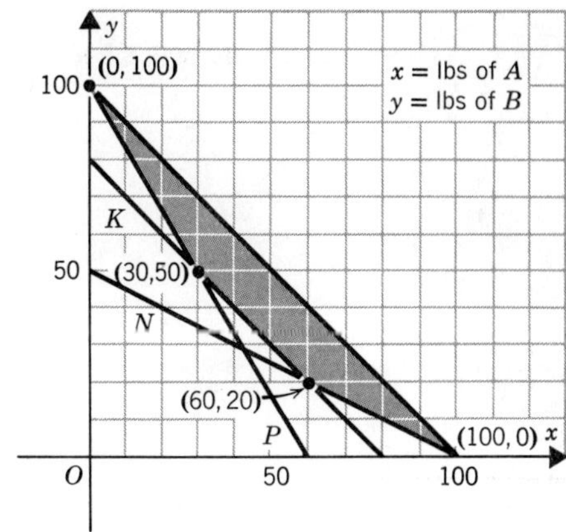
y
(0, 100)
100
x = lbs of A
y = lbs of B
K
50
(30,50)
N
(60, 20)
P
(100, 0)
x
O
50
100

EXERCISE 5.3

1. a. $P = 5$ at $(2,1)$
b. $C = 2$ at $(1,0)$

3. a. $P = 14$ at $(5,1)$
b. $P = 4$ at $(-4,4)$

5. a. $P = 29$ at $(4,7)$
b. $R = 12$ at $(0,9)$

7. a. $F = 55$ at $(5,10)$
b. $G = 0$ at $(1,10)$

9. a. 80 cubic yards from A; 0 from B.
b. When cost for Area A exceeds \$2 per cubic yard, shift to 20 cubic yards from A and 30 from B.

11. 24 standard; 12 deluxe.

EXERCISE 5.4

1. a.

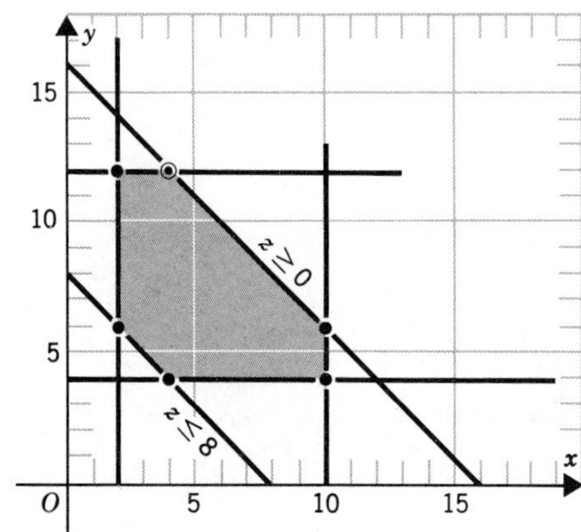

b. $A = 72$ at $(4, 12, 0)$

3. a.

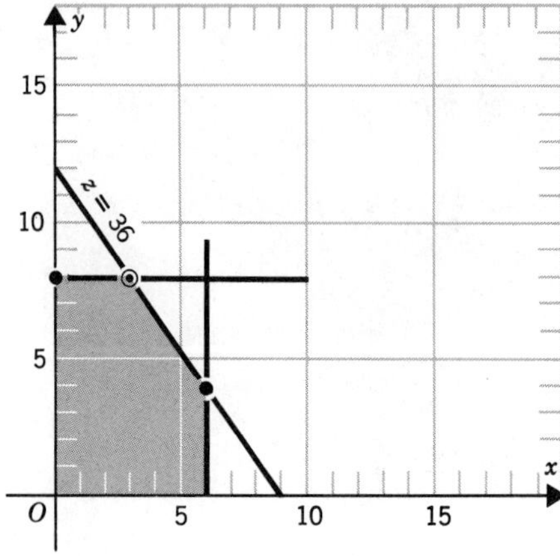

b. $A = 152$ at $(3, 8, 36)$

5.

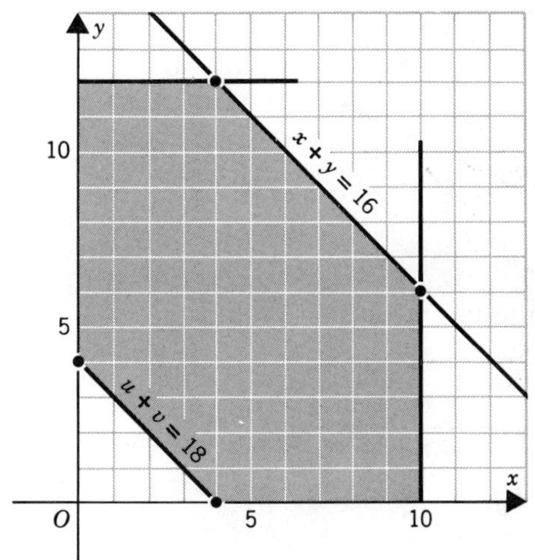

7. a.

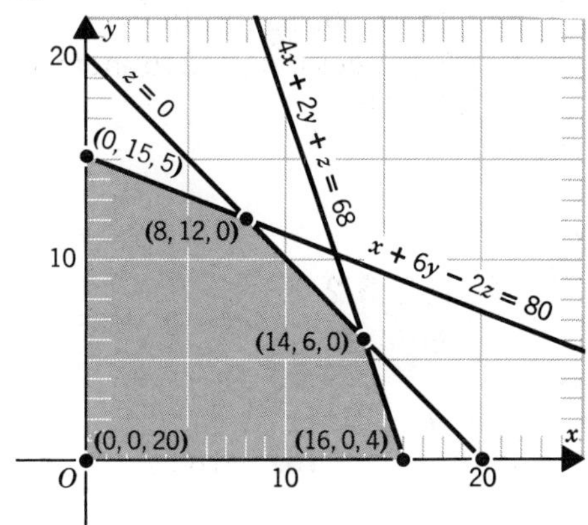

b. $R = 120$ at $(0,0,20)$

c. $C = 48$ at $(8,12,0)$

9. a. $(0,0,0)$, $(0,0,7/2)$, $(0,14/3,0)$ $(7,0,0)$, $(5,3,0)$, $(2,0,3)$

b. Maximum: $C = 39$ at $(5,3,0)$. Minimum: $C = 0$ at $(0,0,0)$.

11. a. $x =$ number of tons Grade A ore
$y =$ number of tons Grade B ore
$z =$ number of tons Grade C ore.

Capacity: $x + y + z = 100$
Lead: $.10x + .04y + .08z \geq 6.8$
Zinc: $.04x + .06y + .05z \geq 4.7$
Manganese: $.02x + .01y + .03z \geq 1.8$.

Minimum Cost: \$510 per hour
60 tons Grade A
30 tons Grade B
10 tons Grade C.

11. b.

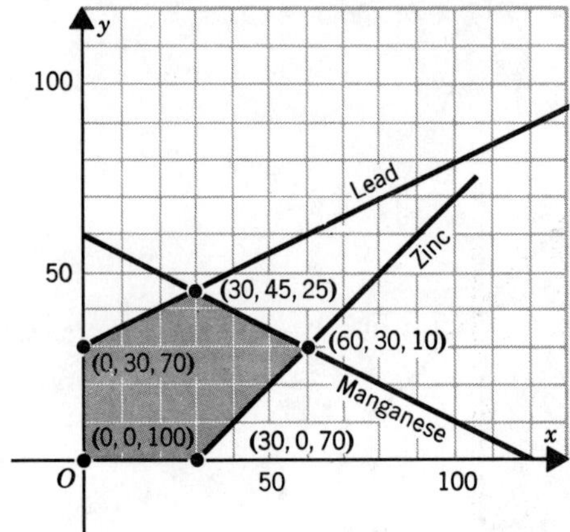

EXERCISE 6.1

1. 36
3. a. 36 b. 36
5. 56
7. 2 desserts; 216 menus
9. 7872
11. 60
13. 13

EXERCISE 6.2

1. a. 12 c. 120 e. 210
3. One
5. 30,240
7. 3024
9. 336
11. a. 120 b. 120
13. 576

EXERCISE 6.3

1. a. 15 c. 56 e. 120
3. a. 10, 9, 8 b. 720 c. 6
5. 210
7. $10^5 = 100{,}000$
9. 1680
11. 8 – 56 different exams possible
13. Approximately 92.8%

EXERCISE 6.4

1. 50
3. 5040
5. a. 720 b. 36 c. 216 d. 72
7. 154
9. 132
11. 95

EXERCISE 6.5

1. a. 70 c. 560
2. a. 35
3. a. 27,720 c. 51,975
5. a. 135 c. 7420
7. a. 70 b. 35
9. 184,800
11. 70
13. Three: 40. Four: 135
15. a. 720 b. 11

EXERCISE 7.1

1.

	O_1	O_2	O_3	O_4	O_5	O_6
Manager:	Hilton	Milton	Hilton	Tilton	Milton	Tilton
Assistant Manager:	Milton	Hilton	Tilton	Hilton	Tilton	Milton

3. The outcomes are different sets of three people each. There are $C(10,3) = 120$ of them.
5. a. One part defective b. 7280
All parts good
At least two parts defective;
etc.
7. One: .001; three: .003.
9. a. .5 b. .75
11. a. The sample space consists of all the possible sets of two balls each.
b. Both red; both white; both green; one red and one white; etc. There are $4 + C(4,2) = 10$ possibilities.
13. a. $S = \{HH, HT, TH, TT\}$. b. .25

EXERCISE 7.2

1. 31/365
3. a. 1/15 b. 4/15
5. a. 1/16 b. 1/4
7. a. None: .33; one: .41; two: .20; three: .05.
b. One c. 10%, same as sample.
d. .59; decrease to zero when sample size = lot size.
9. 11/120
11. Mary Carter: 1/6
Frank Able: 2/9

EXERCISE 7.3

1. .4
3. .48
5. 1/35
7. a. .15 b. .08 c. .4
9. a. 1/8 b. 3/8
11. .57
13. a. .48 b. .64

EXERCISE 7.4

1. a. .12 b. .58 c. .42
3. .5
5. 2/9
7. a. .5 c. .2 e. Yes

9. 5/22

11. a. .12 b. .42

13. .91

15. Events essentially independent – no brand name loyalty.

EXERCISE 7.5

1. a. approximately .35 c. approximately .078

3. a. approximately .28; b. approximately .953

5. a. 1/14 c. 13/14

7. a. 12/65 b. 31/91

9. a. 27/64 c. 189/256

11. a. .0024 b. .163

13. a. 1/1024 b. 106/1024

15. a. .1 b. .0656 c. 7

EXERCISE 7.6

1. a. 1/3 b. 3/7

3. a. $p(A|R) = 5/14$; $p(C|R) = 3/14$

5. a. 1/3 b. 2/3

7. a. 6/13 b. 1/8

9. a. A b. $p(I|B) = .21$; $p(R|B) = .33$; $p(D|B) = .46$

11. approximately .48

13. $p(A|D) = 5/16$; $p(B|D) = 3/8$; $p(C|D) = 5/16$.

15. a. .205 b. $p(M|D) = .13$; $p(BC|D) = .37$; $p(WC|D) = .47$.

EXERCISE 7.7

1. \$1

3. 7 points

5. \$1

7. a. \$0.25 b. \$12

9. .488

11. a. \$9.40 b. 2660

EXERCISE 7.8

1. a. $p(A) = 3/5$; $p(B) = 2/5$ c. $p(A) = 5/6$

2. b. B is favored over A by odds of 4 to 1.
d. The odds against A are 9 to 1.

3. a. .1 b. .25

5. 5 to 1 against

7. a. $p(A) = 3/8$; $p(B) = 1/4$; $p(C) = 5/16$; $p(D) = 1/16$
b. Yes; $p(A) + p(B) + p(C) + p(D) = 1$

9. a. 5/36 b. 1/6

EXERCISE 8.1

1. a.

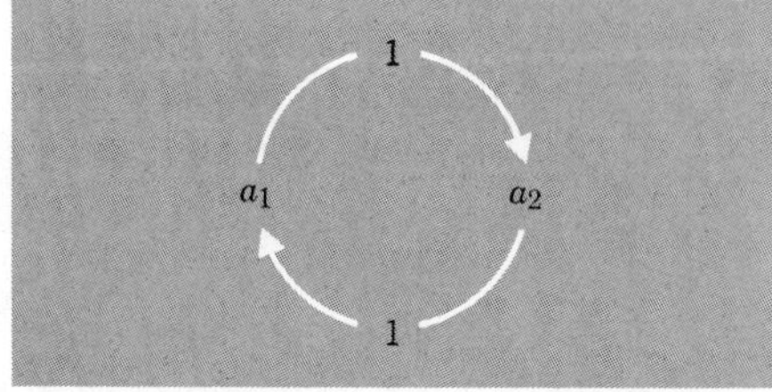

c.

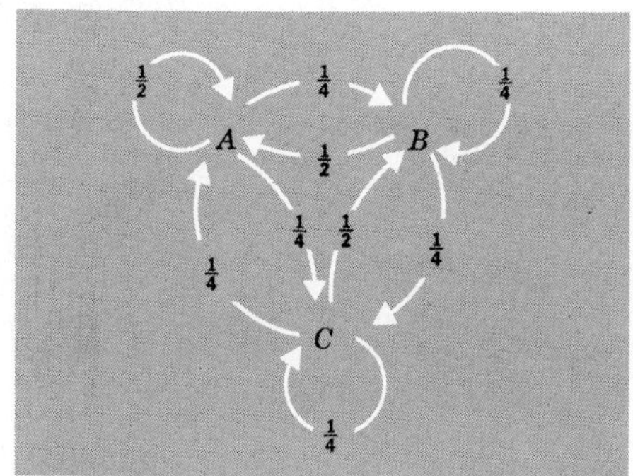

3. a.

$$\begin{array}{c} \\ A \\ B \end{array} \begin{array}{c} \begin{array}{cc} A & B \end{array} \\ \begin{bmatrix} \frac{2}{3} & \frac{1}{3} \\ \frac{1}{2} & \frac{1}{2} \end{bmatrix} \end{array}$$

b.

$$\begin{array}{c} \\ a_1 \\ a_2 \\ a_3 \end{array} \begin{array}{c} \begin{array}{ccc} a_1 & a_2 & a_3 \end{array} \\ \begin{bmatrix} 0 & \frac{1}{2} & \frac{1}{2} \\ 0 & 0 & 1 \\ \frac{1}{2} & 0 & \frac{1}{2} \end{bmatrix} \end{array}$$

5. a.

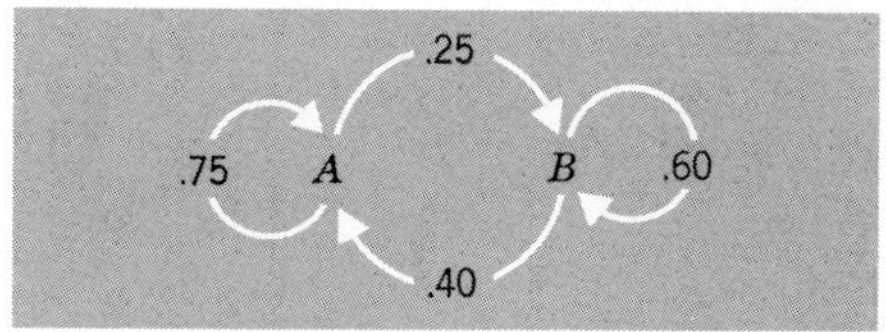

b.

$$\begin{array}{c} \\ A \\ B \end{array} \begin{array}{c} \begin{array}{cc} A & B \end{array} \\ \begin{bmatrix} .75 & .25 \\ .40 & .60 \end{bmatrix} \end{array}$$

7. a.

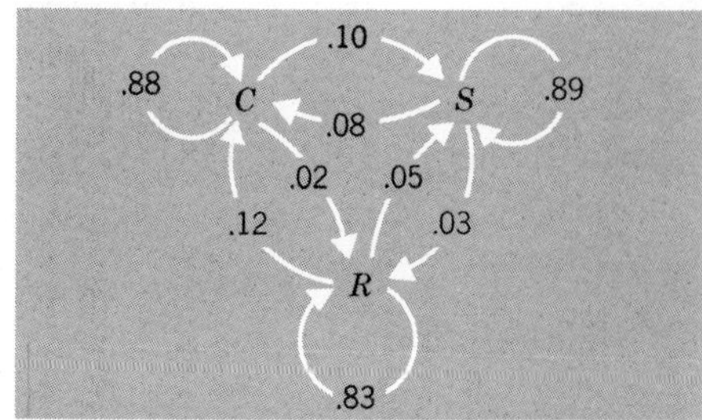

b.

$$\begin{array}{c} \\ C \\ S \\ R \end{array} \begin{array}{c} \begin{array}{ccc} C & S & R \end{array} \\ \begin{bmatrix} .88 & .10 & .02 \\ .08 & .89 & .03 \\ .12 & .05 & .83 \end{bmatrix} \end{array}$$

9. a.

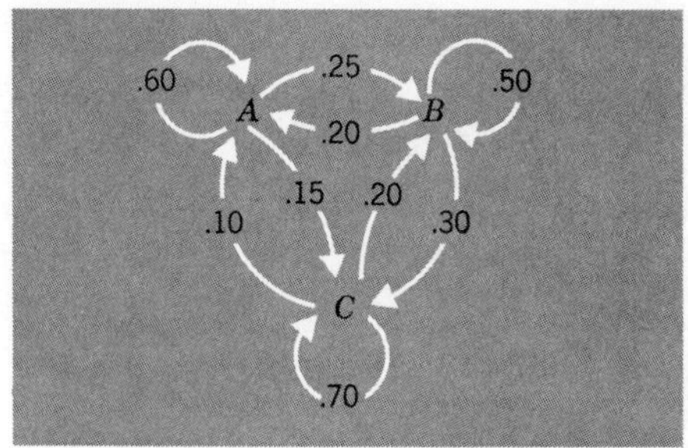

b.

$$\begin{array}{c} \\ A \\ B \\ C \end{array} \begin{array}{c} \begin{array}{ccc} A & B & C \end{array} \\ \begin{bmatrix} .60 & .25 & .15 \\ .20 & .50 & .30 \\ .10 & .20 & .70 \end{bmatrix} \end{array}$$

11. a.

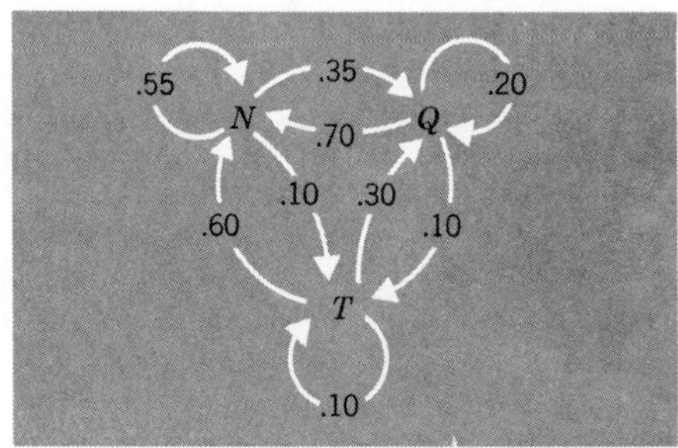

b.

$$\begin{array}{c} \\ NC \\ QC \\ TC \end{array} \begin{array}{c} \begin{array}{ccc} NC & QC & TC \end{array} \\ \begin{bmatrix} .55 & .35 & .10 \\ .70 & .20 & .10 \\ .60 & .30 & .10 \end{bmatrix} \end{array}$$

EXERCISE **8.2**

1. a. [.55, .45] b. [.565, .435] c. [.571, .429]
3. a. .25; .34 b. [8/13, 5/13]
5. City: 42.6%; suburbs: 44.4%; country: 13%
7. [.30, .32, .38] long term: [.255, .306, .438]
9. a. Quick: 30%; thorough: 10% b. .271

EXERCISE **8.3**

1. *a*, *d*, *e*, and *g* are absorbing.

2. a. $R = (.75)$ $Q = (.25)$
$(I - Q)^{-1} = (1.33)$
$(I - Q)^{-1}R = (1)$

e.

$$R = \begin{matrix} \\ a_1 \\ a_3 \end{matrix} \begin{matrix} a_2 \\ \begin{bmatrix} .5 \\ .5 \end{bmatrix} \end{matrix} \qquad Q = \begin{bmatrix} .5 & 0 \\ 0 & .5 \end{bmatrix}$$

$$(I-Q)^{-1} = \begin{bmatrix} 2 & 0 \\ 0 & 2 \end{bmatrix}$$

$$(I-Q)^{-1}R = \begin{bmatrix} 1 \\ 1 \end{bmatrix}$$

3.

$$\begin{array}{c|cc|ccc} & 1 & 5 & 2 & 3 & 4 \\ \hline 1 & 1 & 0 & 0 & 0 & 0 \\ 5 & 0 & 1 & 0 & 0 & 0 \\ \hline 2 & \frac{2}{3} & 0 & 0 & \frac{1}{3} & 0 \\ 3 & 0 & 0 & \frac{2}{3} & 0 & \frac{1}{3} \\ 4 & 0 & \frac{1}{3} & 0 & \frac{2}{3} & 0 \end{array}$$

$$(I-Q)^{-1} = \begin{array}{c} \\ 2 \\ 3 \\ 4 \end{array} \begin{array}{c} \begin{array}{ccc} 2 & 3 & 4 \end{array} \\ \begin{bmatrix} \frac{7}{5} & \frac{3}{5} & \frac{1}{5} \\ \frac{6}{5} & \frac{9}{5} & \frac{3}{5} \\ \frac{4}{5} & \frac{6}{5} & \frac{7}{5} \end{bmatrix} \end{array}$$

$$(I-Q)^{-1}R = \begin{array}{c} \\ 2 \\ 3 \\ 4 \end{array} \begin{array}{c} \begin{array}{cc} 1 & 5 \end{array} \\ \begin{bmatrix} \frac{14}{15} & \frac{1}{15} \\ \frac{4}{5} & \frac{1}{5} \\ \frac{8}{15} & \frac{7}{15} \end{bmatrix} \end{array}$$

The probability of being absorbed into state 5, starting in state 3 is 1/5.

5. a.

$$\begin{array}{c|cc|ccc} & L & R & J & M & S \\ \hline L & 1 & 0 & 0 & 0 & 0 \\ R & 0 & 1 & 0 & 0 & 0 \\ \hline J & .06 & .02 & .80 & .12 & 0 \\ M & .05 & .08 & .02 & .75 & .10 \\ S & .02 & .20 & 0 & .03 & .75 \end{array}$$

c.

$$\begin{array}{c} \\ J \\ M \\ S \end{array} \begin{array}{c} \begin{array}{ccc} J & M & S \end{array} \\ \begin{bmatrix} 5.27 & 2.65 & 1.06 \\ .44 & 4.42 & 1.77 \\ .05 & .53 & 4.22 \end{bmatrix} \end{array}$$

EXERCISE 8.4

1. a. 9/13 b. 38/65 ≈ .58

3. .402

5. .634

7. $p(A) = p(B) = p(C) = p(D) = 1/4$

9. approximately $(1 - r)$ or .5

EXERCISE 8.5

1. $(S,SS), (S,CS), (S,CC), (C,SS), (C,CS), (C,CC)$

3.

$$\begin{array}{l} \\ (S,SS) \\ (C,CC) \\ \\ (S,CS) \\ (S,CC) \\ (C,SS) \\ (C,CS) \end{array} \begin{array}{c} \begin{array}{cccccc} (S,SS) & (C,CC) & (S,CS) & (S,CC) & (C,SS) & (C,CS) \end{array} \\ \left[\begin{array}{cc|cccc} 1 & 0 & 0 & 0 & 0 & 0 \\ 0 & 1 & 0 & 0 & 0 & 0 \\ \hline \frac{1}{4} & 0 & \frac{1}{4} & 0 & \frac{1}{4} & \frac{1}{4} \\ 0 & 0 & 0 & 0 & 0 & 1 \\ 0 & 0 & 1 & 0 & 0 & 0 \\ 0 & \frac{1}{4} & \frac{1}{4} & \frac{1}{4} & 0 & \frac{1}{4} \end{array}\right] \end{array} \begin{array}{l} \text{Absorbing} \\ \text{states} \end{array}$$

5. (S,CS): 5; (S,CC): 6; (C,SS): 6; (C,CS): 5.

7. Blue collar: .09
   Professional: .18
   White collar: .15

INDEX

Absorbing Markov chain, 341, 353ff
Absorbing state, 353, 354
Addition of matrices, 137ff
  associative law for, 138
  commutative law for, 139
Analytical probability, 259
"And," *see also* Conjunction
  truth table for, 49
A posteriori probability, 307
Applicable domain, 130
A priori probability, 308
Abscissa, 108
Argument, 70ff
  conclusion of, 71
  forms, 76ff
  premises of, 71
  truth table for, 72ff
  validity of, 71
Associative law for addition of matrices, 138
  for intersection of sets, 24
  for multiplication of matrices, 146
  for union of sets, 23
Augmented matrix, 164

Basic counting criterion, 221
Bayes' theorem, 307ff
  statement of, 310
Bernoulli trial, 301ff
  definition, 301
Biconditional connective, 55
  truth table for, 55
Binary relation, 98
Break-even point, 127

Cartesian coordinate system, 105ff
Cell of a partition, 247ff
Circular permutation, 231
Coefficient matrix, 163
Column vector, 136
Combination(s), 234
Complement of an event, 264
  of a set, 17
  relative, 17
Compound event, 262, 268ff
Compound set, 23
Compound statement, 46ff, 62ff
Conclusion, 53, 59, 71
Conditional connective, 53ff
  contrapositive of, 57
  converse of, 57
  definition of, 55
  forms of, 55ff
  inverse of, 57
  truth table for, 55
  variations of, 55ff
  verbal expressions for, 60
Condition, necessary, 59
  sufficient, 59
Conditional probability, 275ff
  definition, 276
Conjunction, 47
  truth table for, 49
Connective(s), 46ff
  "and," 47, 49
  biconditional, 55
  conditional, 53
  conjunction, 47, 49
  disjunction, 47, 50
  logical, 46
  negation, 47, 51
  "not," 47, 51
  "or," 47, 50
  symbols for, 47
  truth tables for, 48ff
Consequence, 59
  valid, 71
Contrapositive, 57
Converse, 57
Coordinates, 107
Corner point, 203
Counting, 220ff
  combinations, 234
  criterion for, 221
  elements of a set, 32
  fundamental principle of, 221
  ordered partitions, 248
  permutations, 228
  tree diagram for, 222
  unordered partitions, 249
Counting pattern, selection of, 239
Cross-partition, 38
Current value, 124

De Morgan's laws, 24
Dependent variable, 97, 98, 108
Depreciation, straight-line, 123
Deterministic process, 219, 259
Diagram, tree, 222
  Venn, 26ff
Dimension of a matrix, 136
Directed distance, 105
Direct proof, 78ff
Disagreements in listing, number of, 251
Discounting, loan, 124
Discrete event, 260
Disjoint sets, 18
Disjunction, definition, 47, 50
  exclusive, 50
  truth table for, 50
Distributive law, for matrices, 147
  for sets, 24
Domain, applicable, 130
  of a function, 97, 99
  of a relation, 97, 99

Element(s), of event, 262
  of a matrix, 135
  of a set, 6ff

Empty set, 16
Equality, of matrices, 136
of sets, 10
Equally probable outcomes, 264, 266, 270
Equation(s), linear, 118
systems of, 150ff
Equilibrium condition, 348
of Markov chain process, 348
Equivalence, logical, 67
Equivalent rate of interest, 125
Equivalent statements, 67, 79
Equivalent system of equations, 158
Event(s), complement of, 264
compound, 262, 268ff
definition, 261
discrete, 260
impossible, 263
independent, 286ff
mutually exclusive, 263, 287, 291
odds against, 335
odds for, 334
probability of, 263, 275
simple, 262
union of, 289
Exclusive disjunction, 50
Existential quantifier, 89
Expected value, 320ff
definition, 320
Experiment(s), definition, 261
outcomes, 261
sample space for, 261

Factorial, 228
Falsity, logical, 69
Feasible solution, 204
Finite probability, 220
Finite set, 7
Fixed probability vector, 347
Formal proof, 80ff
Function(s), 99ff
definition, 99
domain of, 97, 99
graph of, 108ff
linear, 115ff
objective, 202, 204
range of, 97, 99
Fundamental matrix of Markov chain, 356
Fundamental principle of counting, 220ff

Gambler's ruin, 365ff
Gauss elimination method, 158
Genetics, Markov processes in, 369ff
Graph, description, 103ff
line, 103
of a function, 108ff
of a relation, 108ff

Half-plane, 191
Hypothesis, 59

Identity laws for sets, 24
Identity matrix, 174
Implication, 56
Impossible event, 263
Independent events, 286ff
definition, 286
Independent trials, 301
Independent variable, 97, 98, 107
Indirect proof, 83ff
Inequalities, algebra of, 185
definition, 185, 186
expressions for, 185
linear, 187ff
properties of, 186
strict, 184, 193
tautologies for, 186
weak, 184, 193
Initial probability vector, 347
Intercept, x-axis, 122 - (6)
y-axis, 118
Interest, equivalent rate of, 125
ordinary, 124
rate of, 124
simple, 124
Intersection, associative law for, 24
commutative law for, 24
definition, 15
Inverse, of a conditional, 57
of a matrix, 174

Joint probability, 282

Line, equation of, 118, 120
graph of, 118, 121
slope of, 116
Linear correlation, 128ff
Linear equation(s), graph of, 118
system of, 150ff
Linear function, 115ff
Linear inequalities, 187ff
graphing, 191ff
systems of, 195ff
Linear programming, 201ff
Loan discounting, 124
Logic, 43ff
algebra of, 45
Logical absurdity, 69
Logical connective, 46
Logical equivalence, 67
Logical falsity, 69
Logically equivalent statements, 67, 79
Logical possibility, 48, 63ff
Logical truth, 68

Markov chain, 339ff
absorbing, 341
definition, 340
fundamental matrix of, 356
regular, 348
transition diagram for, 341
transition matrix for, 342
Mathematical model, 2
Matrix, addition of, 137
augmented, 164
column, 136
definition, 135
dimension of, 136
elementary operations on, 166
equality of, 136
fundamental, 356
identity, 174
inverse, 174
Multiplication of, 143ff
multiplication by scalar, 138

of coefficients, 163
reduced, 165
row, 136
singular, 175
transition, 342
zero, 138
Maximum value of objective function, 204
Members, set, 7
Minimum value of objective function, 204
Model, mathematical, 2
Mutually exclusive events, 263, 287, 291

Necessary condition, 59
Negation, 47
of quantified proposition, 90
truth table for, 51
"Not," *see also* Negation
truth table for, 51
Null set, 16
Number of elements in set, 32ff

Objective function, 202, 204
Odds, computation of, 334
against an event, 335
for an event, 334
Operations, set, 14ff
Optimum solution, 201, 204, 206
"Or," *see also* Disjunction
truth table for, 50
Ordered pair, 98, 107
Ordered partition, 247
number of, 248
Ordinate, 108
Outcome, definition, 261
equally probable, 264, 266, 270

Parameter, 128
Partition(s), counting, 248
cross-, 38
definition, 37
ordered, 247
unordered, 247
Permutation(s), 227ff
circular, 231
number of, 228
Point on graph, 106, 107
Possibilities, logical, 48, 63ff
Premise, 71
Principal, 125
Probability, 259ff
analytical, 259
a posteriori, 307
a priori, 308
assignment of, 260
binomial, 301
conditional, 275ff
joint, 282
of an event, 263, 275
of a simple event, 266
properties of, 263
statistical, 260
transition, 339, 341
Probability vector, 345ff
Product, of matrices, 143ff
of vectors, 142
Profit, 127
Programming, linear, 201ff

Projection(s), 128ff
Proof, direct, 78ff
formal, 80ff
indirect, 83ff
truth table for, 72ff
Proper subset, 11, 12
Proposition, 46
existential, 89
universal, 87

Quantified proposition, 87ff
negation of, 90
Quantifier, existential, 89
universal, 88

Random process, 259
Random variable, 319ff
definition, 320
Random walk, 354, 361ff
duration of, 367
Range, of a function, 97, 99
of a relation, 97, 99
Rate of change, 116
Rectangular coordinate system, 105ff
Reduced matrix, 165
Regular Markov process, 348
Regular transition matrix, 348
Relation(s), 95ff
binary, 98
domain of, 97
range of, 97
Relationship, set, 9
Relative complement, 17
Relative likelihood, 220, 259
Repeated trials, 296ff
Roster method, 7
Row operation, 165
Row vector, 136

Sample space, 261
Sampling, 236
with replacement, 238
Scalar, multiplication of matrix by, 138
Set(s), complement of, 17
compound, 23
disjoint, 18
elements of, 6ff
empty, 16
equal, 10ff
finite, 7
infinite, 7
intersection of, 15
members of, 7
notation, 8
null, 16
number of elements in, 32ff
operations, 14ff
partitions, 37ff
proper subset of, 11, 12
properties of operations, 24
relations between, 9ff
relative complement of, 17
specifications for, 7
solution, 156, 172
subsets of, 9
truth, 96
union of, 14

universal, 16
Venn diagram for, 26ff
Set notation, roster, 7
set-builder, 8
Simple event, 262
Simple statement, 46
Singular matrix, 175
Slope, 116
Solution set, 156
Space, sample, 261
Square matrix, inverse of, 175
Statement(s), biconditional, 55
characteristics of, 45
compound, 46, 62ff
conditional, 53ff
contrapositive, 57
converse, 57
equivalent, 67
false, 69
negation of, 47
open, 46
properties of, 45
simple, 46
true, 68
States, of Markov chain, 339
absorbing, 353
Statistical probability, 260
Study suggestions, 3
Subset, definition, 10
number of elements in, 32ff
proper, 11
Venn diagram of, 26
Sufficient condition, 59
System of equations, 150ff
consistent, 172
inconsistent, 173
solutions for, 155ff
System of inequalities, 195ff
corner point, 203
graph of, 195
solutions for, 195, 214
Stochastic process, 259, 301ff, 339
Markov chain, 339
transition diagram for, 341

Tautology, 68

Transition diagram, 341
Transition matrix, 342
Transition probability, 339, 341
Tree diagram, 222
Trials, repeated, 296ff
Bernoulli, 301
Truth set, 96
Truth table(s), 48ff
for biconditional, 55
for compound statements, 62ff
for conditional, 55
for conjunction, 49
for disjunction, 50
for negation, 51
for validity of argument, 72

Union of sets, 14
Universal quantifier, 88
Universal set, 16
Unordered partition, 247
number of, 249
Urn model, 266

Valid argument, 71ff
Valid conclusion, 70ff
Variable, 96ff
dependent, 97
independent, 97
random, 320
Vector(s), column, 136
fixed, 347
probability, 345, 346
product of, 142
row, 136
Venn diagram, 26ff
Vertex, 203, 214

X-axis, 106
X-intercept, 122 - (6)

Y-axis, 106
Y-intercept, 118

Zero matrix, 138